Sixth Edition

Fire and Emergency Services Instructor

Gloria Bizjak
Maryland Fire and Rescue Institute
Contract Writer

Barbara Adams
Project Manager

Validated by the International Fire Service Training Association
Published by Fire Protection Publications, Oklahoma State University

RECYCLABLE

Cover photo courtesy of Randy Wilden, Utah Fire and Rescue Academy.

The International Fire Service Training Association

The International Fire Service Training Association (IFSTA) was established in 1934 as a "nonprofit educational association of fire fighting personnel who are dedicated to upgrading fire fighting techniques and safety through training." To carry out the mission of IFSTA, Fire Protection Publications was established as an entity of Oklahoma State University. Fire Protection Publications' primary function is to publish and disseminate training texts as proposed and validated by IFSTA. As a secondary function, Fire Protection Publications researches, acquires, produces, and markets high-quality learning and teaching aids as consistent with IFSTA's mission.

The IFSTA Validation Conference is held the second full week in July. Committees of technical experts meet and work at the conference addressing the current standards of the National Fire Protection Association and other standard-making groups as applicable. The Validation Conference brings together individuals from several related and allied fields, such as:

- Key fire department executives and training officers
- Educators from colleges and universities
- Representatives from governmental agencies
- Delegates of firefighter associations and industrial organizations

Committee members are not paid nor are they reimbursed for their expenses by IFSTA or Fire Protection Publications. They participate because of commitment to the fire service and its future through training. Being on a committee is prestigious in the fire service community, and committee members are acknowledged leaders in their fields. This unique feature provides a close relationship between the International Fire Service Training Association and fire protection agencies which helps to correlate the efforts of all concerned.

IFSTA manuals are now the official teaching texts of most of the states and provinces of North America. Additionally, numerous U.S. and Canadian government agencies as well as other English-speaking countries have officially accepted the IFSTA manuals.

ISBN 0-87939-167-7

Library of Congress LC# 99-62199

Printed in the United States of America 10 9 8 7 6

If you need additional information concerning the International Fire Service Training Association (IFSTA) or Fire Protection Publications, contact:

Customer Service, Fire Protection Publications, Oklahoma State University
930 North Willis, Stillwater, OK 74078-8045
800-654-4055 Fax: 405-744-8204

For assistance with training materials, to recommend material for inclusion in an IFSTA manual, or to ask questions or comment on manual content, contact:

Editorial Department, Fire Protection Publications, Oklahoma State University
930 North Willis, Stillwater, OK 74078-8045
405-744-4111 Fax: 405-744-4112 E-mail: editors@osufpp.org

Table of Contents

Preface

The fire and emergency services profession continuously needs competent, experienced, and knowledgeable instructors. These instructors teach initial training courses and continuing education programs to learners who need to meet the obligations of certification, recertification, and professional qualifications. Fire and emergency services training presents many challenges, all of which can be met by well-prepared instructors. Instructors provide learners with knowledge and skill resources that enable them to perform their jobs. Good instructors are aware of changes in standards, safety issues, equipment, and operating procedures and instruct learners about them.

This sixth edition of **Fire and Emergency Services Instructor** (formerly called **Fire Service Instructor**) has been revised to guide the instructor candidate in the many important areas of instructional methods, including the techniques of teaching, the processes of developing training programs, and the tasks of managing and supervising these programs. This text meets the requirements of NFPA 1041, *Standard for Fire Service Instructor Professional Qualifications*, Levels I, II, and III. The new edition encompasses all aspects of fire and emergency services and can be adapted to instructor training in fire, rescue, emergency medical services, police, and other related emergency service areas.

A dedicated and hard-working committee of the International Fire Service Training Association (IFSTA) compiled and validated this **Fire and Emergency Services Instructor** text. Fire Protection Publications of Oklahoma State University gratefully acknowledges and thanks these committee members for their continuing input and diligent work in updating and revising the previous text to this new edition.

IFSTA Instructor Validation Committee

Chair

Russell Strickland
Maryland Fire and Rescue Institute
College Park, Maryland

Vice Chair

Terry L. Heyns
Lake Superior State University
Sault Sainte Marie, Michigan

Secretary

Peter Sells
Toronto Fire Services
Toronto, Ontario, Canada

Committee Members

Charles Anaya
Stillwater Fire Department
Stillwater, Oklahoma

Frank Cotton
Memphis Fire Training
Memphis, Tennessee

Cindy Fuller-Soule
North Carolina Office of State Fire Marshal
Boone, North Carolina

Jerome F. Harvey
Lead Fire Department
Lead, South Dakota

(Continued)

Committee Members (Continued)

Ronald T. Hiraki
Seattle Fire Department
Seattle, Washington

Lee Ireland
Lee Ireland Leadership Seminars
Egg Harbor Township, New Jersey

Alan E. Joos
Utah Fire and Rescue Academy
Provo, Utah

Dennis Katz
Tualatin Valley Fire and Rescue
Aloha, Oregon

Ed Kirtley
Guymon Fire Department
Guymon, Oklahoma

Jarett Metheny
Midwest City Fire Department
Midwest City, Oklahoma

Dana Reed
San Jose Fire Department
San Jose, California

Gregory E. Russell
DOD Louis F. Garland Fire Academy
Goodfellow Air Force Base, Texas

Brad Smith
Sunpro Fire Records Software
Ellensburg, Washington

Thomas E. Solberg
Lee's Summit Fire Department
Lee's Summit, Missouri

Jeff Tucker
Gainesville Fire and Rescue Department
Gainesville, Florida

Chuck Wilson
Vista Fire Department
Vista, California

Special recognition is given to Gloria Bizjak, primary author of the manual, as well as to Terry Heyns, Peter Sells, and Ed Kirtley who each authored a chapter. In addition to the contributions of committee members, several other key individuals assisted in writing or reviewing chapters and in advising or guiding our developers in special and technical areas. We would like to acknowledge the following individuals for their contributions and thank them for their contributions and support:

Richard D. Armstrong
Baltimore City Fire Department
Baltimore, Maryland

Joseph L. Byrnes
Anne Arundel County EMS/Fire/Rescue
Millersville, Maryland

William E. Hathaway
Emergency Health Services Department
University of Maryland Baltimore County
Catonsville, Maryland

Daryl Sensenig
Anne Arundel County EMS/Fire/Rescue
Millersville, Maryland

The following individuals and organizations contributed information, photographs, and other assistance that made the completion of this manual possible:

Larry Ansted, Dan Gross, and Robert Wright
Maryland Fire and Rescue Institute
College Park, Maryland

Jerome Harvey
Lead Fire Department
Lead, South Dakota

Terry L. Heyns
Lake Superior State University
Sault Sainte Marie, Michigan

Ron Hiraki
Seattle Fire Department
Seattle, Washington

Alan E. Joos and Bill Partridge
Utah Fire and Rescue Academy
Provo, Utah

Jarett Metheny
Midwest City Fire Department
Midwest City, Oklahoma

Dana Reed
San Jose Fire Department
San Jose, California

Peter Sells
Toronto Fire Services
Toronto, Ontario, Canada

St. John Ambulance Brigade
Toronto, Ontario, Canada

Jeff Tucker
Gainesville Fire and Rescue Department
Gainesville, Florida

Last, but certainly not least, gratitude is also extended to the members of the Fire Protection Publications Instructor Project Team whose contributions made the final publication of this manual possible.

Instructor Project Team

Project Manager
Barbara Adams, Associate Editor

Editor
Barbara Adams, Associate Editor

Technical Reviewers
Richard Hall, International Fire Service Accreditation Congress Manager
Carl Goodson, Senior Publications Editor

Proofreader
Marsha Sneed, Associate Editor

Production Coordinator
Don Davis, Coordinator, Publications Production

Illustrators and Layout Designers
Ann Moffat, Graphic Design Analyst
Desa Porter, Senior Graphic Designer
Connie Nicholson, Senior Graphic Designer

Graphic Design Technicion
Ben Brock

Production Assistant
Shelley Hollrah

Contract Artist
Mac Crank

Chapter 1

Challenges of Fire and Emergency Services Instruction

This text provides information and guides prospective instructors on becoming effective instructors. This chapter introduces the challenges of teaching the diverse group of people who come to the learning environment with a need to learn something that is relevant to and useful in their jobs and everyday lives. The challenge for instructors is to respond to learner needs while meeting the requirements of individual organizations or jurisdictions, professional standards, and other federal, state/provincial, and local laws. As these requirements and laws are changed to require safer operations and improved or more efficient services and skills, organizations will need to develop and revise programs that prepare personnel to apply new knowledge and skills on the job. With the constant changes in jobs, communities, and individual lives, people must be committed to lifelong learning — continually pursuing knowledge to adapt to these changes in life and work. Instructors are an integral part of the lifelong learning process. The decision to become an instructor who will be involved in program planning and instruction is a commitment to the responsibilities of fulfilling the needs of the learners, their organizations, and the laws and standards that direct change and training.

Instructors have many obligations. This chapter discusses those obligations and provides information and direction in such areas as the roles of the instructor, the characteristics of an effective and efficient instructor, how to effectively communicate with learners, the various challenges for instructors, the importance of and resources for professional development, and last, the importance of instruction to fire and emergency services organizations. References and supplemental readings and job performance requirements are given at the end of the chapter.

The Roles of the Instructor

[NFPA 1041: 2-5.5, 3-2.3, 3-2.5, 3-5.4, 4-2.3, 4-2.7]

Instructors belong to a noble and honored profession. Consider the meaning of the term: A *profession* is a calling that requires specialized knowledge and long, intense preparation that includes (1) learning scientific, historical, or scholarly principles of specific skills and methods; (2) maintaining high standards of achievement and conduct; and (3) committing to continued study and work — all with the prime purpose of providing a public service. Fire and emergency services instructors would agree that what they do falls into the category of a profession.

Communities have historically held the profession of instructor, or teacher, in high esteem — especially those who maintain a professional attitude about their profession. Instructors as professionals have a set of performance standards and ethics that are critiqued by their peers, similar to the professions of law and medicine. Instructors are professionals who meet a standard that is not based on working for pay or volunteering time but on a high level of performance. They are individuals who have developed authority and practical experience in an area of knowledge and skill.

This chapter emphasizes how the high esteem that instructors have earned is due not only to just what instructors know but also to who instructors are — the characteristics that justify such high regard from the community. The following sections describe the roles and characteristics of professional fire and emergency service instructors.

The Instructor Job

Instructors have the challenge of meeting many obligations in their role of fire and emergency service instructor. Their primary role is to plan and conduct training that meets performance standards and measures learner knowledge and skill with validity and consistency. Included in this task are the responsibilities of evaluating the training program and providing feedback to both supervisors and learners on the results of program evaluations (Figure 1.1). Feedback in the form of reports to supervisors and managers on program results and student performance is important for future program planning and equipment purchases. Formal, informational feedback in the form of grades, assessments, or performance evaluations is important to learners who will use the information as a guide for changing study habits, improving testing skills, preparing for lessons, and most importantly, for transferring learning to job performance (Figure 1.2).

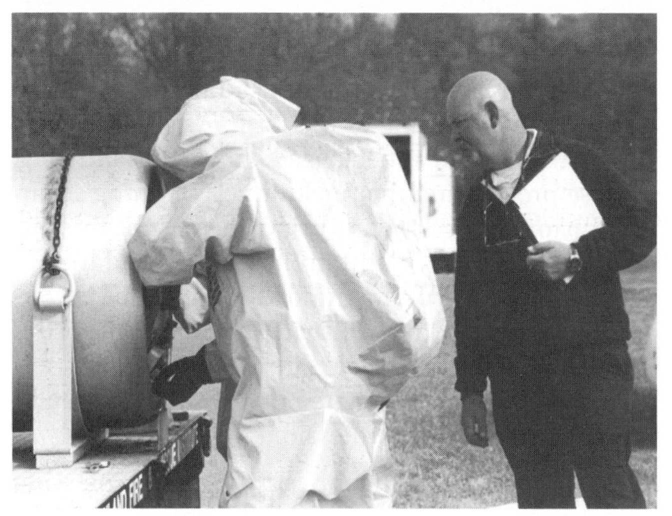

Figure 1.1 Using performance tests is a way for the instructor to receive feedback. *Courtesy of Larry Ansted, Maryland Fire and Rescue Institute.*

The Instructor's Role in an Organization's Success

The fire and emergency services field is a dynamic profession, is powerful, and is always changing. Change is a necessary part of this field as standards are revised, equipment becomes more technical, protocols are updated, and personnel retire or are promoted or replaced. Change requires constant reevaluation of training programs and personnel performance. Instructors have multiple roles in this changing environment. They research, observe, and participate in the planning, development, and implementation of programs by their organizations and by other organizations in similar service and safety areas. The instructor may be in the position to plan interorganizational programs and training sessions that broaden knowledge and skills of many levels of fire and emergency services personnel, enhance cooperation and understanding among personnel and organizations, and ensure current standards are adopted and applied.

In addition to planning, developing, and teaching programs, instructors must meet the expectations of both learners and their organizations. These duties include being effective communicators and good listeners, being able to present new knowledge and skills in a positive manner, and acting as role models and advisors. These multiple challenges are not always easy to manage, yet the instructor must be able to perform in all these areas in order to meet the expectations of both management and program participants.

Figure 1.2 Skills learned in the classroom must be transferable to the job. *Courtesy of Dan Gross, Maryland Fire and Rescue Institute.*

Canon of Ethics for the Fire and Emergency Services Instructor

The preservation of human life, the freedom from injury, and the protection of valuable property depend upon the careful and thorough execution of instructor duties. Instructors recognize the value of their work, whether performed on a full- or part-time basis, as a volunteer or as a career. In recognizing the importance of performing correctly, instructors pledge themselves to the highest standards of professional and ethical conduct, which includes a commitment to perform the following:

- Achieve and maintain professional competency.
- Advance instructor professional competency through networking and mentoring.
- Teach only those subjects that you are qualified to teach.
- Prepare for each presentation because a life depends upon it.
- Evaluate program results honestly.
- Work to continually improve programs and presentations.
- Use only current and accurate material, information, and statistics.
- Perform instructor duties with integrity.
- Respect the work of other instructors through the courtesies of crediting their ideas and materials appropriately and through compliance with copyright laws.
- Recognize when your conduct does not meet this canon of ethics, and resolve to improve.

The diversity of learner needs and experiences may be the instructor's greatest challenge. Each individual brings something different to the learning environment and expects something significant from the learning experience. From the time they enter school, learners' views of themselves and the world are shaped by their instructors. Every learner remembers a teacher who influenced his or her life, interest in learning, and job performance. The performance of instructors contributes to the overall success of their organizations.

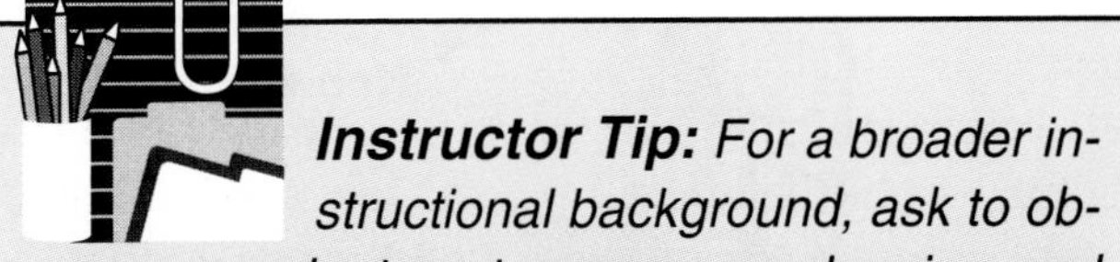

***Instructor Tip:** For a broader instructional background, ask to observe or volunteer to serve on planning and development committees. These experiences provide insight into training program needs and requirements, planning and budget processes, and decision-making methods.*

The Instructor as a Role Model

Fire and emergency service instructors have opportunities to positively influence the actions and ideas of personnel, beginning with new recruits or cadets and continuing through all the ranks and levels of fire and emergency services organizations. By their leadership abilities and knowledge, attitudes, actions, and examples, instructors are modeling and influencing performance. This influence goes beyond the classroom. As fire and emergency services personnel practice what they learn, they show the extent of their instructors' knowledge and experience, positive values, interest, and motivation.

Fire and emergency service instructors can positively affect learner self-esteem and create a desire to learn and a determination to succeed. An instructor's influence — that good or bad impression that affects learners' attitudes and actions — is lasting. An important part of being an instructor is to favorably influence and impress learners (Figure 1.3).

Instructors must act as role models. Actions indicate attitudes towards training. Interest and motivation must always reflect professionalism and commitment to learners and the organization. Instructors communicate the performance expectations of the organization to personnel through training situations and learning experiences, support the mission of the organization, and ensure the success of the organization by training personnel through quality instruction. The organization cannot meet its mission without well-trained personnel who are products of effective, committed instructors.

By now, it should be evident that as an effective, committed instructor, there are many roles and

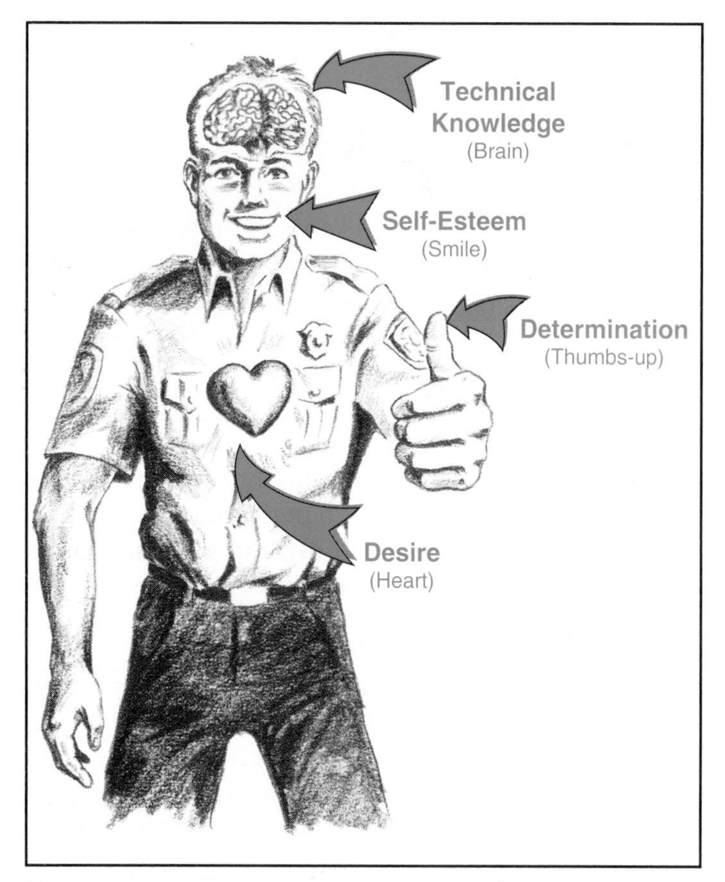

Figure 1.3 The instructor must give the learner more than just technical knowledge.

obligations to fulfill among learners, peers, organizations, and communities — roles that place an emphasis on instructors as role models to others. In the role of instructor, he or she is often the first person to influence individuals in a learning situation. Instructors are "on stage" and are watched by learners who want to find out whether they actually do what they say. As a role model for change, the instructor's responsibility toward the organization is to introduce and advocate new procedures and methods of the profession. Sometimes the instructor role as organization advocate may conflict with the role of learner advocate. In cases of conflict, instructors must communicate learner needs to supervisors for consideration while at the same time support organization requirements for learner consideration. Learners must see that instructors follow and support organizational requirements, but they also need to see that instructors understand and communicate learner concerns as well. This difficult role of communicating conflicts and advocating needs of both sides to enable understanding and develop a clear vision of ideas puts instructors in a unique position: a role model who fosters cooperation and helps support the organization's mission.

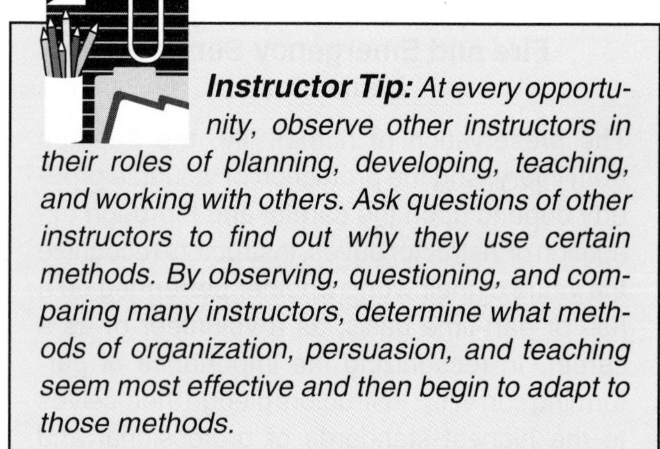

Instructor Tip: *At every opportunity, observe other instructors in their roles of planning, developing, teaching, and working with others. Ask questions of other instructors to find out why they use certain methods. By observing, questioning, and comparing many instructors, determine what methods of organization, persuasion, and teaching seem most effective and then begin to adapt to those methods.*

Communicating and Listening

In this unique position of role model, it is essential that instructors are effective communicators. Communication skills include much more than simply speaking in the classroom; communication is creating dialogue with and among learners so they may learn and share their needs and experiences, ideas and issues, debates and arguments, and attitudes and beliefs. Sometimes the results of these communications must be relayed to management for consideration and resolution. Listening is a key part of communication. An important communication process is through the evaluation reports and feedback among instructors, learners, and supervisors. The instructor's interest and skill in listening and transmitting ideas among all levels of personnel are important assets.

Characteristics of Effective Instructors

[NFPA 1041: 2-4.2]

Are there certain characteristics that instructors must have or develop in order to be effective? In past experiences as learners, instructors themselves had instructors who were able to maintain interest and stimulate motivation. Also, there were also likely some instructors who were not effective in teaching or providing guidance through knowledge and skills. Consider the characteristics of those instructors who were most effective in stimulating and facilitating learning experiences. Some of the qualities that come to mind are discussed in the following sections.

Leadership Abilities

Many believe that good instructors are natural leaders. But what makes a good instructor, and what does one do in his or her leadership position that causes others to follow? Good instructors lead or guide learners through the requirements, knowledge, and skills of a class while ensuring that the needs of each learner are met. Good instructors provide appropriate learning opportunities, examples, and ideas and encourage learners to discuss, think, and draw conclusions. A good leader also knows when to follow. Good instructors as good leaders enable learners to take the lead. Instructors will also find themselves in follower positions as well. Though they may not have had a hand in influencing and guiding policy-making decisions, they must always follow and promote organizational and administrative policies.

Instructor Attributes

The attributes of effective instructors cannot be scientifically analyzed, but certain personal qualities and behaviors contribute to their success and effectiveness. Think of instructors who have been influential and what they did to gain respect and admiration. Would they be considered role models? Will their best characteristics be imitated? Which characteristics discussed in the following paragraphs did these instructors have and what characteristics might be added (Figure 1.4)?

Ability to Understand and Work Well with People

Instructors must have a desire to understand the needs of learners, peers, and supervisors. Relationships are built on mutual respect, rapport, and confidence. Instructors must work well with others. They must be fair and impartial to all learners, and they must be open-minded and willing to hear, consider, and discuss the ideas of others (Figure 1.5).

Figure 1.4 The characteristics of the instructor enhance the effectiveness of the instruction.

Desire to Teach

Instructors can improve teaching skills through experience, study, and professional development, but unless they have a desire to teach, no amount of knowledge and experience can make them good instructors. Desire affects performance and motivation, which contribute to success in the classroom and on the job.

Competence in the Subject, Technical Skills, and Educational Methods

Instructors must have the background knowledge and experience to teach the subject and the skills, and they must have the ability to transfer that knowledge and experience to others. Learning is a lifelong process. It does not stop with the end of the class or the degree received in the program. Instructors must continually seek to increase their knowledge and skills in technical subject matter and educational methods. They must also be open-minded and consider or attempt to learn and understand alternative methods and ideas of others. It is important to add to and maintain skill ability and knowledge so that they may be passed on to learners. Instructors cannot effectively or convincingly teach or accept new skills or policies if they themselves have not learned about them or kept up with changes. Through continuing education and professional development, instructors can renew professional and career abilities, maintain skills, and improve instructional methods so they may continue to interest and motivate learners and coworkers. A responsible instructor seeks opportunities to pursue continuous

Figure 1.5 Instructors must have a desire to understand learner needs. *Courtesy of Jarett Metheny, Midwest City (OK) Fire Department.*

Figure 1.6 Skill and knowledge of the profession are essential, but good instructors must also know how to teach. *Courtesy of Dana Reed, San Jose (CA) Fire Department.*

improvement and maintain certification requirements rather than depend on a supervisor to track his or her progress (Figure 1.6).

Enthusiasm

The characteristic of enthusiasm is contagious. Instructors generate interest and enthusiasm in both learners and administrators alike when they show a high degree of interest and enthusiasm in a subject, in teaching it, in learners and the outcomes of their progress, and in the success of the program. Class becomes fun and exciting, and willingness to participate is high. Administrators will likely be supportive of instructor needs as a result.

Motivation

Instructors must have the motivation or the desire and determination to achieve goals and to stimulate that desire in their learners. To motivate others, instructors must give them a clear idea of what must be performed and how. Communicate to learners the end results, and show them how to get there. Show learners why they need this information. Make the steps easy, and allow for mistakes as learners practice and improve. Encourage learners as they try, reinforce correct tries, and redirect incorrect ones without embarrassing them. Instructors must show learners that they are motivated to help each individual learn by giving them every opportunity to do so.

Ingenuity and Creativity

An effective instructor understands that a technique suitable for one group of learners may not be suitable for another. Different audiences may need to learn or be approached through different instructional methods. Instructors can demonstrate ingenuity and creativity by developing or using various training aids and supplemental materials and by discovering and using innovative means of presenting information to meet the needs of every learner.

Empathy

Empathy is the ability to understand the feelings and attitudes of another. Instructors must be able to put themselves "in the shoes" of others and understand the learners' points of view, problems, or crises. Empathetic instructors have a sincere desire to help individuals learn, are not condescending or punitive, and do not act superior or threatening. Having empathy is especially important when working with learners who are having difficulty learning.

Mediation Skills

The instructor needs to act as a negotiator and mediator between learners and the organization and between the learners themselves. Instructors may have to mediate in situations involving disputes in the class, on evaluations and tests, with many types of personalities and responsibilities, and on a variety of other issues that may arise in the course of a program. As a mediator, an instructor listens to both sides and suggests solutions and may have to assist the sides in negotiating a solution. Remember, instructors work for both management and learners, so they are advocates for both parties. The instructor must work to create win-win situations and relationships among all parties.

Qualities to Emphasize

Instructors are constantly observed and evaluated by learners. The impressions that learners form of their instructors affect responses, enthusiasm, and motivation. If learners develop a negative attitude, learning is inhibited. As a past or current learner, instructors may recall classes in which their own instructors created unfavorable impressions. The following sections compare some positive and negative characteristics that affect learner impressions of instructors. As potential instructors develop instructional skills, they should avoid the negative pitfalls and emphasize the positive characteristics (Figure 1.7).

Honesty Versus Bluffing

Instructors must always be truthful and honest. No one knows all the answers, and learners realize this. Instructors are allowed to say *"I don't know, but I'll find out."* Learners want and expect honesty and want an instructor who is willing to find answers to their questions. Instructors do not need to be embarrassed if they cannot answer a question during class. But instructors who bluff their way through a question quickly lose credibility. Following are some guidelines for effectively handling this type of situation:

- If a question is appropriate to the lesson, promise to find the answer and do so.
- If a question is of interest to only the person asking, tell the individual where to find the information.
- If there is no exact answer to a question, refer the learner to related information.
- If a question refers to material that is in a later lesson, answer briefly and explain that it will be covered later. Make a note to refer to the question when the material is presented.
- If a question is pertinent to the lesson and of interest to the class, initiate a discussion and ask learners for their opinions.

Sincerity Versus Sarcasm

Attitudes and responses that show a sincere interest in helping learners learn are important characteristics. Learners react, respond, and cooperate more positively and willingly with instructors who demonstrate a concern with helping them reach program objectives. An instructor who uses sarcasm toward a learner places the entire class on the defensive. The emotional reactions of the learner and the class block effective communication. Instructors must not retaliate even if learners are sarcastic toward them. Evaluate the

Figure 1.7 Instructors must avoid negative teaching pitfalls and emphasize the positive.

situation, consider the circumstances, and deal with it in a mature and reasonable manner.

Solutions Versus Complaints

Few instructors work under ideal conditions, but complaining about the situation, especially to learners, accomplishes nothing and creates a negative impression. Learners have greater respect for instructors who present interesting classes with enthusiasm and optimism than for those who chronically complain. An instructor may have little control over the learning environment, so class time should not be taken by apologizing or making excuses for a situation. When preparing to teach a program, steps such as preplanning lessons, checking equipment and supplies ahead of time, arranging for appropriate assistance and props well in advance of the lesson, and always having alternate plans as backups will minimize problems. Even if the best of plans and arrangements are made and something still goes wrong, instructors can use other resources, creativity, their imaginations, or alternate plans. Make the best of a problem situation with a positive attitude for now, and plan to resolve the problem before the next session.

Inspiration Versus Intimidation

The instructor who views the trainer's role as a drill sergeant or an autocrat who gets results through threats and fear will not stimulate learners to reach their potentials. The instructor who gets the best results is the one who, as teacher and leader, inspires learners to learn, participate, and keep trying by using patient, positive, and reasonable methods of guidance and reinforcement. Learners cannot learn in an environment of fear, stress, and intimidation. Intimidating tactics, or bullying, is usually an expression of fear or frustration on the part of the one who intimidates. People react to intimidators with fear, contempt, and rebellion. The intimidating instructor earns no respect and stimulates no interest or motivation from learners.

Positive Humor Versus Offensive Humor

Humor can add emphasis to and create interest in a subject. It can also make learning fun and memorable and release tension in the classroom. Instructors must use it appropriately but cautiously. Some people have the knack or personality for using humor while others who attempt to be humorous often fail. Instructors must recognize which type of person they are and act accordingly. Often there are learners in the class who can see humor in a tough learning situation and can help other students realize the fun in the topic. Humor can lighten a difficult subject and make it easier to learn. Instructors should allow learners to express their personalities, but they must monitor the expressions to ensure that they do not offend others. Offensive or inappropriate humor, jokes, and stories that belittle or degrade anyone have no place in any training session. Constant humor can make learning ineffective, but absolutely no humor makes any learning situation dull. Instructors must find the appropriate balance between humor and the serious aspects of fire and emergency services training.

Communication

[NFPA 1041: 2-4.3, 2-4.4, 2-4.5, 3-4.2]

The effort people make to talk with each other and express ideas, give and get information, share knowledge, add gestures to expressions, or show pictures are all ways in which they communicate. Whether through

> ### *Case Study: Instructor Behaviors, Personality, and Effectiveness*
>
> You can recall instructors who were "good," whose classes you enjoyed, and who made you feel like you learned something. You can also likely recall instructors whose classes or teaching techniques were so boring you dreaded attending, but you had to because it was a certification or promotion requirement. What characteristics made the difference in the two types of instructors? Certain characteristics — such as attitude, behavior, and personality, as well as knowledge and ability to convey it — contribute to effective instruction. List and discuss characteristics that you think instructors should practice and exhibit that help in presenting instruction effectively. Listing characteristics is easy. Now, discuss ways in which instructors can achieve effectiveness. For example, if instructors must be good listeners, how can they perform good listening techniques? If instructors must be knowledgeable, how should they gain and maintain knowledge or remain current in their fields?

some simple or complex method, communication is a two-way process of sending and receiving a message, and there are many ways to communicate a message. A problem arises with communications when the message sent is not clear to the receiver. What causes misunderstandings? What steps can instructors take to ensure messages are received correctly? The following sections provide some positive guidelines for developing communication skills.

Instructor Tip: Check personal communication techniques by videotaping different presentations. Look for posture, gestures, expressions, and body movements; listen for voice inflections, loudness, quality, and enthusiasm. Do personal techniques help or hinder the presentations? What do others do in teaching or meeting situations? What attributes seemed effective? Which ones could be imitated?

Communicating Effectively

The ability to communicate effectively is an important quality of a good instructor. As mentioned in The Roles of the Instructor section, an instructor must have knowledge and skill as well as the ability to communicate them. Effective communication is more than just talking: It is a two-way process of telling and listening interactively. Effective communication takes practice and involves sharing ideas, attitudes, and abilities.

Instructors must provide opportunities for two-way communication and encourage all parties to participate in exchanging ideas for an effective communication process. The instructor, as facilitator of the communication process, also acts as negotiator, referee, and mediator as learners discuss, debate, disagree, and discover learning points in their efforts to build and arrive at logical conclusions. Through instructor facilitation, learners can understand the ideas of others — a vital point in the communication process. Learners must clearly understand information and be willing to ask questions and provide thoughtful responses.

The Communications Model

Instructors must practice and perfect communication skills to become effective and professional communicators. Teaching that is simply telling is a one-way action. Learning requires two-way action between senders and receivers, which includes feedback between both parties that creates understanding of the factors that may affect, influence, or inhibit learning. Individuals send and receive messages in many forms through numerous instructional media, but all communication forms exchange information to some extent. Regardless of the form used, the communication process consists of, and is influenced by, six essential elements:

- ***Sender*** — Information or idea transmitter
- ***Message*** — Information or idea
- ***Instructional medium*** — Method for transmitting the information or idea
- ***Receiver*** — Information or idea recipient
- ***Feedback*** — Receiver and sender responses that clarify and ensure understanding
- ***Environment*** — Surrounding factors that may distract or enhance, inhibit or promote, and encourage or discourage communication

People learn more effectively from interactions with other people than by passively watching videos, listening to lectures, or reading books or notes on a chalkboard/writing board. They learn by becoming involved in shared experiences, knowledge, and skills. To enhance and enable learning, instructors use a variety of tools to facilitate, guide, and coordinate the communication process. For example, to transmit information to learners, an instructor may present a lesson supported by various types of visuals, reference texts, journals, and other resources. To provide feedback, the instructor involves learners in discussions, activities, and evaluations that allow them to practice, apply, and judge their understanding of the knowledge and skills presented in the lesson. These multiple processes enable both instructors and learners to ask questions, practice skills, correct techniques, and verify levels of understanding and ability.

Senders and Receivers

The communications process begins when a person (the sender) encodes and sends a message to another (the receiver). *Encoding* is any process in which a person transforms a behavior or experience into words, pictures, symbols, or gestures from which another person can extract a meaning. The sender develops a message and selects a medium for transmitting it so that the receiver can understand or decode it. *Decoding* occurs when the receiver extracts meaning from the sender's words, pictures, symbols, or gestures. A basic communication model consists of these three components: sender, medium, and receiver (Figure 1.8a).

Instructor Tip: Recall a class in which an instructor primarily lectured and expected listening and extensive note-taking. Compare that class with one in which an instructor generated discussions, expected expressions of opinion, and encouraged questions. Which class had more participation, interest, application of information, and learning? Ensure that in every lesson there is a two-way exchange of information, not just a one-way message that consists of the instructor's viewpoint.

Communication is an active process that is more than just a message from a sender. Effective communication creates an understanding between sender and receiver. Each person in a conversation has a turn to act as both sender and receiver (Figure 1.8b). Feedback between receiver and sender gives the communicators an idea of how each is perceiving the other. It is important to realize that each may perceive the message differently because of factors in their backgrounds that act as filters to understanding content and intent. Background and experience, or what people already know, help them interpret new information they see and hear. Because everyone's experience is different, people will not always interpret a message in the same way. Different perceptions create unintentional barriers to communication and understanding. Senders must consider the attitudes, experiences, and personal backgrounds of receivers. Instructors must develop

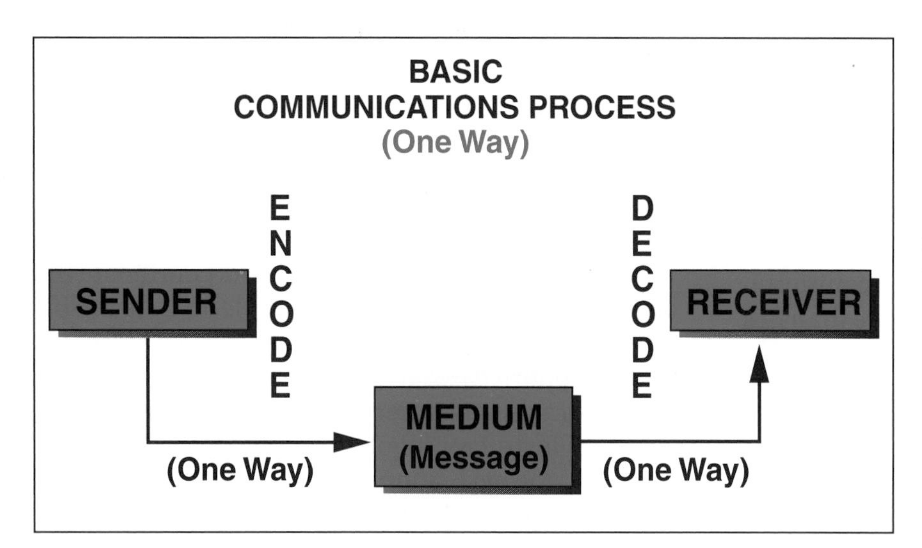

Figure 1.8a The basic components of a one-way communication model. *Courtesy of Maryland Fire and Rescue Institute.*

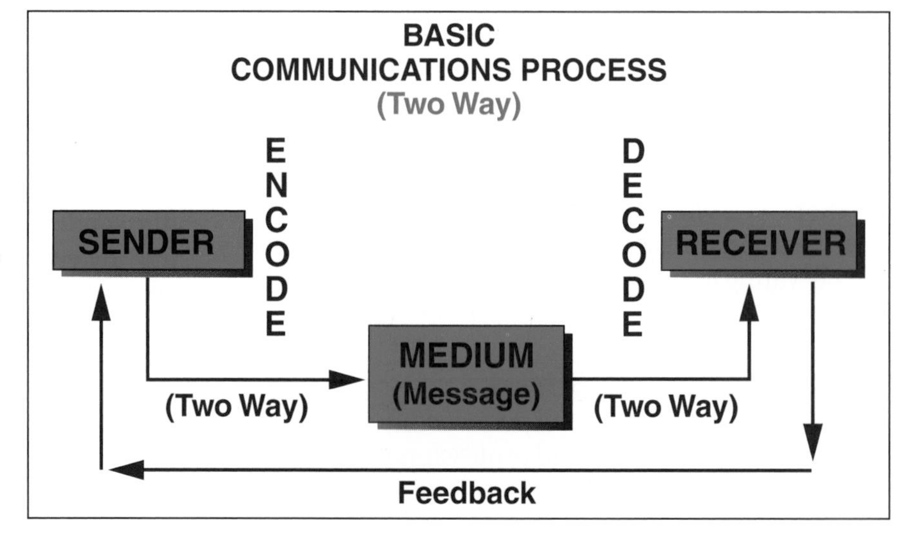

Figure 1.8b The effective communication process is a two-way exchange. *Courtesy of Maryland Fire and Rescue Institute.*

thoughts into understandable messages, send and receive feedback through appropriate instructional media, and consider the following filters that affect the messages of both parties (Figure 1.9):

- *Semantics* — Semantics is the study of the meaning in words and symbols. It refers to language, word meanings, and meaning changes due to context. Meanings are affected by individual background, knowledge, and experience.
- *Emotions* — Emotions are complex and often intense reactions accompanied by strong feelings. They can be powerful influences and can shift attention away from the intent of a message. Emotions may be expressed through nonverbal cues (gestures, voice tones, etc.).
- *Attitudes* — Attitudes are values and beliefs backed by emotions. Positive attitudes aid acceptance of messages; negative attitudes contribute to resistance to messages.
- *Role expectations* — Role expectations are learned ideas of how we expect ourselves and others to act based on our culture, social status, and job. Some individuals may not accept ideas from individuals perceived to be in "lower" roles or may resist ideas from individuals perceived to be in "higher" roles.
- *Nonverbal cues* — Nonverbal cues are messages without words that are often transmitted by gesture, posture or body language, eye contact, facial expression, tone of voice, and appearance.

The Message

The message communicated between the sender and receiver must convey the intended meaning and be in a form that is understood. This process is sometimes more difficult than it appears. The sender's message must clearly convey what is wanted or expected of the receiver. And the messages sent must be useful and necessary to the receiver. Useless information only clutters and confuses the communication process. The receiver must clearly understand what is expected and know that what is conveyed is important. If the message is unclear, the entire communication process falters.

There are many barriers to receiving and understanding a message clearly. Besides having different backgrounds, experiences, and knowledge levels, people may speak different languages and may misinterpret terminology. Instructors can help learners

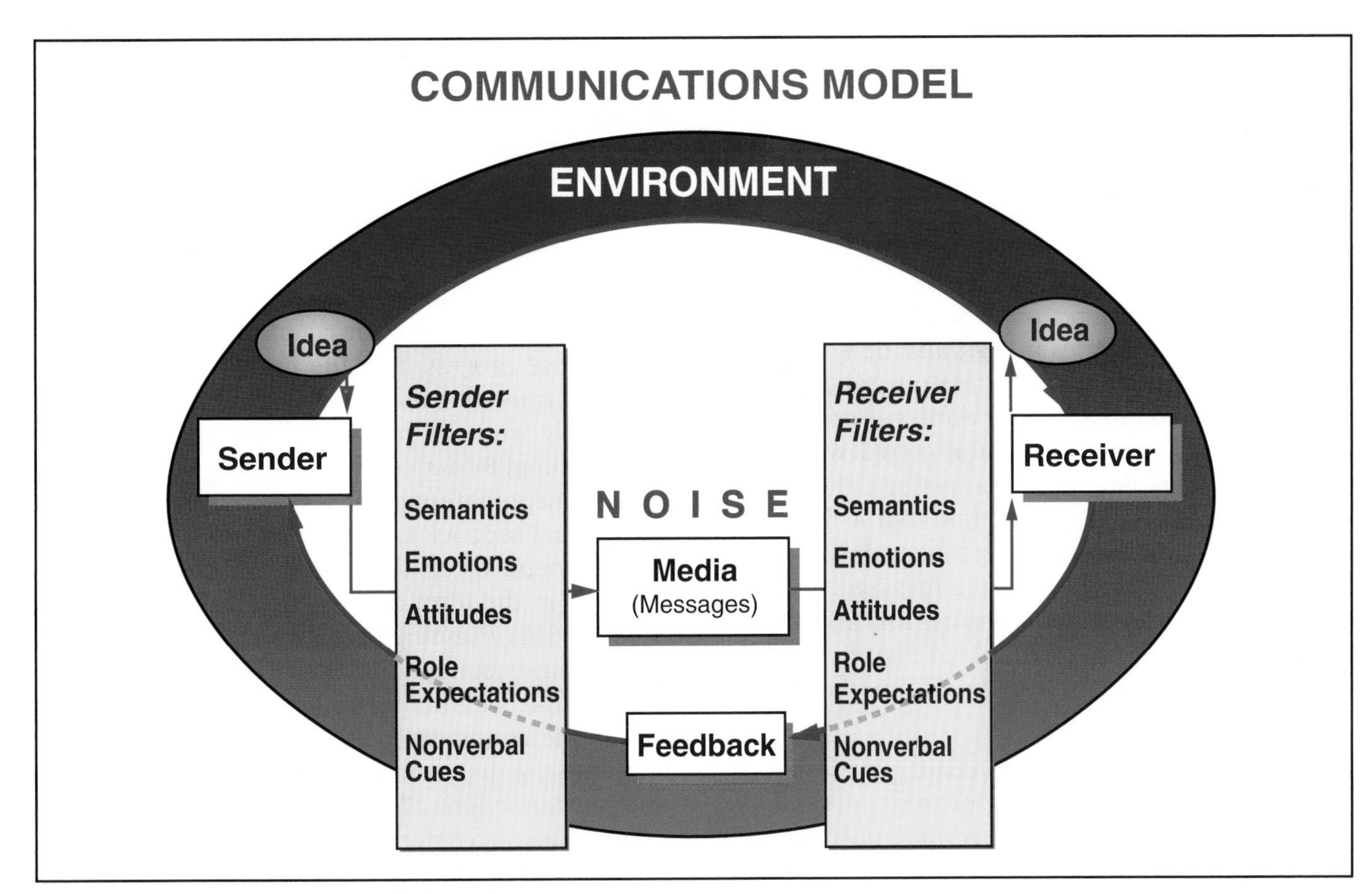

Figure 1.9 The complete communications model. *Courtesy of Maryland Fire and Rescue Institute.*

receive messages by using simple words in brief, uncomplicated messages. This type of communication is important in fire and emergency situations when excitement and activity levels are high, which may result in misunderstanding. The media through which messages are communicated are key considerations when instructors convey information to learners.

The Instructional Medium

The *medium* of communication is the method or format through which the message is transmitted between the sender and receiver. The selection of a medium for transmitting information depends upon many factors such as available time, purpose, language barriers, or the information being delivered. For example, it may be acceptable on the emergency scene to send a verbal, face-to-face message or to transmit a message by radio rather than try to send a written message. Other examples of instructional media include the following:

- *Physical action* or body language of an individual such as hand signals or gestures, facial expressions, or body postures
- *Visual displays* in a classroom such as bulletin boards, static displays, and models
- *Sounds* such as tone or volume of voices, alarms, buzzers, or music
- *Touch* such as gripping or manipulating a tool or piece of equipment

Combinations of instructional media can be used to send messages — *gestures* emphasize *words* that point to a *display* describing its function or purpose; computer *monitors* and *keyboards* enable individuals to manipulate *words, figures,* or *icons* to rearrange paragraphs in a *text* or figures in a *graphic* design. Often what an individual communicates nonverbally is not always received as intended, but words can add to how a message is sent, received, and interpreted. Words are usually the major ingredients of a message. The sender has a choice in selecting a method that is the most effective to convey the message; the result should be that the receiver understands the message. Using a variety of instructional media ensures that all learners receive the message in a format that they can understand — one that stimulates and maintains their interest in the communication.

The written form of communicating messages — written words, graphic displays, or common symbols — is a common and important instructional medium. In this form of communication, the receiver must visually identify the message, interpret the meaning, and respond or provide feedback that the message was understood. Examples of written communication must convey the information that the sender wants the receiver to understand, and it must be in a form appropriate for the message. Examples of written forms of communication include the following:

- Letters or other correspondence
- Instructional texts, handout materials, and words on slides or transparencies
- Symbols used in mathematical calculations, hazardous materials identification, and traffic control
- Training aids such as graphs, charts, and diagrams

Listening

Listening is an important part of the communication process. But how do instructors ensure careful and complete listening? First, what is listening? What is its definition? It isn't easy to define, and many researchers have been trying for years to create a definition that best describes the term. Andrew Wolvin and Carolyn Coakley (1985) developed the following definition: *"Listening is the process of receiving, attending to, and assigning meaning to auditory stimuli."* In plainer words, *listening* is a process of steps that gives information that listeners try to understand. Through some kind of *vocal sound* or *auditory stimulus,* individuals *hear* or *receive* information, *think about* it or *attend to* it (determine how important it is and decide whether to remember or forget it), compare it with what they already know, and *give* or *assign* it some meaning so that what they heard makes some sort of sense and they can recall it for later use.

The act of listening is just as important to the communications process as the act of speaking. Listening is an active process, and the individual who is not speaking must remain actively involved by listening attentively with the purpose of understanding and participating. By actively participating in listening, a person can prepare feedback in response to what he or she hears. Feedback may be a question, a comment, an argument, or an agreement and may be stated orally or in the form of a gesture, but the response validates that listening (as well as thinking in an attempt to understand) took place. The speaker is able to get some measurement of understanding and can plan to summarize and complete the discussion, add to or continue the discussion, or take the discussion in a different direction. Through listener responses, the speaker can also determine whether different tactics should be used in order to gain, stimulate, or redirect the listener's attention and interest.

The act of speaking to a group does not guarantee that the message was heard and understood as it was intended. As individuals listen, they try to link the new information to something they already know so they can recall it later. Researchers have shown that the average person takes in only a fraction of what is heard. The passive listener does not receive much information. For an individual to remember what is heard, that person must actively participate in the communications process and make attempts to understand what is heard. Several factors, discussed in the following sections, affect listening. Speakers must be aware of them and attempt to reduce distracting factors.

The Environment

One factor that may contribute to the listening process is the environment. Hearing is not the same as listening. Individuals hear many sounds and noises in their environment and may not pay attention to the sound to which they are supposed to be listening. Distracting sounds in and around the learning environment may gain the attention of listeners more than the sounds or messages of the speaker. Those individuals who "tune out" the speaker to attend to another noise or conversation miss the speaker's message or misunderstand its meaning. Understanding the message requires that the listener take an active role in the communication process — the listener must attend to what the speaker says.

A speaker can monitor the attention or active listening cues of the receivers. Answers to the following questions provide the speaker with feedback on how the message is being received — on how the receivers are listening:

- Do they have alert and inquisitive facial expressions?
- Do they have attentive body positions and postures?
- Do they respond with questions and comments?

The Speaker

Speakers themselves may distract listeners. To help listeners understand intent, speakers must give cues and context that make words and meanings clear. The instructor as speaker must realize that listeners are trying to interpret meanings. The instructor must make context clear, make comparisons to known things, and take the listener back to areas when words and meanings are unclear and try again. Words have different meanings to different people, so a speaker must attempt to make meanings clear, and the listener must attempt to understand the speaker's intent. When listening, the receiver must do more than just hear the words. The receiver must attempt to understand the words, interpret meanings, compare words and meanings to what is already known and understood, and apply the meanings to everyday knowledge and abilities. Chapter 7, Instructional Delivery, gives more information on presentation methods.

The Receiver

Sometimes it is the receiver who is a distraction in the listening process. Adults have many responsibilities with demands of jobs, families, friends, and personal needs. Often learners come to class wanting to be there but with many other things on their minds, and they are not ready to listen. Learners come to class for the purpose of improving themselves and/or their job prospects, but they also come with varying skills, knowledge, language and reading abilities, and attention spans. They want to be there but may interpret the speaker in many ways and at many levels of understanding. A challenge for instructors is to meet the needs of all learners while attempting to communicate with them at the same level. Instructors who plan to deliver simple messages with clear context and explanations will be effective and reach the majority of listeners. Feedback is also important: Listen carefully for learner comments. As instructors try to attend to all class activities, it is easy to become distracted and miss the importance that learners are placing on their questions, comments, and responses.

Speaking Techniques

A speaker can use a number of techniques to gain the attention of receivers and ensure they listen. An effective speaker is a good communicator, so instructors must learn, practice, and perfect speaking skills (Figure 1.10). See Chapter 7, Instructional Delivery, for some effective speaking techniques.

Positive techniques that deliver a clear message gain class interest and attention and stimulate active listening participation. Instructors must learn to recognize their own distracting techniques and work to eliminate them. Individuals who wish to be instructors should recall the positive and negative events of their own learning experiences. Avoid the negative and concentrate on enhancing the techniques that will be effective while speaking and communicating. Chapter 7, Instructional Delivery, gives more information on speaking.

date on current management styles in the workforce or on information and skills in the training arena. How did people perceive that kind of individual? It is likely that people complained about and had little or no respect for the person. Instructors can prevent that negative attitude toward themselves by continuing their efforts of lifelong learning through professional development. If individual states/provinces or organizations do not offer resources, check with local colleges and libraries. Professional journals and books report the latest research, provide resources for further study, and offer excellent articles from experienced professionals. Many colleges and journals also offer continuing education programs by correspondence and through distance learning technology. See Appendix A, Instructor Resources.

Schedule Training

Many organizations are "downsizing" and requiring that fewer personnel perform more duties for which they must be trained. Many fire and emergency services organizations are conducting more public service programs in addition to necessary daily duties and meetings. As organizations get busier, there is often less time for training, and it can become a low priority, but organizations must keep up with required training as well as attempt to fulfill desired training needs. Organizations must have a mission statement that describes goals and includes training as an integral part of successful operations. Instructors can perform needs analyses that can guide management in planning and scheduling appropriate and required training. See Chapter 11, Management and Supervision of Training, for more information.

Manage Funding and Resources

It seems that when federal, state/provincial, and local funds are short, training programs are the first to lose their funding, yet required training must continue. Reduction in revenues may have an impact on the resources available for training programs, and instructors must create new ways to achieve training goals. Some creative methods of meeting goals and maintaining the training program include (1) developing cooperative relationships with other training organizations, with industry, and with colleges; (2) seeking funding from new sources such as grants, business and industry sponsors, and donations; and (3) raising funds by providing training to the private sector. See Chapter 11, Management and Supervision of Training.

Recruit Qualified Instructors

Some organizations have a high turnover of personnel, and it is often difficult to fill a vacant position in a timely manner with someone who has the experience, ability, and qualifications of a quality instructor. Instructors often work closely with other instructors and usually know who is effective and fits the needs of the organization. This type of "inside" knowledge or sharing of the qualities and qualifications of other individuals is called *networking*. It is a valuable asset when developing cooperative programs and looking for new personnel.

Assets within the organizations are often overlooked. Watch for instructor qualities among internal personnel and consider developing those individuals into quality instructors. For small or nongrowing organizations, looking for, developing, and using potential internal resources are essential elements for maintaining a quality organization. See Chapter 11, Management and Supervision of Training, for more details.

Professional Development

[NFPA 1041: 2-4.3, 2-4.4, 3-4.2, 4-3.3, 4-3.4]

Because the fire and emergency services profession is dynamic and constantly changing, instructors must always be aware of new improvements or developments. Competent instructors look for opportunities to learn and improve their professional, technical, and instructional skills (Figure 1.11). Continuing education contributes to lifelong learning and expands the instructor's professional "tool box." There are several ways in which instructors may continue to learn and expand their professional abilities and credibility. Some of these ways are as follows:

- Maintain instructor skills by teaching classes and seminars or by facilitating discussion groups and workshops.
- Belong to a professional organization.
- Network with other instructors through local, state/provincial, or national/international instructor associations and with instructors in other organizations.
- Earn professional certification as an instructor.
- Maintain management skills such as those used in fire command and supervision by attending professional development seminars and reading professional journals.
- Maintain technical skills such as those used in fireground operations, hazardous materials inci-

Figure 1.11 Instructors must remain current on changes and advances in their profession. Some examples are self-study and attending workshops and conferences. *(left) Courtesy of Jarett Metheny, Midwest City (OK) Fire Department. (right) Courtesy of Larry Ansted, Maryland Fire and Rescue Institute.*

dents, rescue situations, and emergency medical service (EMS) operations through drills, workshops, and other continuing-education programs.

- Develop the ability to criticize constructively as a learning tool rather than as a personal indictment about another's abilities or personality.
- Develop an open-minded willingness to hear and consider the ideas of others before condemning or criticizing them.

Organization or certification standards may require that instructors pursue a minimum number of hours of continuing education, but it is the responsibility of instructors to maintain those requirements. Responsible instructors who foster the listed qualities are valuable assets to the organizations that hire them and to the trainees who must learn the requirements of the organization and the skills of the profession through them (Figure 1.12).

Networking

Networking with other fire and emergency services professionals and instructors is a valuable resource. Joining a professional association gives an instructor the opportunity to meet other professionals. These associations exist at local, state/provincial,

Figure 1.12 Instructors must take pride in and achieve personal satisfaction from their jobs. *Courtesy of Toronto Fire Services.*

national, and international levels. Refer to Appendix A, Instructor Resources, for a list of several well-known associations. The list is not all-inclusive and is not intended to endorse any association. There are many organizations and associations that provide a wealth of information on professional and training issues.

Through these associations, instructors can find information on a variety of topics, including safety during training evolutions. Associations can provide the latest information based on experience and research through formats such as articles, seminars, conferences, workshops, web sites, and electronic mail (e-mail).

International Organizations

Organizations such as the International Association of Firefighters (IAFF) and the International Association of Fire Chiefs (IAFC) are important sources of safety information. Many international government organizations provide resources and information on safety issues. Some of these organizations include the World Health Organization (WHO), the International Red Cross (IRC), and the International Civil Aviation Organization (ICAO). WHO recommends specific protocols to follow during emergency medical operations; the IRC helps people prevent, prepare for, and cope with emergencies including those involving blood supplies, disasters, tissue transplants, social services, or health and safety. The ICAO addresses aircraft crash and rescue protocols. These resource organizations can provide extensive information that will enrich an instructor's knowledge and provide credible documentation for following certain safety procedures during fire and emergency services evolutions such as live fire training and confined space training.

Importance of Instruction to the Organization

[NFPA 1041: 4-2.3, 4-2.7, 4-5.3, 4-5.4]

A unique aspect of the fire and emergency services profession is that the consequences of wrong actions can be devastating. Fire and emergency services professionals often face potentially life-threatening conditions and must be able to react to those conditions quickly, effectively, and safely. The safety of these professionals, and those whom they rescue, depends on the quality of instruction that they receive through their organizations. Providing quality training and instruction are critical qualities for successfully meeting the mission of all fire and emergency services organizations. Success lies in the ability of personnel to meet customer needs in an effective, efficient, and safe manner. Personnel must be well-trained in all the skills required for the job to meet these needs. The instructor's job is to prepare employees to meet the requirements of the organization. Effective training is accomplished when instructors take the following actions:

- Ensure the safety of personnel who must meet the requirements and demands of the job.
- Provide effective services to customers and the community through organization personnel.
- Motivate personnel to achieve their best while engaged in the mission of the organization.
- Provide challenges for personnel to overcome in training by teaching them how to choose the right people with the right tools and equipment for the given situations and to perform to the organization's requirements.
- Reduce organizational and personal liability through well-planned programs that follow standards and develop skilled personnel who are less likely to injure customers and other personnel or damage customer property or organization equipment. See Chapter 3, Legal Considerations, for more information on liability.
- Promote creation and maintenance of training programs by developing policy recommendations to support new and ongoing training programs.
- Provide the organization's administrators with unbiased reports and well-supported evaluations, conclusions, and recommendations for training programs.
- Encourage organizational self-assessment by developing evaluation plans that measure training objectives and meet organizational policies.
- Create evaluation plans for the purpose of course improvement that include learner input for evaluating instructors, course components, and facilities.

References and Supplemental Readings

Cross, K. Patricia. *Adults as Learners.* San Francisco: Jossey-Bass Publishers, 1981.

Davies, I. K. *Instructional Technique.* New York: McGraw-Hill, 1981.

Ireland, L. "Negotiating Skills for Instructors." *The Voice,* International Society of Fire Service Instructors (ISFSI), April, 1993.

Ireland, L. "Using Humor in the Classroom." *The Voice*, International Society of Fire Service Instructors (ISFSI), August, September/October, 1994.

Knowles, Malcolm. *The Modern Practice of Adult Education.* Chicago: Follett, 1980.

Merriam, Sharan B. and Rosemary S. Caffarella. *Learning in Adulthood.* San Francisco: Jossey-Bass Publishers, 1991.

Morse, Henry. *Teaching in the Fire Service.* Albany, NY: Delmar Publishers, 1988.

Powers, Bob. *Instructor Excellence.* San Francisco: Jossey-Bass Publishers, 1992.

Wolvin, Andrew D. and Carolyn Gwynn Coakley, *Listening.* Dubuque, IA: Wm. C. Brown Publishers, 1985.

Job Performance Requirements

This chapter provides information that addresses the following job performance requirements of NFPA 1041, *Standard for Fire Service Instructor Professional Qualifications* (1996 edition). Colored portions of the standard are specifically covered in this chapter.

Chapter 2 Instructor I

2-4.2 Organize the classroom, laboratory or outdoor learning environment, given a facility and an assignment, so that lighting, distractions, climate control or weather, noise control, seating, audiovisual equipment, teaching aids, and safety, are considered.

2-4.3 Present prepared lessons, given a prepared lesson plan that specifies the presentation method(s), so that the method(s) indicated in the plan are used and the stated objectives or learning outcomes are achieved.

2-4.4 Adjust presentation, given a lesson plan and changing circumstances in the class environment, so that class continuity and the objectives or learning outcomes are achieved.

2-4.5 Adjust to differences in learning styles, abilities and behaviors, given the instructional environment, so that lesson objectives are accomplished, disruptive behavior is addressed, and a safe learning environment is maintained.

2-5.5 Provide evaluation feedback to students, given evaluation data, so that the feedback is timely, specific enough for the student to make efforts to modify behavior, objective, clear, and relevant; include suggestions based on the data.

Chapter 3 Instructor II

3-2.3 Formulate budget needs, given training goals, agency budget policy, and current resources, so that the resources required to meet training goals are identified and documented.

3-2.5 Coordinate training record keeping, given training forms, department policy, and training activity, so that all agency and legal requirements are met.

3-2.6 Evaluate instructors, given an evaluation form, department policy, and job performance requirements, so that the evaluation identifies areas of strengths and weaknesses, recommends changes in instructional style and communication methods, and provides opportunity for instructor feedback to the evaluator.

3-4.2 Conduct a class using a lesson plan that the instructor has prepared and that involves the utilization of multiple teaching methods and techniques, given a topic and a target audience, so that the lesson objectives are achieved.

3-4.3 Supervise other instructors and students during high hazard training, given a training scenario with increased hazard exposure, so that applicable safety standards and practices are followed, and instructional goals are met.

3-5.4 Analyze student evaluation instruments, given test data, objectives and agency policies, so that validity is determined and necessary changes are accomplished.

Chapter 4 Instructor III

4-2.3 Develop recommendations for policies to support the training program, given agency policies and procedures and the training program goals, so that the training and agency goals are achieved.

4-2.7 Present evaluation findings, conclusions, and recommendations to agency administrator, given data summaries and target audience, so that recommendations are unbiased, supported, and reflect agency goals, policies, and procedures.

4-3.3 Design programs or curriculums, given needs analysis and agency goals, so that the agency goals are supported, the knowledge and skills are job related, the design is performance based, adult learning principles are utilized, and the program meets time and budget constraints.

4-3.4 Modify an existing curriculum, given the curriculum, audience characteristics, learning objectives, instructional resources and agency training requirements, so that the curriculum meets the requirements of the agency, and the learning objectives are achieved.

4-5.3 Develop course evaluation plan, given course objectives and agency policies, so that objectives are measured and agency policies are followed.

4-5.4 Create a program evaluation plan, given agency policies and procedures, so that instructors, course components, and facilities are evaluated and student input is obtained for course improvement.

Chapter 2

Safety: The Instructor's Role

Safety is a key issue and an important responsibility of all those involved in fire and emergency services — instructors and trainees, management and staff, and office and field personnel. Everyone at all levels of fire and emergency services must be aware of safety requirements. The most important aspect of safety is that everyone must support, practice, and enforce all requirements. Risks to fire and emergency services personnel can be reduced if everyone involved takes a proactive role in safety. Instructors, in particular, must ensure that all training meets federal, state/provincial, and local regulatory obligations and supports the goals of the organization.

This chapter gives information and ideas on the role of the instructor and the importance of teaching safety, supporting and enforcing safety policies, and teaching program participants to look for and correct potential hazards. Methods for preventing and managing accidents are given along with resources for safety guidelines and regulations. With these issues come specific responsibilities for instructors. Among these responsibilities is the importance of knowing and following the organization's guidelines and policies and the regulations of federal, state or provincial, and other local agencies. References and supplemental readings and job performance requirements are given at the end of the chapter.

The Safety Concept

[NFPA 1041: 2-4.2, 2-4.2.1, 2-4.5.1, 2-4.5.2, 3-4.3, 3-4.3.1]

Fire and emergency services personnel make life and death decisions on almost every response. Fire and emergency responses often involve operating equipment in risky environments such as inclement weather, fire situations, and hazardous materials incidents. Decisions and actions are based on training that was modeled and reinforced by instructors who were responsible for demonstrating procedures and guiding practice in safe learning conditions. When safety is emphasized and learned in training, it tends to be remembered and practiced on the job. The instructor is the initial role model for safety and must take that role seriously. Instructors cannot just mention safety guidelines and expect they will be followed. Instructors must demonstrate and reinforce these guidelines.

Following safety guidelines or plans developed or adopted by training organizations has a significant impact on reducing injuries and deaths of fire and emergency services personnel in training and on the emergency scene. Fire departments conduct emergency operations under an emergency management system that provides an organized structure for operational coordination and effectiveness as well as for safety aspects of emergency situations. Instructors must be familiar with established incident management and safety procedures and integrate them into training exercises.

Instructor Tip: *Be sure you are aware of your organization's safety policies and guidelines for each emergency services training program that you teach. If you don't have a copy, get one. If you are not sure how to implement these policies and guidelines, talk with other instructors. If a policy is in place and you, as an instructor, do not follow its guidelines, you may be held liable for injuries to program participants.*

Reinforcing Safety

Safety is an issue that must be emphasized, and instructors must plan and devote appropriate time in every session to cover all areas of safety. These areas include requirements or procedures for assuring safety, safeguards and equipment used for preventing accidents, and procedures for managing and investigating accidents.

An instructor is in a position to influence trainees and other personnel in using proper safety procedures. Because instructors constantly interact with personnel in planning and presenting programs, they set the stage and provide role models for implementing, enforcing, and following safety requirements (Figure 2.1).

Accident prevention and safety procedures begin before personnel arrive at an emergency scene where conditions may contribute to personnel injuries. Safety awareness and practice begin in training sessions. In classroom lessons, instructors must describe possible hazards and explain cautions, proper equipment use, and safety rules before the participants begin to practice a skill (Figure 2.2). When *planning* manipulative training programs, instructors must look for potential hazards in the sessions and eliminate them. Instructors must also plan to address the precautions to take to prevent injury while training, train learners to recognize job hazards, and teach them how to control or eliminate these hazards. These steps help ensure an acceptable level of risk in a risky profession and prevent injuries.

When *conducting* manipulative training sessions, instructors must appoint an individual who acts as a safety officer or monitor. The job of this individual is to ensure that all safety guidelines are followed, that appropriate personal protective equipment is used, and that equipment is used properly. Another important step to take before starting every manipulative skill session is to review the safety rules and equipment-use guidelines. To ensure that all learners understand these rules and guidelines, provide them in writing, and read the written rules and guidelines aloud as the learners read them. Some organizations require that participants in a training program sign a paper stating that they have read and understood these rules and guidelines.

Fire and emergency services programs often train nonfire or nonemergency services personnel. Though these participants may not be required to perform all aspects of the curriculum on their jobs, instructors must enforce all safety procedures as these personnel participate in the program. It is still important that they follow safety guidelines even though all training procedures may not pertain to or affect them. Safety must be stressed and made a high priority from the beginning of the program (Figure 2.3). In the position of role model, instructors will demonstrate safe procedures that learners

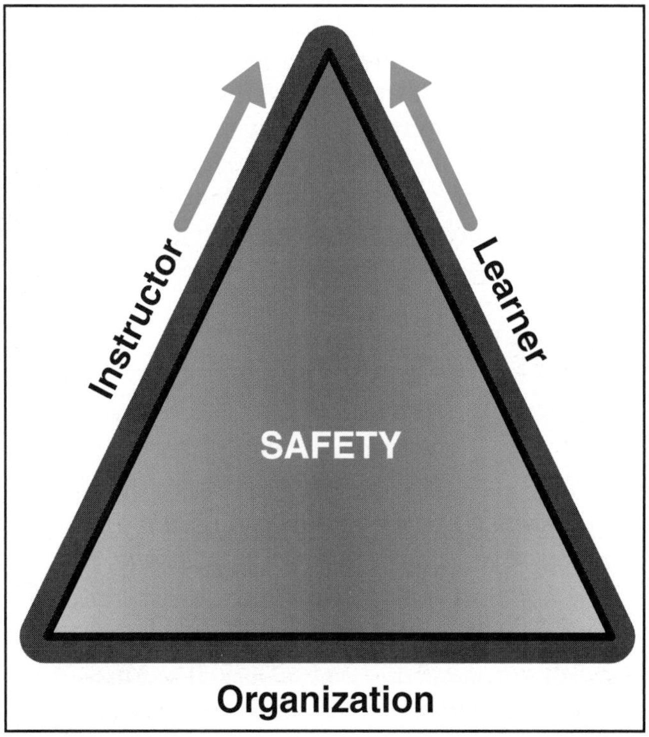

Instructor
Learner
SAFETY
Organization

Figure 2.1 Instructors can have a positive impact on safety in the fire and emergency services.

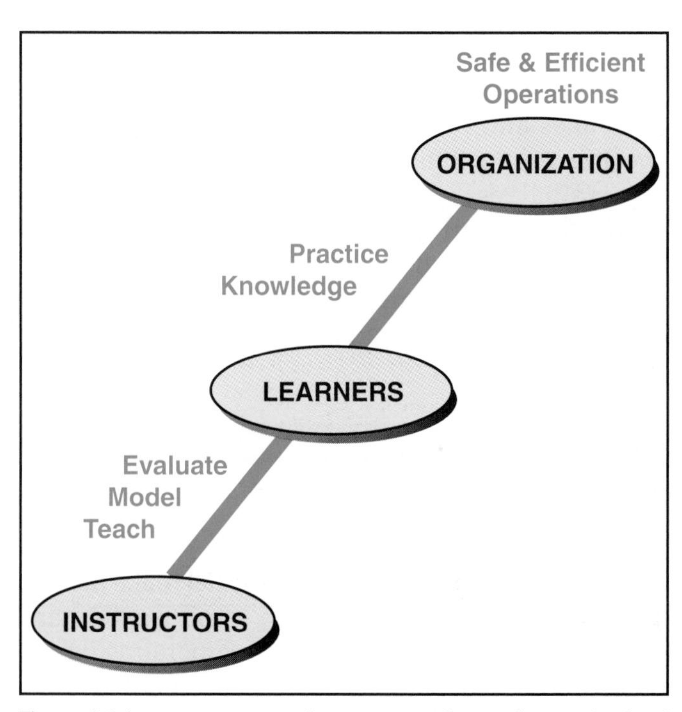

Figure 2.2 Instructors prepare learners to perform safe organizational operations.

will imitate and incorporate into their skills so they become habits (Figure 2.4).

Safety Plan

All organizations must have safety policies and procedures to follow in the event of an accident during classroom and drill ground exercises. As instructors begin all programs and manipulative skill sessions, they should first discuss and explain the organization's safety policy. In some organizations, the chief or supervisor presents the safety policies to new trainees or personnel, which emphasizes management's support and demonstrates that instructors are authorized to enforce the policies. In case of an accident, instructors must know the procedures to follow and explain them to learners. Learners need to know what signals are given if an accident occurs, what to do when the signal is given, what steps to take to help the injured, and whether training stops or proceeds. Parts of the safety plan include what personnel to contact, what duties to perform or assign, and what types of documentation to complete and maintain.

Instructors must also make learners aware that investigation procedures do not cast blame but determine why the accident occurred. The purpose of an investigation is to research and examine the events that caused the accident so future accidents can be prevented and training procedures can be reevaluated and revised for safety effectiveness. Often, personnel and training participants who are involved in or witness an accident are afraid to give information, thinking that they may "get someone in trouble." But the situation cannot be resolved nor can future situations be prevented if valuable information is withheld. The instructor's role, then, is to explain all points of the organization's safety policy, ensure that all safety precautions are followed, describe all steps that must be taken in the event of an accident, and emphasize the positive purpose of an investigation (see Accident Investigation section).

Figure 2.3 Safety must always be a primary concern. Following the manufacturer's operating instructions is important. In this photo, the shift officer reviews the operation of the pressure-relief valve. *Courtesy of Terry L. Heyns.*

Preventing and Managing Accidents

[NFPA 1041: 2-4.2, 2-4.2.1, 2-4.5, 2-4.5.1, 3-4.3, 3-4.3.1]

The first step to safe training programs is accident prevention. The learning environment, whether inside or outside, must meet safety codes and be free of structural problems and natural or man-made hazards that can potentially cause injury. Preventing accidents means preventing the loss of personnel and time, not to mention the time required for the myriad

Figure 2.4 Safety procedures must become habits to fire and emergency responders to prevent injuries, property losses, and deaths. *Courtesy of Larry Ansted, Maryland Fire and Rescue Institute.*

of paperwork that must be completed to report an incident for review, evaluation, and determination of methods for preventing a recurrence in a similar incident.

Managing accidents is the next step for safe training. Accident management includes identifying accident factors, investigating accidents that do happen, and analyzing those accidents to look for causes and trends. Then recommendations are made for accident prevention.

Instructor Tip: *Before starting class sessions, check the environment for safety hazards, and alleviate them. At inside training sites, look for unstable or broken chairs, tables, and desks; look for loose carpeting and cords that individuals can trip over, frayed wires, overloaded sockets, and plugs without grounds. At outside training sites, look for debris, rocks, holes, or uneven areas that could cause falls and injuries; look for unstable building parts, such as walls, chimneys, and flooring, that could collapse. What other hazards might be present in both environments?*

Prevention

An *accident* is a sequence of unplanned or uncontrolled events that produces unintended injuries, deaths, or property damage. Accidents are usually the result of unsafe acts by persons who are unaware or uninformed of potential hazards, are ignorant of the safety policies, or fail to follow safety procedures. Accidents may also be caused by conditions in the physical environment that were not examined or considered as potential hazards. The majority of accidents are predictable and preventable (Figure 2.5). The occurrence of an accident indicates that someone failed to plan, someone performed in an unsafe manner, or an unsafe condition existed. Instructors can reduce the potential for accidents by planning for safety and maintaining control over the event. Accident prevention steps include carefully planning for training events, having appropriately trained personnel assist in supervising the event, informing learners of safety policies and potential hazards, and modeling and reinforcing safety policies and procedures.

Accident Factors

After the accident, it is often easy to clearly see the cause and wonder why anyone would create or tolerate unsafe mechanical or physical conditions or take a risk and perform an unsafe act. There are many factors that cause individuals to disregard risks, all of which can be controlled. Some accident factors include the following:

- ***Management*** — Oversight, omission, or mismanagement, which affect other factors
- ***Situation*** — Facilities, tools, equipment, or vehicles unavailable, in disrepair, or not used
- ***Environment*** — Noise, vibration, temperature extremes, poor lighting, and moisture that affects concentration or ability to function
- ***Human*** — Individuals and their experiences, ignorance or innocence, and self-assurance

While all factors are significant and contribute to every other factor, the human factor plays a major part in accidents. Analyses of industrial accidents find that accidents do not distribute themselves by chance but happen frequently to some people and infrequently to others. A reason for this occurrence is the human factor. Some individuals fail to learn from experience or instruction how to control known hazards, and they become involved in accidents more frequently (Figure 2.6).

Three human factors contribute to accidents. Before allowing participation, an instructor must determine whether any of the following apply to the individual learner:

- ***Improper attitude*** — The instructor must attempt to readjust faulty attitudes or personalities so that the individual does not create or become involved in an accident. Is the individual
  - — Irresponsible or reckless?
  - — Inconsiderate or uncooperative?
  - — Fearful or phobic of the situation?
  - — Egotistic or jealous?
  - — Intolerant or impatient?
  - — Excitable or oversensitive?
  - — Obsessive or absentminded?
- ***Lack of knowledge or skill*** — The instructor can correct these defects by providing proper training and an appropriate opportunity to practice under supervision. Is the individual

Figure 2.6 Accidents can be caused by human factors. *Courtesy of Dan Gross, Maryland Fire and Rescue Institute.*

Figure 2.5 Most accidents are predictable and preventable.

Figure 2.7 *Before* operating equipment, individuals must understand the operating instructions so that potential for personal injury, as well as equipment damage, is reduced. *Courtesy of Terry L. Heyns.*

— Sufficiently informed about the training?

— Capable of interpreting the training and convinced of its need?

— Experienced in prerequisite knowledge and skills and capable of decisive actions? (Figure 2.7)

— Properly trained and able to recognize potential hazards?

- ***Physical limitations*** — The instructor may not be able to help learners overcome all physical constraints but can guide individuals to seek professional assistance. In many cases, an instructor can work with individuals to improve their confidence and abilities. Before doing any physical activity, an instructor can lead participants through some simple warm-up activities that include muscle stretching and strengthening exercises. An instructor who is not familiar with how to perform these properly and safely should consult the organization's health unit or physician for guidance. Is the individual

— Able to see and hear well enough for the situation?

— Able to safely perform a task based on height or weight?

— Limited by aerobic capacity or strength?

— Affected by a medical condition, allergy, or illness?

— Limited by a physical or mental condition?

— Affected by substance abuse causing reduced reaction times?

Especially important to instructors is any condition or factor that could lead to the injury or death of a participant. Undoubtedly, good training plays an important role in enhancing safety while responding to, acting on, and returning from any fire or emergency scene. The twofold concern of safety on the fire or emergency scene and safety on the training ground both depend on teaching, enforcing, and reinforcing behaviors that participants must learn, adapt, and practice at all times.

The National Fire Protection Association (NFPA) publishes articles and reports based on collected statistics that show firefighter death and injury trends in both emergency and training situations. An update from NFPA's Fire Analysis and Research Division shows that there were 157 deaths (including 10 in training) in 1977 and 95 deaths (including 8 in training) in 1997. During these 20 years, the total death rate rose to a high of 171 in 1978 (training deaths rose to 17 in 1987) and fell to 75 in 1992 (training deaths fell to 3 in 1979 and 1984). Washburn, LeBlanc, and Fahy (1998) give further details on these trends over the past 20 years in their article, "Firefighter Fatalities." The article gives information on the death and injury figures by type of duty and by cause and nature in all duty areas (fireground, responding/returning, nonfire calls, training, and other) and shows some fluctuation over the years but a gradual downward trend. The authors also discuss causes for the trend and suggest actions for continuing the reduction of deaths and injuries.

For the most part, this positive trend is due to improved engineering: better protective clothing and equipment, safer fire apparatus, better training facilities, and better incident management. But engineering accomplishments have their limits; the remaining improvements must come from behavior changes. Some critical areas need ongoing attention and enforcement. For example, the following topics are critical to fire service training organizations and must be emphasized in their programs:

- ***Driver education*** — Firefighters and emergency responders must pay more attention to and practice safe driving habits, including controlling speed, observing traffic rules, and ensuring that everyone on the apparatus or in the vehicle is wearing a seat belt.
- ***Health and fitness*** — Better health and fitness can result in additional reductions in deaths of fire and emergency services personnel. Screen applicants and current personnel by using NFPA 1582, *Standard on Medical Requirements for Fire Fighters*, continue to screen personnel so they meet fitness requirements throughout their careers, and test health annually.
- ***Personal alert safety system (PASS) devices*** — Train all firefighters and rescue personnel in the use of PASS devices. Properly maintain and test PASS devices just as other protective gear components are tested.
- ***Incident management plan*** — Include in the plan an accountability system to track personnel by location and function so the incident commander knows where personnel are operating and what they are doing so they can be rapidly located if a PASS alarm is activated.

Can training and education programs reduce deaths and injuries among fire and emergency responders? Certainly, training and education programs play an increasing role in the overall efforts to reduce both deaths and injuries. Driver training can address many pertinent and current concerns. Health and safety courses specifically targeting critical issues are already available. More research and course development are needed to continue the downward trends of death and injury rates. Instructors can contribute to downward trends by being aware of the importance of stressing and enforcing safety in every training program, especially when conducting live fire or other practical training evolutions (see Chapter 8, Practical Training Evolutions).

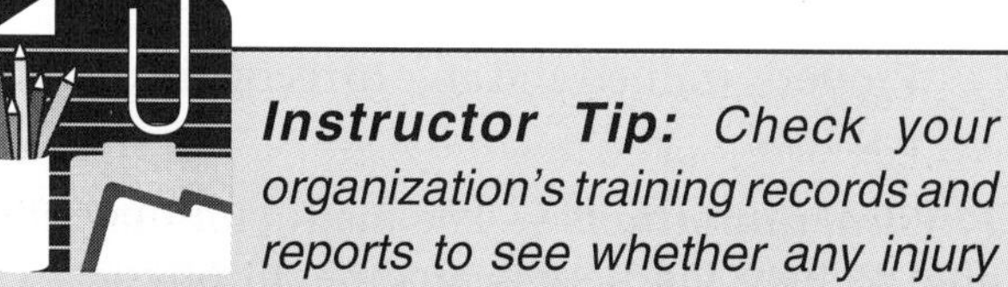

Instructor Tip: *Check your organization's training records and reports to see whether any injury trends occurred during training over the past five to ten years. From the information given, can you determine whether an injury was due to faulty equipment or to improper behavior? Based on the information or conclusions, were new safety guidelines implemented? What steps (that you may have been unaware of before) will you now take to teach learners appropriate safety behaviors?*

Accident Investigation

Accidents usually occur through a logical and predictable sequence of events that must be investigated so organizations may take steps to prevent their reoccurrence. Investigations are *fact-finding* but not *fault-finding* procedures. There will always be certain pressures involved in accident investigations, but they must be conducted with objective determination and free of personal feelings toward management or those involved in the incident. The purposes of accident investigation include the following:

- Avoid loss of human resources and equipment.
- Ensure better cost effectiveness in the use of personnel and equipment.
- Improve the morale of both the department's personnel and members of the community.
- Determine the change or deviation that caused the accident.
- Determine hazardous conditions to which fire and emergency services personnel may be exposed.
- Direct the attention of management or officers to the causes of accidents.
- Examine facts as if they will have a legal impact on accident cases

Instructors usually work with safety officers in conducting investigations of accidents that occur during training. The instructor and the safety officer will likely be on the scene of the incident and are most able to provide detailed information.

Accident Analysis

After investigators collect accident data, they review and analyze it to look for causes and make recommendations for prevention. Included in the process of accident review and analysis is the examination of current and past records for significant areas that may have been overlooked in recognizing a potential accident and preventing its occurrence. Keeping records of activities becomes an important responsibility of instructors and fire and emergency services training organizations. Careful record keeping often reveals minor situations that indicate the need for corrective action. For example, a scratch is not disabling, but if it was received from a piece of equipment that was placed or used in an awkward position or without protective equipment, the incident points to a need for corrective action. Investigators reviewing records will note the number of scratches reported in similar situations and follow the record review with research, interviews, and discussions that will lead to recommendations for appropriate changes.

While reviewing records for types and frequency of injuries, investigators can list and categorize injuries in order to prioritize hazards and take appropriate corrective action. Consider the following two factors when prioritizing hazards (Figure 2.8):

- Performance frequency of the hazardous activity and how it directly relates to the accident or injury frequency
- Relative severity of the potential loss

For example, a frequency performed task may have the risk of causing a minor hand injury if the task is not performed correctly. It may be a higher priority to address this hand-injury hazard than to address an infrequently performed task that has the potential for a more serious injury. Ideally, all hazards are addressed as high priorities for training, but hazard prioritization is often a reality and a necessity imposed by limited training resources and a common-sense approach to eliminating injuries.

Analyzing circumstances or conditions surrounding accidents can enable an investigator to perform the following actions:

- Identify and locate principal sources of accidents by determining the materials, machines, or tools most frequently involved with accidents and job-producing injuries.
- Disclose the nature and size of the accident problem in different operations.
- Indicate the need for engineering revisions by identifying the unsafe conditions of various types of equipment.
- Identify problems in operating procedures and processes contributing to accidents.

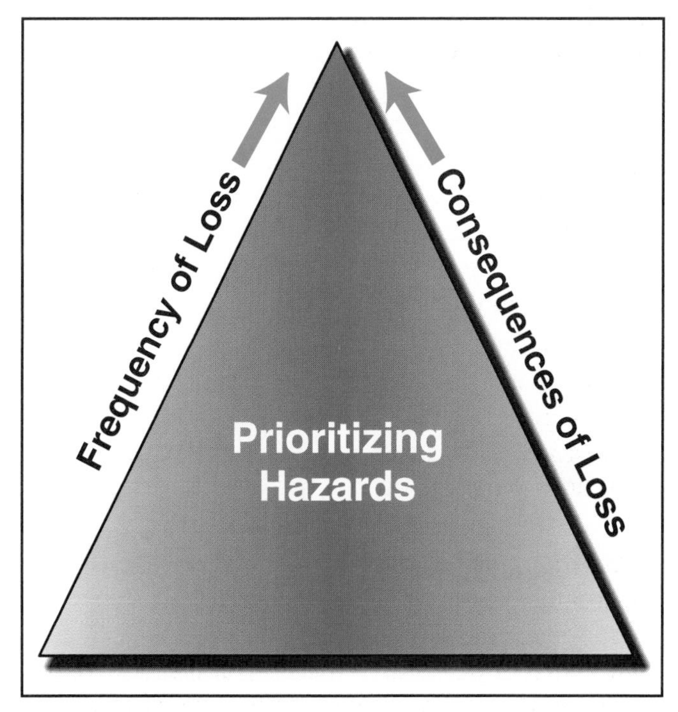

Figure 2.8 Prioritizing hazards: Consider two factors: frequency of loss and consequences of loss.

- Disclose unsafe practices that might suggest or direct additional training.
- Identify personnel placement where inability or physical handicap contributes to accidents.
- Enable supervisors to use time, personnel, and resources more effectively by providing information about hazards and unsafe practices in their operations.
- Permit an objective evaluation of the safety program's progress.

Accident analysis can also reveal the need for additional training in specific procedures. Individuals may act with good intentions yet not understand the consequences of their actions. In many cases, individuals who perform unsafe acts are not aware of the safety factor until an accident occurs. Under these conditions, individuals can become involved in accidents that result in injuries, deaths, or property damage. The instructor's role is to ensure that all training is conducted under the safest conditions possible (Figure 2.9). The instructor looks at all training exercises with safety in mind from the beginning to the end and also makes learners aware of the need to look for accident potentials.

> ***Instructor Tip:*** *Be sure you are familiar with every piece of equipment you use in training. Locate and read the manufacturer's instructions completely, train the program participants in the appropriate operations, demonstrate the safety and maintenance features, and supervise the participants as they begin operations. Do not guess at operations or try to operate equipment without learning how to use it appropriately. Improper use leads to broken or malfunctioning equipment and increases the potential for injury.*

Resources for Safety Guides and Regulations

[NFPA 1041: 3-4.3, 3-4.3.1]

Instructors may find it difficult to keep up with all the new safety regulations and guidelines and the revisions to old ones that government agencies and organizations create, but they can contact several reliable resources for information. Knowing the resources and how to contact them (or being on their mailing lists) are valuable assets for any instructor.

Figure 2.9 Training instructors are responsible for ensuring safety on the training ground. Instructors must observe operations and stop and correct learners to prevent injuries. *Courtesy of Larry Ansted, Maryland Fire and Rescue Institute.*

Government Agencies

Several United States government agencies are responsible for developing, regulating, and ensuring safe workplace policies. The two primary agencies concerned with safety in the workplace are the National Institute for Occupational Safety and Health (NIOSH) and the Occupational Safety and Health Administration (OSHA). Both agencies were created by the Occupational Safety and Health Act of 1970. The responsibilities of NIOSH include investigating, researching, and evaluating safety and health hazards in the workplace. The responsibilities of OSHA include setting and enforcing workplace safety and health standards. OSHA has the authority to issue citations and fines to enforce safety and health standards.

There are both federal and state OSHA regulatory agencies that enforce worker protection standards for employees. In some states, called federal OSHA states, the federal government enforces OSHA regulations and covers private and federal firefighters in nonstate-plan states. Many states have their own occupational safety regulations that are enforced by a state agency. See Appendix B for a list of state-plan states and nonstate-plan states. Coverage for volunteers varies from state to state. State plans must equal or exceed federal requirements. It is important for instructors to know which regulatory organization has jurisdiction over their operations. Occupational safety and health regulations significantly influence fire and emergency training. These regulations vary among jurisdictions, therefore, instructors must know and use the applicable regulations to ensure safety. To determine specific regulations that apply to instructional situations, contact state training organizations or other state agencies.

Safety Resources and References

When instructors develop programs or plan lessons, where do they find reliable and accurate information? How do they know that the information is current? Many resources and references are available, and many professional organizations and associations can ensure that instructors have credible sources.

State/Provincial or Local Safety Agencies

States, provinces, and local safety and occupational agencies often have review and enforcement functions. For example, states and provinces may fall under the guidelines of an enforcement organization whose rules and regulations must be observed during a training evolution or exercise.

In countries with a federal structure, states, provinces, or local agencies may have to follow regulations that may differ from, expand upon, or exceed national rules. They may also have to follow additional regulations that are not covered at the national or federal level. Instructors and safety officers must know the regulations of all federal, state or provincial, and local agencies, how they differ or contradict each other, and which must be followed to what extent.

Instructors should not limit their inquiry to regulations for fire and emergency services. Instructors must obtain copies and follow all applicable regulations. State/provincial and local agencies may have safety regulations written for other industries or organizations that apply to fire and emergency services training. For example, regulations for construction workers on a roof may be applicable to firefighters training for vertical ventilation. In addition to enforcement inspectors, many regulatory agencies have consultants or educators who will review pertinent safety regulations with instructors who are planning a training program.

National Fire Protection Association

NFPA is one of many recognized organizations that develops safety standards and guidelines that many training organizations adopt. Government and other agencies can adopt NFPA standards as their guidelines for safety compliance.

Sample List of NFPA Standards

- ***NFPA 1041, Standard for Fire Service Instructor Professional Qualifications —*** Identifies the professional levels of competence required of fire service instructors; includes requirements for conducting high-hazard training using applicable safety standards and practices
- ***NFPA 1402, Guide to Building Fire Service Training Centers —*** Lists guidelines to follow when building training facilities, including burn buildings, smoke buildings, and combination buildings, and when conducting outside drill ground activities
- ***NFPA 1403, Standard on Live Fire Training Evolutions*** — Lists guidelines to follow during live fire training evolutions, including information on structures, fuel materials, safety, instructors, and reports and records; also provides guidelines for conducting outside evolutions (not in structures), including information on site preparation, safety, instructor-to-student ratio, and postburn activities
- ***NFPA 1561, Standard on Fire Department Incident Management System —*** Contains the minimum requirements for an incident management system to manage all emergency incidents; also used to meet the requirements of NFPA 1403 1-2
- ***NFPA 1500, Standard on Fire Department Occupational Safety and Health Program*** — Contains information on the guidelines for fire departments to follow in order to ensure the health and safety of firefighters and help prevent accidents and health problems
- ***NFPA 1521, Standard for Fire Department Safety Officer —*** Contains minimum requirements for the assignment, duties, and responsibilities of a health and safety officer and an incident safety officer for a fire department or other fire service organization

Instructors must be aware of and familiar with NFPA standards that relate to safety and guide the performance of live fire training evolutions and other hazardous training. Standards are reviewed and revised periodically, and often new standards are created from existing ones. Some of these standards are briefly described in the list on page 31, but instructors must check with their organizations for the most current standards and to see which ones apply to their organization, personnel, and instructional responsibilities.

American Society for Testing and Manufacturing (ASTM)

ASTM is an organization that has developed and published numerous technical standards that are used by fire and emergency service organizations. It also develops and conducts training in the performance, use, and application of ASTM standards, many of which apply to equipment used by fire and emergency services personnel. Its technical committees research and develop standards to cover areas such as hazardous materials, protective equipment, footwear, occupational health and safety, building materials, search and rescue, and many other areas that affect personnel in fire fighting and rescue operations. Product safety has been a major factor in the reduction of training injuries, but this result is only possible when training programs emphasize and follow all safety guidelines.

Other Sources

As discussed in Chapter 1, Challenges of Fire and Emergency Services Instruction, instructor associations and organizations provide information on safety and networking opportunities. Instructors do not need to feel alone when faced with developing a program or preparing for a class. Instructors have many resources through a variety of local, state/provincial, national, and international associations and organizations. These organizations will often also provide information and sources at the request of instructors who are not members but who need information for their programs.

References and Supplemental Readings

Bachtler, Joseph R. and Thomas F. Brennan (eds.). *The Fire Chief's Handbook* (5th ed.). Saddle Brook, NJ: Fire Engineering Books and Videos, A Division of Pennwell Publishing Company, 1995.

Crapo, William F. "The Treadmill." *Fire Engineering,* August, 1997, pp. 105-110.

Case Study: Safety on the Training Ground

Instructors are gathering and inspecting the property and the equipment they will use for a combined training exercise with police, fire, and rescue personnel. The scenario describes a partial collapse of a seven-story building under construction in a busy downtown area that is being renovated. Inclement weather had weakened poured concrete supports, and the top floor collapsed, blowing out lower supporting structures, bricks and blocks, and wooden framework. There are multiple casualties including construction workers, bystanders and passersby, and occupants of nearby vehicles and shops.

To simulate a collapsed building and a disaster scene (and to have space to gather personnel and units), the instructors are using an abandoned and decomposing warehouse in an old, little-used industrial section of town near the riverfront. The training organization has permission from the city and the site owners to use the property.

Answer the following questions:

1. Think of similar areas in your jurisdiction or in cities you may have visited. List and discuss some hazards that the instructors may expect to find in the described location.
2. What types of injuries could occur to participants if these hazards are not cleared away?
3. Regarding some of the hazards you may think of, such as waste product containers or used tires, what special arrangements must the instructors or their organization make to dispose of these items? What agencies in your jurisdiction would you contact?
4. If you were to plan a similar exercise in your jurisdiction, what possible alternative locations could you use? Are these locations safer? Are they more convenient? In what ways?
5. Does your organization have safety guidelines for selecting training sites? Have you had the opportunity to use them? After reading this chapter, do you think the guidelines are reasonable? Why or why not?

Krieger, Gary R. and John F. Montgomery (eds.). *Accident Prevention Manual for Business and Industry* (11th ed.). Itasca, IL: National Safety Council, 1997.

National Safety Council. *Public Employee Safety & Health Management.* Itasca, IL, 1990.

Washburn, Arthur E., Paul R. LeBlanc, and Rita F. Fahy. "Firefighter Fatalities." *NFPA Journal,* July/August 1998, pp. 50–62.

Job Performance Requirements

This chapter provides information that addresses the following job performance requirements of NFPA 1041, *Standard for Fire Service Instructor Professional Qualifications* (1996 edition). Colored portions of the standard are specifically covered in this chapter.

Chapter 2 Instructor I

2-4.2 Organize the classroom, laboratory or outdoor learning environment, given a facility and an assignment, so that lighting, distractions, climate control or weather, noise control, seating, audiovisual equipment, teaching aids, and safety, are considered.

2-4.2.1 *Prerequisite Knowledge:* Classroom management and safety, advantages and limitations of audiovisual equipment and teaching aids, classroom arrangement, and methods and techniques of instruction.

2-4.5 Adjust to differences in learning styles, abilities and behaviors, given the instructional environment, so that lesson objectives are accomplished, disruptive behavior is addressed, and a safe learning environment is maintained.

2-4.5.1 *Prerequisite Knowledge:* Motivation techniques, learning styles, types of learning disabilities and methods for dealing with them, methods of dealing with disruptive and unsafe behavior.

2-4.5.2 *Prerequisite Skills:* Basic coaching and motivational techniques, adaptation of lesson plans or materials to specific instructional situations.

Chapter 3 Instructor II

3-4.3 Supervise other instructors and students during high hazard training, given a training scenario with increased hazard exposure, so that applicable safety standards and practices are followed, and instructional goals are met.

3-4.3.1 *Prerequisite Knowledge:* Safety rules, regulations and practices, the incident command system used by the agency, and leadership techniques.

Chapter 3

Legal Considerations

Instructors and their organizations are held legally responsible for their actions. Therefore, instructors must be knowledgeable about the law and follow and enforce standards and policies. When performing routine duties, providing information, and demonstrating skills in training, instructors must perform diligently and correctly. Incorrect performance or failing to perform normal duties on the job and in training situations can have a legal impact that can result in loss of assets and reputations of individuals and organizations. Instructors may also be liable for the actions of former learners who perform improperly on the job because of improper instruction while training (Figure 3.1).

Instruction, then, is not simply teaching and demonstrating what one knows to interested learners. Instructors must also be aware of legal terms and issues and comply with organizational regulations when involved in planning, preparing for, delivering, and evaluating instruction.

This chapter provides an overview of legal issues that are important considerations for instructors who have a basic responsibility for the end performance of learners. In addition, the chapter reviews the origins of the law, lists pertinent laws and policies, explains types and categories of laws, explains torts and liability, discusses legal protections, discusses substance abuse, and reviews copyright and permission issues. References and supplemental readings and job performance requirements are given at the end of the chapter.

Overview of the Law

[NFPA 1041: 2-5.2, 2-5.2.1, 3-2.5, 3-2.5.1]

Most instructors have little, if any, legal background and like most people do not think about the legal system until a legal issue must be addressed or resolved. If the issue is beyond a simple traffic violation, an individual may wish to consult an attorney. Instructors must become aware of the laws, ordinances, and regulations that have an influence on their organizations and training programs. The following sections provide some general background information and definitions on the origin of laws, types of laws, and categories of law (public and private). This information helps in understanding where an instructor's legal responsibility lies within an organization and where the organization's legal responsibility lies within the legal system (Figure 3.2).

Origin of Laws

Most laws are based on federal, state, or provincial constitutions that provide principles of law for a governing system. Constitutions guarantee that people are protected from government abuse, and they protect the rights of minorities from arbitrary actions of those in power. Constitutions provide the basic law to which all others must conform. This law cannot be repealed or overruled by statutes, regulations, or ordinances.

Types of Laws

The term *law* is used broadly and commonly to cover many legal concepts. There are three types of laws: *legislative, administrative,* and *judiciary.* The terms apply to the three different jurisdictions that create the laws.

Legislative Law

Legislative laws are made by federal, state/provincial, county/parish, and city/township legislative bodies. Legislative bodies have broad power to enact any statute as long as it has some reasonable relationship to protecting the health or general welfare of the public.

Figure 3.1 Instructors have a responsibility toward the fire and emergency services and the individuals with whom they work.

Figure 3.2 Instructors must be aware of the legal aspects of fire and emergency services programs. *Courtesy of Jarett Metheny, Midwest City (OK) Fire Department.*

This broad power is often called *police power,* and with it comes the authority to regulate.

It is important that instructors become familiar with the legislative laws in their states/provinces and jurisdictions. Some examples of United States (U.S.) legislative laws are as follows:

- ***Title VII of the Civil Rights Act*** — Covers equal opportunity employment and education practices to prevent discrimination and prohibits sexual harassment
- ***Americans with Disabilities Act (ADA)*** — Provides for reasonable accommodation for learners with disabilities
- ***Privacy Act or Buckley Amendment*** — Prevents disclosure of personal or learner information to unauthorized parties
- ***Affirmative action policies*** — Required by federal statutes and regulations that are designed to remedy discriminatory practices in hiring minority group members

States/provinces create statutes that become statutory laws. *Statutory laws* regulate matters from subtle to complex such as state/provincial taxes, industry, commerce, and professions. Examples of state/provincial statutes are as follows:

- Driving laws for fire and emergency services vehicle operation
- Protection policies such as worker's compensation acts and Good Samaritan laws
- Professional certification requirements such as those for physicians, nurses, paramedics, and lawyers.

Municipal corporations, such as those formed by counties/parishes, cities, or townships, often have their own local needs and create laws (ordinances) that cover matters beyond federal or state/provincial laws such as local leash, loitering, or litter laws. An *ordinance* is a local law that applies to persons and things in a jurisdiction. It is an act created by a city council or local government body and has the same force and effect as a statute. Jurisdictions enact ordinances to regulate local needs such as zoning, highway speed, parking, trash disposal, and other community issues.

Administrative Law

Administrative or regulatory agencies, such as the Occupational Safety and Health Administration (OSHA), create administrative laws or regulations. A *regulation* is a rule or similar directive issued by an administrative agency. These agencies have authorization to issue and enforce their directives. Agencies that are authorized to issue regulations follow certain steps such as giving prior notification of the action in a public record and providing an invitation and opportunity for public comment. Regulatory agencies work for the executive branches of government and create laws that regulate areas such as safe operations of certain machinery (for example, breathing apparatus, oxygen delivery systems, and forklifts) in the workplace. Certification of legal, educational, medical, and allied health personnel are also regulated by administrative agencies such as the following:

- ***Board of Examiners*** — Professional certifications
- ***Board of Bar Overseers*** — Lawyer certification
- ***Board of Education*** — Teacher certification

These agencies have authority to enforce their regulations just as states/provinces and cities/townships have the authority to enforce their laws and ordinances. Specific regulatory agencies, such as the Federal Communications Commission (FCC), have authority to control and supervise particular activities of public interest. The FCC administers laws that regulate access to communication facilities such as fire and emergency services communications channels.

Judiciary Law

Judiciary law is made by judges and is called *judicial legislation* or *judge-made-law.* This type of law is established by judicial precedent and decisions rather than by statutory or administrative laws. A more familiar term for this type of law is *common law.* U.S. common law is based on precedent from English common law. Common law is based on principles rather than rules and does not consist of absolute, fixed, and inflexible rules but on broad principles of justice, reason, and common sense. The principles are determined by the social needs of the community and change as needs change with the progress of society. This type of law is also called *unwritten law* because it is not written in statutes. Each case is based on previous similar cases and generally guides society in its actions. When a new case is decided, a precedent is set, and successive similar cases are based on this precedent. In common law, precedents are regarded as the major sources of law. Precedents may involve novel questions of common law or be interpretations of statutes. Interpretation of a common law is sometimes unclear until a complainant brings an issue to court and it is decided by a judge. Examples of common law are those that cover issues of patient consent and negligence.

Public/Private Laws

Laws guide society's actions and conduct so everyone feels safe and has just recourse for wrongs. There are two categories of law: *public* and *private.* See Table 3.1. A person can be charged under both categories because they are separate systems. This chapter concentrates on the issues of private law, specifically torts (see Torts section).

Laws and Policies

[NFPA 1041: 2-5.2, 2-5.2.1, 3-2.5, 3-2.5.1, 4-2.7]

Instructors are not typically lawyers and are usually limited in legal knowledge, but they must be aware of laws that affect training; they must do more than plan lessons and teach. Their actions in the classroom affect trainees' actions on the job and in relationships with employers. Many instructors are employed by training organizations, which means that all employment-related regulations apply to them. Instructors must be aware of state/provincial, county/parish, local, and organization laws and regulations that affect learners and training.

How does an instructor become aware of which laws affect instruction and management of learners? Most organizations' policies are based on federal, state/provincial, and local laws that require compliance and are not, as some instructors believe, created by the whim of an autocratic employer. Employing organizations should make their policies clear when hiring and orienting instructors, long before any legal issue occurs and legal action is required. Usually, the employing organization informs and orients their instructors to the laws that must be followed, and it likely hands out organizational and training regulations so that instructors may guide learner actions. It is the instructor's responsibility to ask for, study and understand, and adhere to any laws and regulations that affect training programs. Some of these laws and policies are described in the following sections.

U.S. Federal Laws

There are several U.S. federal laws that can have an impact on employment and instruction. Brief descriptions of some of the most significant U.S. federal laws are given in the following sections.

Civil Rights Act

The Civil Rights Act of 1964 is a federal act that was passed to amend statutes passed after the Civil War. Its purpose is to provide stronger protection for rights guaranteed by the U.S. Constitution such as preventing discrimination in public accommodations.

Title VII—equal employment opportunity (EEO)—of the Civil Rights Act of 1964 outlaws employment practices that discriminate based on race, color, reli-

Table 3.1
Public and Private Law Systems

Public Law	Private Law
Issues concerning the Constitution:	***Issues concerning individuals, companies, businesses:***
• Federal	• Contracts
• State/Province	• Domestic disputes
• County/Parish	— Child custody
• City	— Death with no will
Administrative:	***Torts*** (requires proof with a preponderance of evidence):
• Regulatory Agencies	• Civil wrongs between individuals
Criminal (requires proof beyond reasonable doubt):	• Intent is to compensate victim
• Penalties for accused (jail, fines, death)	
• No compensation to victims for loss	

gion, sex, or national origin. This act created the Equal Employment Opportunity Commission (EEOC). See the Equal Employment Opportunity Laws section for details on these laws.

Title VII also protects female employees from sexual harassment, but recent lawsuits on discrimination on the basis of gender have applied to both men and women and include both verbal and physical harassment. *Harassment* occurs when a superior purposely exercises authority in a manner that is unnecessarily oppressive and implies malicious and discriminatory actions. The initial intent to protect women from sexual harassment was due to the fact that women held few, if any, positions of workplace authority over men. But in recent years, women have risen to supervisory, management, and authority levels and are in positions to intimidate and harass male subordinates. Often, the authority figure is in a position to withhold pay and promotion, make working conditions difficult, or fire or demote the employee.

Americans with Disabilities Act

The Americans with Disabilities Act of 1980, which is similar to the 1973 Rehabilitation Act, defines a *disabled person* as one who has a physical or mental impairment that limits one or more "life activities," has a record of such impairment, and is regarded as having the impairment. This act prohibits discrimination against a qualified individual with a disability in application, hiring, advancement, discharge, compensation, job training, and other terms, conditions, and privileges of employment. A *qualified individual with disability* is a person with a disability who, with or without reasonable accommodations, can perform the essential functions of the position. The term *reasonable accommodations* means making existing facilities (rest rooms, telephones, and drinking fountains) readily accessible to and usable by individuals with disabilities. It also includes the following:

- Acquiring or modifying equipment or devices
- Adjusting or modifying examinations or training materials or policies appropriately
- Providing qualified readers or interpreters
- Adjusting work schedules
- Providing other reasonable accommodations for disabled individuals

Employers must provide reasonable accommodations for disabled workers, and training organizations must provide them for disabled learners. The ADA prohibits asking job applicants certain questions. These questions include those of medical history, worker's compensation or health insurance claims, absenteeism due to illness, mental illness, and past treatment for alcoholism.

Privacy Acts

Privacy acts are federal and state statutes that prohibit invasion of a person's right to be left alone or free from unwanted publicity. They also restrict access to personal information such as personnel files and learner grades. Many U.S. state privacy laws have been nationalized by congressional legislation known as the Family Educational Rights and Privacy Act of 1974, which affects federal aid to education and governs educational records — those directly related to learners (see Learners' Rights section). The act guarantees access to records only by the covered learner or eligible parent or guardian and prevents disclosure of personal information without consent. Its guidelines are followed by all training agencies.

Affirmative Action Programs

Affirmative action programs are employment programs required by federal statutes and regulations designed to correct discriminatory practices in hiring minority group members. The purposes of affirmative action programs are as follows:

- Eliminate existing and continuing discrimination.
- Remedy lingering effects of past discrimination.
- Create systems and procedures to prevent future discrimination.

Equal Employment Opportunity Laws

One reason for employees to take classes is to enhance job opportunities, and they have certain rights as employees. Regardless of their race, color, nationality, origin, sex, handicap, or age, both job applicants and employees have specific rights and privileges that are protected by law. EEO laws apply to protected groups of individuals who have experienced past workplace discrimination. Table 3.2 summarizes the major federal EEO laws.

State/Provincial, County/Parish, and Local Laws and Department Regulations

When state/province and local jurisdictions make their own laws, it is for the purpose of ensuring compliance to issues or situations that are specific to their particular needs. A rule about local law-making requires the law to meet or exceed the federal law; it cannot require

Table 3.2
Summary of Major Federal EEO Laws

EEO Law	Description
Equal Pay Act of 1993	Requires equal pay for men and women doing the same job.
Titles VI and VII, 1964 Civil Rights Act, as amended by the 1972 Equal Employment Act	Prohibits discrimination in all employment practices (recruiting, selecting, compensating, classifying, assigning, promoting, disciplining, terminating, and setting eligibility for union membership) based on race, color, sex, religion, or national origin.
Executive Order 11246, as amended by Executive Order 11375 of 1967	Prohibits employment discrimination by organizations having federal contracts of $10,000 or more. Requires affirmative action programs where necessary.
1975 Amendment to the Age Discrimination in Employment Act (1967)	Prohibits hiring or employment discrimination of workers over 40 years of age unless a bona fide occupational qualification (BFOQ) can be established.
1973 Rehabilitation Act and Executive Order 11914 of 1974	Prohibits discrimination of physically or mentally handicapped applicants and employees by federal contractors.
Vietnam Era Veterans Readjustment Assistance Act of 1974	Prohibits federal contractors from discriminating against disabled veterans and Vietnam era veterans. Requires affirmative action in employing veterans.
Americans with Disabilities Act of 1992	Prohibits job discrimination against disabled people. Requires businesses with 25 or more employees to provide *reasonable accommodations* for qualified disabled job applicants and employees.

less than the federal law unless the state/province or jurisdiction is willing to forego federal funding for that specific funded project. For example, there are federal and state clean-air acts. When states create regulations to control pollution output of local organizations, the requirements must equal or exceed the federal requirements.

Laws in each state/province, in each county/parish of that state/province, and in each local jurisdiction of that county/parish are different because they are based on the good and welfare of local citizens and the general environment. State/provincial laws are generic for the state/province, but city laws vary somewhat from city to city and are different from rural or township laws within the state/province. Similarly, organizations in the cities and towns create their own regulations to guide management and employee actions as they perform duties that must also comply with state/provincial and other regulations. Many state/provincial and local laws are based on federal laws and include such laws as hiring and firing policies, affirmative action policies, and worker's compensation insurance. Organizations create their own regulations to cover these and additional areas such as the following:

- Liability and personal insurance requirements
- Professional development and certification requirements
- Substance abuse testing
- Driver testing and record policies
- Criminal record policies

Learners' Rights

Everyone, including learners, has *substantive rights,* which include a right to equal enjoyment of fundamental rights, privileges, and immunities. A right to privacy of learner records and grades and a right to free speech and expression are also included. Some organizations feel that these rights make it difficult to maintain authority and discipline within their training programs. Reasonable regulations that are for the purpose of protecting the learner while in training can usually be supported in a lawsuit, but regulations that limit or take away learners' substantive rights are not legal to begin with. Learners tend to ignore such unreasonable regulations, and instructors are reluctant to enforce them. For example, requiring hard hats on a construction site and requiring full turnout gear on the drill ground are safety regulations. No eating in the classroom is a safety regulation appropriate for a biology or chemistry lab but may be inappropriate for adults who come directly from work to an evening classroom lecture session.

Torts

[NFPA 1041: 3-4.3, 3-4.3.1]

A *tort* is a private or civil wrong that is the result of a breach of legal duty or failure to perform based on society's expectations of conduct. For our purposes, there are two types of torts: *intentional* and *unintentional.* See Table 3.3.

Intentional Torts

Intent is a state of mind and difficult to prove. It must, instead, be inferred from circumstances and fact. An *intentional tort* is a wrong performed by someone who performs an intended act against a law. Intentional torts include *assault, battery, defamation (libel and slander),* and *false imprisonment.*

Table 3.3
Types of Torts

Intentional Torts	Unintentional Torts
Mimic criminal charges including: • Assault • Battery • Defamation — Libel — Slander • False Imprisonment	Includes only one, which affects all aspects of fire and emergency services: • Negligence

Assault

An *assault* is an attempt or a threat to inflict bodily injury on another person, coupled with the apparent ability to do so, where that person is in immediate fear of harm. An assault can occur without actually touching or striking the individual or performing bodily injury.

Battery

Battery is defined as the unlawful touching or application of force to a person and requires physical contact. The legal protection from battery extends to any part of a person or to anything so closely attached as to be considered part of the person. These *attached items* can be clothing, a cane, a bicycle, or an automobile in which the person is the driver or passenger. If the touching is offensive but harmless, it can still be ruled as battery, but the complainant may be awarded only nominal damages (see Defenses to Negligence section).

Defamation

Publication of anything that injures the good name or reputation of or brings disrepute to a person is considered *defamation.* Actually, there is no legal action called defamation. The action itself is called *libel* or *slander.* For either type of defamation, proof of truth is the defense.

Libel. *Libel* is performed by placing false and malicious words in print that defame a living person. It is a serious offense because libel provides a relatively permanent record of the action. Though there may be variations by state/province, libel includes any unauthorized false and malicious publication in print (reports or newspapers), script (radio addresses or formal statements), or pictures (television broadcasts) that exposes a person to public or professional scorn, hatred, contempt, or ridicule or implies some incapacity or lack of qualifications in the individual's office, trade, business, or employment. See Copyright Laws section.

Slander. *Slander* is defamation of a person through spoken words that tend to damage that person's reputation. It is limited to false remarks but requires proof or presence of actual damages.

False Imprisonment

A person who is unjustifiably detained for any appreciable duration can claim *false imprisonment.* The victim believes that he or she is being restrained unwillingly, even if no physical force is used. False imprisonment includes any unlawful intentional restraint that interferes with a person's liberty.

Unintentional Torts

An individual may not intend to perform a wrong against another, but the individual may act improperly or fail to act appropriately because of ignorance or forgetfulness but with no willful intent to cause harm. Society's members have a duty to avoid harming one another. A legal duty exists to exercise a degree of care in any act because a certain regard for and certain actions toward each other are expected such as offering a seat to a pregnant woman on a crowded metro or yielding the right-of-way to a driver. An individual is considered negligent when he or she fails to exercise that degree of care that persons of ordinary prudence would exercise under similar circumstances. Hypothetically, a *reasonably prudent person* is one who exercises those qualities of attention, knowledge, intelligence, and judgment required of society's members for the protection of the interests and welfare of others. The following sections discuss negligence and the four elements of proof.

Negligence

Negligence is a breach of duty where there is a responsibility to perform. It is a failure to act in a *reasonable and prudent manner,* such as failing to perform to the standard established by law. A *standard of care* is a uniform standard of behavior on which the theory of negligence is based. In the training situation, standard behavior requires an instructor to perform as another reasonable person (instructor) of ordinary prudence would in a similar situation. If conduct falls below the standard, the instructor may be liable for damage that results from that conduct (Figure 3.3).

In tort law, risk should be reasonably perceived and avoided, and this is the common-law duty of the instructor toward the student. But the fire and emergency services profession is dangerous and involves a "peculiar risk" that cannot be avoided. If an instructor asks learners to take risks that are dangerous beyond that peculiar risk — the learners' level of ability — then the risk would be considered unreasonable. For example, firefighters must enter smoke-filled buildings; it is a peculiar risk of fire fighting for which they are trained. But instructors would not direct new fire-training learners to go into a smokehouse without first teaching them how to use self-contained breathing apparatus.

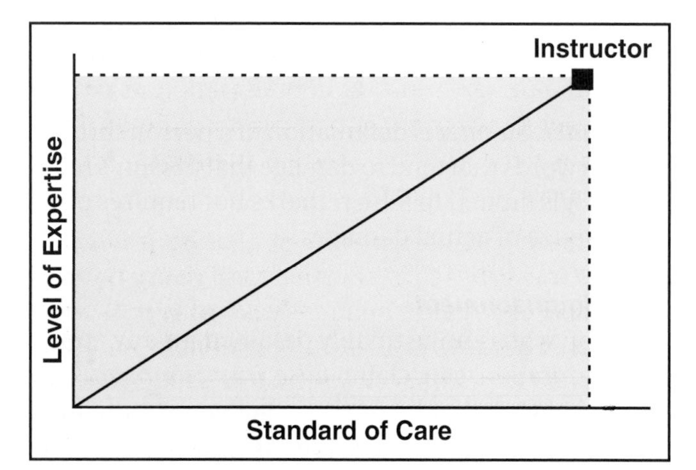

Figure 3.3 The standard of care is related to the expertise a person in a particular situation is expected to possess.

In any training situation, there may be some factors that limit or control how an instructor will or can act. But the instructor has a responsibility to act in a reasonable manner based on known information, current conditions, and available resources. What is *reasonable and prudent* is often established by common sense (common law) or social norm, and the action taken is judged based on what a *reasonable person* would *prudently* do in the same circumstances. As individuals learn more and become more experienced in their lives and careers, their levels of reasonable prudence change, and they are held to higher standards. For example, an adult is held to a higher standard than a child, a medical person is held to a higher standard than a nonmedical person, and an instructor is held to a higher standard than a learner.

What actions are considered reasonable and prudent for instructors? If a person has a greater amount of expertise than another person, the duty of the expert person is greater in proportion to the nonexpert. For example, a fire and emergency services instructor would have judgment, knowledge, and actions similar to other instructors but greater than learners. A reasonable and prudent instructor uses qualified instructors who can foresee risks to assist in teaching learners with less knowledge and judgment. If a potentially hazardous condition exists, instructors must consider the following factors, especially if resources are not available to correct them:

- Potential for and degree of harm that may result from the condition
- Likelihood that harm will occur because of the condition
- Availability of alternative methods and/or appropriate equipment to correct the condition
- Burden of removing or changing the condition

Fire and emergency services instructors owe a duty to learners to provide and enforce a safe standard in training. Instructors must ensure they provide a reasonably safe learning environment and take steps to prevent injury and limit the possibility of liability actions against themselves and their organizations. Instructors also owe a duty to members of the community who depend on properly trained fire and emergency services personnel to provide a reasonable standard of care or service during interactions with them. Instructors who do not meet standard requirements in their training programs may be considered negligent.

Instructors teach all program requirements to the expected standards and emphasize the importance of performing to these standards. Eliminating or abbreviating a portion of a program because it does not apply to a certain group creates a potential for liability for the instructor who failed to teach and for the participants who failed to perform that expected duty. Individuals who fail to perform or perform incorrectly may be found negligent and held liable for their actions.

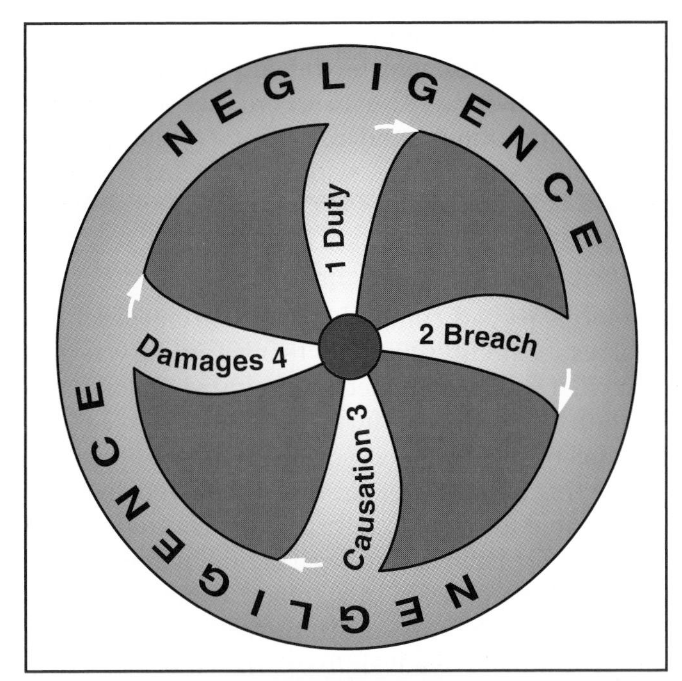

Figure 3.4 To prove negligence, the four elements of proof (duty, breach, damages, and causation) must exist.

Proving Negligence

The relationship between the action or inaction of the instructor and the resulting injury must be proved by the injured party. Consider the following questions:

- Was the instructor supposed to teach a certain point before the learner proceeded?
- Was the learner directed to proceed without proper instruction?
- Did an injury occur because the instructor failed to teach that point?
- Did the party or complainant sustain damage as the result of an injury caused by the instructor's failure to perform?

To prove negligence, there must be *prima facie* evidence, which literally means *first view*. This type of evidence needs no further support to establish the existence, validity, or credibility of the facts. Elements of proof include *duty, breach of duty, damage,* and *proximate cause (causation).* When these four elements are proved, there is negligence (Figure 3.4).

Duty. A *duty* is a human act or obligation that conforms to the laws. Instructors owe a legal duty to learners to teach them job skills based on accepted standards and to protect them from unreasonable risk during training. Fire and emergency services personnel have a duty to act in emergency situations, and the victims of these situations expect a certain level of service from those personnel. Once fire and emergency services personnel take on the responsibility of the duty, they must follow through or complete the actions necessary to fulfill the obligation of that duty. The level of the duty expected is based on an individual's knowledge, skill, and profession.

Breach of duty. When an individual fails to provide the standard of care that a reasonably prudent person would exercise under similar conditions, it is called *breach of duty.* Under the law of negligence, an individual who owes a duty to another must provide the care or service to the expected standard to avoid negligent injury. Breach of duty can occur when an individual fails to perform to the expected standard or when the individual performs incorrectly.

Damage. *Damage* includes loss, injury, or deterioration caused by negligence of one person to another. The victim must have some kind of physical damage. Stress or mental anguish is not considered until after the physical damage has been proven. In plural form, the term *damages* refers to compensation in money for a loss or damage. An individual who suffers a damage has a right to compensation (damages) for the injury caused by another.

Proximate cause. There must also be proof that damage was caused. In law, there are many types of cause, and one that is frequently heard is *proximate cause.* It is a cause that directly produces an effect (injury) and arises out of a wrongdoer's negligence or conduct. Without the action or negligence of the wrongdoer, the injury would not have occurred. Proximate cause may be difficult to prove because there must be proof of responsibility. Most accidents are the result of multiple factors or parties, all of which may be partly responsible. It also may be difficult to show that the procedures used were not in accordance with acceptable organizational standards.

Liability

[NFPA 1041: 3-4.3, 3-4.3.1. 4-2.2.1, 4-5.2, 4-5.2.1]

Liability is a broad, comprehensive term that describes legal responsibility. It implies that if a wrong was performed, it will be corrected by the individual or organization who performed it. Instructors or organizations may be negligent for wrong instruction or for not instructing in some area for which they were liable or responsible for teaching. Instructors can reduce the potential for liability and legal action against themselves by being aware of standard expectations, by teaching to standards, and by providing a safe learning environment relatively free of risks that may cause injury. Instructors are expected to foresee and prevent potential injury events while training personnel for appropriate performance on the job.

Program planners must remember that training programs require participants to learn how to perform some risky activities that they don't yet know how to do. Program planners must foresee instructional problems and ensure that all instructors can appropriately perform and demonstrate the skills in a thorough step-by-step introduction and then properly supervise the practice of those skills. Learners are being trained in a nonemergency environment for skills that will take place in an emergency environment. Although it is a benefit to make the training situation "realistic," training activities are supposed to be controlled. The inability to manage risk during training activities can lead to injured learners and create a liability. Injuries that result from real fire and rescue situations have special defenses that do not apply to those who are responsible for setting up situations that cause injuries in the training environment. Exposing learners to hazardous training environments without having adequate controlling elements could be perceived as negligence in court if any injury occurs.

Each training session has many potential variables in participants, instructors, tools and equipment, scenarios and evolutions, and the environment that affect the safety of the program. When planning a training session, instructors must make every effort to maintain control and foresee potential dangers or injury situations and reduce, if not eliminate, risks to learners and personnel.

Even when every precaution has been taken to reduce risks, program participants may still be injured. Injuries may not be due to the fault of instructor oversight but due to the fault of participants who do not follow safety guidelines. Consider the following questions:

- Can instructors be held liable for actions of learners who irresponsibly act on their own?
- Can individuals be held personally liable for contributing to their own injuries and to the injuries of others?
- To what extent are employers liable for injuries caused by their employees?

The following sections explain the concepts of vicarious liability, define foreseeability, list criteria to help prevent employer liability, and suggest steps that will reduce the possibility of personal liability.

Vicarious Liability

Vicarious liability means that the blame for the actions of one person is placed on another. Employers are responsible for the actions of employees; individuals are responsible for their own negligent actions that cause injury to themselves or others. The instructor is in the middle in this situation, but if the instructor performed as directed or could not perform because there was no directive, then the blame is shared by another (responsible) party — the organization that failed to provide proper directives.

Foreseeability

Foreseeability is a concept used to limit the liability of a party (the instructor) for the consequences of his or her acts to the consequences that a person of ordinary prudence would reasonably expect to occur. Instructors can expect certain conditions in training, can foresee potential problems with those conditions, and can plan to reduce the risks. Foreseeability must extend to the conditions that the instructors anticipate learners to meet once they are on the job as well. While training, instructors must cover all knowledge and skill competencies so that learners are prepared to perform beyond the classroom. An effective instructor keeps up with changing trends, policies and protocols, and new technology and research that affect job performance in all areas of fire and emergency services. Instructors must foresee potential risks that learners may be exposed to on the job.

Practicing foreseeability requires training officers and instructors to regularly conduct an occupational or job analysis that identifies contemporary duties and tasks of the field. Reviews of litigation involving vocational instructors show that courts expect reasonable and prudent instructors to do the following:

- Have a plan or formally developed strategy to prevent injuries.
- Follow the plan.
- Provide for health and safety.
- Give proper instruction.

Instructors can take precautions to minimize the chances of becoming involved in a liability case. Some of these actions are as follows:

- Ensure that all assisting instructors meet the organization's qualifications and requirements.
- Check equipment regularly for safe operating conditions.
- Maintain written objectives, and document each training session.
- Provide learners with a written course description so they understand all requirements.
- Make sure trainees are physically fit and prepared for the tasks.
- Instruct and test all learners in the safe operation of equipment.
- Do not leave learners unattended while they are practicing potentially dangerous skills.
- Respect learner privacy.

Fire and Emergency Services Employer Liability

Laws are constantly changing. They also vary in each state/province and in jurisdictions within a state/province. Fire and emergency services organizations must be aware of legal changes and realize they are vulnerable to lawsuits. They are responsible for actions or inactions of their instructors and employees. Before accepting personnel in a training program, the host training organization must determine who is responsible for injuries to program participants. Often, the people participating in a training program are from different organizations. It is not safe to assume that each individual has insurance coverage. Typically, organizations provide insurance coverage for employees, associates, and volunteers, especially if the individuals are required to perform risky activities as part of their job responsibilities. A part of the program registration process may include requiring insurance coverage by organizations on all participants. Any program participant who feels wronged by an instructor or is injured during a training exercise under the supervision of an instructor may initiate a lawsuit. Is it the responsibility of the instructor, the host training organization, or the individual's organization to cover damages?

In addition, any training activity practiced in public that may cause injury to the public during the training exercise has the potential for liability (such as training drivers on fire and emergency apparatus on public streets before the trainee is certified and licensed to drive the unit). Any consumer who feels wronged or is injured by fire and emergency services personnel, whether during a training exercise or in a real-life emergency, may initiate a lawsuit. The individual named in the suit will in turn fault the instructor, the training organization, and his or her own organization. The instructor may be found liable and required to cover a portion of the damages. Defending a lawsuit is costly in terms of time, money, and reputation, even if the accused is cleared of the wrong. Typically the organization, who has the "deep pocket" or more money to cover damages, is held liable.

Organizations have attempted to prevent lawsuits by having learners or consumers, such as ambulance patients, sign a release form or waiver. Release forms are not legal if they attempt to waive liability or take away someone's rights. Rights are individual liberties that are expressly provided for citizens in state/provincial or federal constitutions, and an organization cannot overrule a constitution and have individuals sign away their rights.

Organizations must have certain criteria in place to reduce the potential of lawsuits, and instructors must be aware of and follow the criteria. Instructors should check with organizational policies on the following issues:

- Risk management programs that make an effort to understand and control risk and protect itself from losses
- Expectations and policies on professional conduct that set guidelines for competency and dealing with such issues as harassment
- Safety policies that follow established local or national standards
- Insurance policies for personnel in the event of injury (worker's compensation insurance and vehicle insurance) or lawsuit (liability insurance)
- Physical fitness requirements that are met by all program participants

A few other issues that can create liability problems are common and often accepted practices. Training organizations and instructors must also address the following issues:

- ***Competitive exercises*** — These activities may be considered recreational and not training. Consider the following questions:
  - — If participants are injured, who is liable for damages?
  - — Who is sponsoring the program?
  - — Who is sponsoring the participant?
  - — Is it a recognized training program, or is it a recreational activity among competing organizations?
  - — Will all organizations take the responsibility of injuries to their personnel?
  - — Will the organizations' insurance companies be willing to cover injuries resulting from these types of programs?
- ***Underage trainees*** — Many organizations recruit or accept trainees in their training programs who are underage and call them *cadets*. Fire and emergency services work is hazardous, and exposing underage individuals to these hazards may be unlawful. Consult with the legal office before accepting underage recruits or cadets into a program that involves them in an activity other than participating in cognitive classroom learning and observing demonstrations in a safe environment.

In addition, instructors must be sure that there are appropriate standby emergency care and rescue teams at the site of all hazardous training exercises. If an injury occurs during training, follow the organization's policies on caring for the injured, halting or continuing training, and reporting the injury. It is critical that the organization complete a thorough investigation on the incident to determine cause and prevent further incidents. If an investigation is not done, a court may determine that the organization demonstrated a lack of concern, which can have unfavorable results in a lawsuit.

Personal Liability

If a court decides that a fire and emergency services employee is liable for a negligent act, it may place a judgment of damages against the employee. The employee may in turn blame the instructor or the employing organization by saying he or she acted, or failed to act, based on what was taught in class or condoned on the job.

Though fire and emergency services instructors may be covered by the employing organization's insurance, they still face situations for which they may be individually liable, and they must take precautions to prevent liability (Figure 3.5). Situations that instructors should be aware of in order to take steps to prevent legal action include the following:

- Negligence in training learners, including exposing them to unnecessary risks or failing to warn them of the potential danger of an exercise
- Injuries to learners after training caused by poor or incorrect instruction (**Note:** The facts of the injury may be difficult to establish but can be proved if necessary safety warnings were not included as part of the instruction.)
- Injuries to a third party because of poor or incorrect instruction
- Misrepresenting qualifications or benefits of training, including making claims for training that cannot be supported (for example, instructors who teach in areas for which they have no qualifications)
- Improperly supervising interns or trainees (**Note:** Organizations must have specific guidelines that instructors must follow for governing the quality of the trainees' work and actions.)
- Poor communications, including libel, slander, or breach of confidentiality
- Improper personal actions, including discrimination and harassment

In the career of any instructor, there will likely be an unforeseen injury to a learner in some training program. Instructors may want to consider personal li-

Figure 3.5 Instructors can reduce their exposures to liability if they take the appropriate precautions of understanding and practicing risk management principles. *Courtesy of Jarett Metheny, Midwest City (OK) Fire Department.*

ability insurance coverage in addition to coverage provided through their employing organizations. An instructor should check the extent of coverage and the conditions under which the employer's insurance is effective before purchasing additional coverage.

Legal Protections

[NFPA 1041: 3-4.3, 3-4.3.1, 4-2.2, 4-2.2.1, 4-2.7, 4-5.2, 4-5.2.1]

Society has provided itself with a few protections and defenses against lawsuits. The protections described in the following sections do not guarantee immunities, and in some cases these protections are limited or have been eliminated. Discussions of damages and precautions an instructor can take are also included.

Immunities

An *immunity* is a right given that exempts someone from a duty or a penalty. It is a benefit granted to someone that is contrary to the general rule or expectation of others. The following sections describe three types of immunities: *charitable, sovereign,* and *Good Samaritan.* Only one (Good Samaritan) is still completely effective for fire and emergency services providers.

Charitable Immunity

Charitable immunity was given to those individuals or organizations (such as Sisters of Charity) who provided free services to those in need. Over the years, charitable services have become a lucrative, for-profit business, so immunity is no longer granted to these organizations.

Sovereign Immunity

Sovereign immunity was part of English Common Law doctrine designed by kings to protect themselves from damage claims of lowly citizens. Sovereigns could not be held liable (although they were responsible) for wrongdoing or failure to act on behalf of society. Society could not use the law to sue the lawmaker, the sovereign. For years nations continued this type of immunity from the statutes that they enforced on their citizens. Sovereign immunity was carried over to American law. In the United States, sovereign immunity gave federal, state, and local government bodies and their employees immunity from liability for any action taken, negligent or otherwise.

In 1946, the United States Government waived its immunity from liability and allowed for the litigation of claims from suits for injury to persons or property. Over the past few decades, many legislative bodies have modified or abolished sovereign immunity, and today there is a continuing trend against it. Negligence liability exists in all but a few jurisdictions. Most jurisdictions have set limits on their liability for damage. Instructors should note that the sovereign or governmental agency, as well as the individual employee, may now be held liable in court.

Case Study: Legal Issues with Training Groups

In one training situation, the training organization did not provide standards or policies to the instructor, and as a result, the instructor provided incorrect or no specific instructions, policies, or guidelines to the class group. In another training situation, the instructor was provided the proper standards and policies and passed them on to the training group, but an individual learner ignored them and was injured and caused injury to another learner as a result of her own improper actions.

Answer the following questions:

1. Who is to blame for the absence of standards and policies? Who is to blame for the injuries? Are the actions of the learner a type of liability? If so, what type is it?
2. Can instructors be held liable for actions of learners who irresponsibly act on their own? Can individuals be held personally liable for contributing to their own injuries and to the injuries of others?
3. To what extent are employers liable for injuries caused by their employees?

Discuss these questions with others in your organization and draw conclusions. Consult *Black's Law Dictionary* or Gifis' *Law Dictionary* for information. Compare your responses with your organization's policies. Determine what instructors must do to protect themselves from potential liability.

Good Samaritan Immunity

Good Samaritan laws provide protection for those who provide care for others in need as long as they provide care at their level of skill and ability. Good Samaritan immunity is still in effect and effective. The United States has no federal Good Samaritan statute; however, each state and all U.S. possessions have their own statutes with their own variations. Good Samaritan laws are a defense against ordinary negligence, but they do not provide protection from events outside of patient care (such as injuries resulting in an accident en route to an emergency scene) nor does it cover gross negligence.

Gross negligence is a lawyer's term meaning willful and wanton disregard. It is usually left to the court to determine gross negligence by deciding that the individual intentionally failed to perform an obvious duty in reckless disregard of the consequences to the life or property of another.

Defenses to Negligence

In addition to immunities, there are several other defenses against torts of negligence. These include the following:

- ***Failure to prove all elements of negligence*** — Each element requires proof with a preponderance of evidence. If proof in any element is missing, the suit is invalid.
- ***Contributory negligence*** — If an individual contributes to his or her own injury by performing below the standard established for protection, that person is just as liable for the damage as the other contributing party. Any amount of contributory negligence may eliminate recovery by the injured individual. Some states have adopted a comparative negligence rule that assigns a portion of the blame between the injured party and the contributing party based on the extent of negligence or damage caused by each.
- ***Assumption of risk*** — Individuals assume a risk when they knowingly expose themselves to a hazard or danger created by another. In such an event, the other person is not responsible for damage to the individual who assumes the risk. In the fire and emergency services profession, firefighters and other rescue personnel expressly agree to assume the risks of the ordinary hazards expected in their profession. They contribute to negligence when they fail to exercise due care. Assumption of risk occurs regardless of the care used because the risk they take is based on consent.
- ***Statute of limitations*** — There is a limit to an individual's memory and the amount of time to which an individual is required to suffer mental anguish while waiting for an injured party to pursue action. The *statute of limitations* is any law that fixes a time period for enforcing rights; it is a certain time allowed by a statute for bringing litigation. After that time period has passed (with the exception of murder), injured parties are barred from pursuing a claim.

Damages

When damages are awarded for negligence, they are based on proven losses, and a judge or jury decides the form and amount of compensation. The types of damages are described as follows:

- ***Actual or real compensatory damages*** — These may include reimbursement or some form of restitution for loss of pay, to cover medical expenses, for loss of a relationship with another, and for pain and suffering.
- ***Exemplary or punitive damages*** — For intentional torts and for cases of gross negligence, these damages are meant to punish the party who caused the injury.
- ***Nominal or technical damages*** — In some cases the court finds in favor of the complainant but only awards a nominal amount, one dollar for instance. This type of award means that the court recognizes that a legal injury was sustained, but it was slight, and no recoverable loss can be established. The injured party wins the case but only on a technical issue.

Instructor Precautions

Even when instructors follow all the organization's policies and prescribed standards while performing instructional duties, a mishap or misunderstanding can occur and someone can perform incorrectly or fail to perform as directed. Ideally, an instructor takes all appropriate steps to prevent a court appearance, and these steps include the following:

- Always appear professional in appearance on the job and in the training situation. Generally, appearance gives a good impression and implies respect for the responsibilities of the profession.
- Maintain skill level. Instructors are held accountable for skills. Maintaining current certifications through appropriate and credible refresher programs is critical. The defense of *"I didn't know"* is not acceptable.

- Treat other instructors and coworkers, all learners, and any community members in all situations with courtesy and respect. Attitude is a major factor in how others judge people and their abilities.
- Seek advice and guidance from a higher authority when in doubt. Do not attempt to make decisions beyond your knowledge or authority.
- Document all issues of discrepancy, complaint, and injury accurately and in detail with dates, times, conversations, suggested resolutions, outcomes, and follow-up plans.

Instructors must remember to abide by the following restrictions:

- Do not exceed your skill level when training learners or working with other instructors.
- Do not ignore, short cut, or exceed protocols or policies.
- Do not presume anything. Ensure in as many ways possible (verbal, written, and graphically) that learners and instructors understand the intent and outcome of all directives or instructions as well as the consequences if procedures are not followed.
- Do not joke about serious situations or belittle the actions of others in any learning or service situation.
- Never disclose personal information (except to appropriate authorities) about learners, other personnel, or any victim or patient who required emergency services. Follow the organization's policy on disclosing information to insurance companies, hospital personnel, legal representatives, news reporters, or other persons who want information on issues that are going through litigation.

Substance Abuse

[NFPA 1041: 3-2.5, 3-2.5.1]

Substance abuse is a sensitive issue in today's fire and emergency services. Organizations have found the need to develop regulations to guide their actions in handling employees with substance-abuse problems. Instructors are responsible for knowing their organizations' policies. The nature and responsibility of the fire and emergency services profession require mentally alert and physically able and capable employees. The profession cannot risk mistakes made by personnel who are incompetent because of substance abuse.

Instructors who work closely with program participants may notice indications of substance abuse. Instructors have moral and ethical obligations to that learner's employer and to the public whom that learner will later serve while on the job. Reporting suspicions of substance abuse is not invading that person's privacy; it is preventing potential injuries to others. Instructors should be aware of the employee assistance programs offered through their organizations. Often, a part of new employee and learner orientation is informing people of the organization's benefits, which includes information on these types of assistance programs.

Copyright Laws

[NFPA 1041: 3-2.5, 3-2.5.1]

Copyright laws protect the works of artists and authors and give them exclusive rights to publish their works or determine who may publish them. Since the Copyright Act of 1976 was passed, almost all copyright law in the United States is governed by federal statute. Copyright protection for most works is for the life of the author plus 50 years. Unauthorized use of copyrighted materials is considered infringement on authors' rights, who have a right to recover damages or gain profits from the use of their works.

Instructors and learners often copy materials from texts and journals for use in class. Is this copyright infringement? It depends on how the material is used. The Fair Use Doctrine grants the privilege of copying materials to those other than the owner of the copyright without consent if the material is used in a reasonable manner. Section 107 of the Copyright Act lists the privileges or factors to be considered when determining if the use of copyright materials is "fair use" (Figure 3.6). A few of the copyright guidelines are as follows:

- Teachers may make single copies of the following for scholarly research or when preparing to teach a class:
  - — Chapter from a book
  - — Article from a periodical or newspaper
  - — Short story, essay, or poem
  - — Chart, graph, diagram, drawing, cartoon, or picture from a book, periodical, or newspaper
- Teachers may make multiple copies of items for classroom use and discussion by their learners provided that the copied material is brief and the idea to copy an item is spontaneous because it is currently appropriate to the day's lesson. Copying must be for the particular class being taught at the time

Figure 3.6 Instructors must know where they can find information on copyright laws. *Courtesy of Robert Wright, Maryland Fire and Rescue Institute.*

and not *cumulative* or copied repeatedly for subsequent classes.

- Copying shall not substitute for buying books, publisher's reprints, or periodicals.
- Learners cannot be charged for copied materials beyond the cost of photocopying.

There are many emerging issues with the use of materials obtained from the Internet. Instructors must use proper citations when they download and use these materials.

Permission for Pictures

[NFPA 1041: 3-2.5, 3-2.5.1]

Part of learning is seeing the "real thing," and one way instructors include reality in their instructional methods is through pictures taken at the emergency scene. When fire and emergency services personnel take photographs or shoot motion picture films at a scene, they are, in a sense, invading the privacy of the individuals or victims involved in the incident. *Invasion of privacy* is the wrongful intrusion into a person's private activities by the government or by other individuals. Individuals have the right to control the use of pictures of themselves and their property. An instructor or organization who uses these pictures or films can be sued for invasion of privacy or libel if permission is not obtained. Tort law protects the private affairs of individuals from unwarranted exploitation or publicity that causes mental suffering or humiliation to the average person. But the right to be left alone is not always superior to the rights of the public. The level of privacy varies for individuals. The right to be left alone may exist to a lesser degree for a public figure (such as a well-known politician, film or music celebrity, or criminal) who makes a living in front of a camera or is in the public eye because the public has a rightful interest in these types of individuals.

The legality and success of lawsuits against the right of privacy depend on several factors such as where the photograph or film was shot and who was photographed or filmed. Some events are newsworthy and are photographed or filmed for the public interest; however, using films or photographs a month after the incident has lost its public appeal and is no longer newsworthy steps over the line of privacy.

Taking photographs or shooting films is allowable if the setting is a public place rather than an individual's private home. The law also allows for photographing and filming publicly famous or notorious individuals who are out in public. It may still be considered an invasion of privacy if any individual, ordinary citizen, celebrity, or criminal is photographed or filmed on a public highway stripped of clothing and dignity having emergency procedures performed while being rescued from a serious accident.

Because of potentially serious legal and professional consequences, organizations must always make arrangements ahead of time to obtain permission in writing from individuals to take photographs or shoot films of an event and use them after the event. Organizations must also explain the purpose of intended use so individuals can make an informed decision or restrict use to a specific form such as training only. Individuals have the right and must be given the opportunity to preview the pictures or films and make a decision before they are used. Individuals may also require that the organization maintain their anonymity while showing the film or photograph. This is usually done by blocking faces with graphic overlays or by showing them in shadows and by altering voices on films by audio distortions.

References and Supplemental Readings

Gifis, Steven H. *Law Dictionary* (4th ed.). NY: Barron's Educational Series, Inc., 1984.

Black, Henry Campbell. *Black's Law Dictionary*. St. Paul, MN: West Publishing Company, 1991.

Brannigan, Vincent. "Training Can Pose Special Liability Risks." *Fire Chief*, January, 1995.

Job Performance Requirements

This chapter provides information that addresses the following job performance requirements of NFPA 1041, *Standard for Fire Service Instructor Professional Qualifications* (1996 edition). Colored portions of the standard are specifically covered in this chapter.

Chapter 2 Instructor I

2-5.2 Administer oral, written, and performance tests, given the lesson plan, evaluation instruments, and the evaluation procedures of the agency, so that the testing is conducted according to procedures and the security of the materials is maintained.

2-5.2.1 *Prerequisite Knowledge:* Test administration, agency policies, laws affecting records and disclosure of training information, purposes of evaluation and testing, and performance skills evaluation.

Chapter 3 Instructor II

3-2.5 Coordinate training record keeping, given training forms, department policy, and training activity, so that all agency and legal requirements are met.

3-2.5.1 *Prerequisite Knowledge:* Record keeping processes, departmental policies, laws affecting records and disclosure of training information, professional standards applicable to training records and disclosure of training information, professional standards applicable to training records, databases used for record keeping.

3-4.3 Supervise other instructors and learners during high hazard training, given a training scenario with increased hazard exposure, so that applicable safety standards and practices are followed, and instructional goals are met.

3-4.3.1 *Prerequisite Knowledge:* Safety rules, regulations and practices, the incident command system used by the agency, and leadership techniques.

Chapter 4 Instructor III

4-2.2 Administer a training record system, given agency policy and type of training activity to be documented, so that the information captured is concise, meets all agency and legal requirements and can be readily accessed.

4-2.2.1 *Prerequisite Knowledge:* Agency policy, record keeping systems, professional standards addressing training records, legal requirements affecting record keeping, and disclosure of information.

4-2.7 Present evaluation findings, conclusions, and recommendations to agency administrator, given date summaries and target audience, so that recommendations are unbiased, supported, and reflect agency goals, policies, and procedures.

4-5.2 Develop a system for the acquisition, storage, and dissemination of evaluation results, given agency goals and policies, so that the goals are supported and those impacted by the information receive feedback consistent with agency policies, federal, state, and local laws.

4-5.2.1 *Prerequisite Knowledge:* Record keeping systems, agency goals, data acquisition techniques, applicable laws, and methods of providing feedback.

Chapter 4

The Psychology of Learning

Pedagogy is defined as the principles, methods, and profession of teaching and instruction. A *pedagogue* is traditionally a teacher of children and youth. Traditional educational methods had children and youth sitting in rows of chairs taking notes while listening as a teacher delivered information, often faster than they could write it, and memorizing facts so they could give them back on some type of written exam. When slides or films were added, the lecture format became "illustrated," but there was often little if any guidance as to the purpose of these types of illustrations. Learners thought films were a break from the routine (and sometimes instructors used them as such) or a form of entertainment. Tests given on class material were a measure of intelligence on the subject. If learners did not do well on tests, teachers considered them unable to learn or blamed them for not studying, and no steps were taken to examine or consider the reason for failure.

These traditional methods of teaching and measuring learning continued for many adults in training classes because adults who are teachers today often rely, for the most part, on these same traditional, familiar methods. Without training in how to teach, instructors will use the methods by which they learned, even if they are incorrect. Adults learn differently from children. The differences are a result of adults' experiences from which they have learned to think and draw conclusions. Adults must be taught with methods different than those for children — methods that allow them to think and draw conclusions rather than dictate what they must think, know, and do.

Dr. Malcolm Knowles (1970), professor, researcher, consultant, and author, was among the first theorists to use the term *andragogy (an-dra-**go**-je)*. Originally coined by a German teacher in 1833, it refers to the art of teaching adults. It describes the characteristics of adult learners and a set of assumptions for most effectively teaching adults. The theory of andragogy is now widely accepted and includes the following assumptions (Lee, 1998):

- ***Self-concept*** — Because of little experience, children need to be directed. Adults, with their extensive experience, have a need to be self-directed by the trainer or training program rather than forced into dependent roles, like children.
- ***Experience***—Adults have accumulated quite a store of experiences that serves as resources for them and on which they have a broad base to relate new information.
- ***Readiness to learn***—Adults become ready to learn whatever they need to know or do in order to meet job requirements or social roles.
- ***Learning orientation*** — Children's orientation to learning is subject-centered, and they master subject content to be promoted. Adults' orientation is problem-centered; they have specific purposes for learning and want skills or knowledge they can apply to real-life problems.
- ***Motivation*** — Adults have internal incentives or motivators and are more motivated to learn by such factors as increased self-esteem than they are by external rewards such as pay raises and promotions.

Though Knowles' many books describe and promote the andragogy *versus* pedagogy theory, many teachers found that children sometimes learned better with andragogal methods, and they discovered that pedagogical methods were appropriate for adults in certain learning situations. Knowles' research indicated that children have as much need for life-, task-, and problem-centered learning as adults do. Adults have a universal characteristic of their *experience,* which is their main resource.

Other educational experts have contributed their theories and suggested practices that they have found successful in teaching adults. Dr. Benjamin Bloom (1984), an icon of education known for his work in developing a taxonomy of objectives, proposed the theory that if instructors gave learners appropriate time and opportunity, they could learn anything. Most adults come to class interested in the subject and eager to learn something, but they may not be given appropriate time and opportunity to master course requirements. As a result, both the adult and the instructor feel unsuccessful and discouraged. Instructors feel they have done their jobs, but may wonder what more they could have done to help those unsuccessful learners.

The universal goal of instruction is to *"send learners away from instruction with at least as favorable an attitude toward the subjects taught as they had when they first arrived."* This positive theory on instruction is from psychologist, Dr. Robert Mager (1984), a contemporary educator who is a world-renowned expert on the design, development, and implementation of instruction.

How can instructors ensure that learners remain interested and feel successful? How can instructors apply teaching methods that enable adults to gain success in a training program? This chapter answers these questions by describing learning processes, domains, styles, and laws. What influences and motivates learning, how individuals learn and remember, and the factors that affect learning are included. The chapter also describes approaches to learning, learner characteristics, and types of learner performance and suggests strategies for managing learner performance. References and supplemental readings and job performance requirements are given at the end of the chapter.

Learning Processes

[NFPA 1041: 2-3.2, 2-3,2.1, 2-4.3, 2-4.3.1, 2-4.3.2, 2-4.5, 2-4.5.1, 2-4.5.2, 3-4.2, 3-4.2.1, 4-3.3, 4-3.3.1, 4-3.4, 4-3.4.1]

Learning is an active process; learners progress through a series of mental steps. Some learners stumble on those steps if instruction is not clear, if information is given in an unfamiliar format, or if information is presented in a style that is difficult to learn. Instructors must understand the learning processes and principles and know how to use teaching methods that gain interest, stimulate motivation, and ensure successful learning.

To understand these processes, instructors may ask: What is teaching? What methods are effective? What is learning? What occurs mentally when a learner is given new ideas? What is motivation? What motivates learners? Simply defining terms answers a few of these questions.

- ***Teaching*** — Method of giving instruction through various forms of communicating knowledge and demonstrating skill. If teaching is successful, it causes an observable change in learner behavior. Teaching must provide activities and include opportunities for learners to demonstrate knowledge and skill and to receive feedback on progress toward the expected behavior change or performance.
- ***Learning*** — Relatively permanent change in behavior that occurs as a result of learning new information, skills, or attitudes through some form of instruction. Learning is enhanced if it is reinforced through practice, which must be frequent and intense in order to promote understanding and ability.
- ***Motivation*** — Arousal and maintenance of behavior directed toward a goal and normally occurs in someone who is interested in achieving some goal. The individual maintains interest when each step in a series of small successes places the individual closer to a desired goal. Motivation may come from within the individual, or it may be fostered by an instructor who designs and delivers instruction that stimulates those interests.

The processes of teaching, learning, and motivation are linked and complement each other. Instructors can motivate learners and help them learn by using a variety of teaching methods that expose every learner to information through the learner's preferred learning style. Instructors can stimulate motivation by making classes relevant to life and work and by using activities that provide opportunities to practice and apply knowledge and skill.

The mental processes of learning that occur within a person are not directly observable, but the effects of learning are seen through changes in behavior. Learning occurs in many ways. It can be purposeful such as the type of knowledge and skill that learners are exposed to in firefighter or emergency medical services (EMS) training. It can be incidental such as the knowledge that smoke means that a fire is burning or that an injured child requires a different approach than an injured adult. Learning occurs through each of the senses and by exposure to different experiences. Regardless of how learning takes place, learners are most likely to learn in conditions where attention is focused on something specific and the effort to learn is deliberate (Figure 4.1). When individuals want to learn, are motivated enough to focus on learning, and are provided knowledge and opportunity to practice the new behavior, they learn rapidly and efficiently. When instructors prepare to teach, they select from three learning domains and many teaching styles.

Learning Domains

[NFPA 1041: 2-4.3, 2-4.3.1, 2-4.3.2, 2-4.5, 2-4.5.1, 2-4.5.2, 3-3.2, 3-3.2.1, 3-3.3, 3-3.3.1, 3-3.3.2, 3-4.2, 3-4.2.1, 3-4.2.2]

The three domains of learning are *cognitive (knowledge), psychomotor (skills),* and *affective (attitude).* The domains are not independent areas of learning. Rather, they are interrelated areas in which learning occurs. When combined in instructional methods, they enable learners to perform a behavior or a job skill. Having an understanding of the domains helps instructors present effective instruction. Through the cognitive domain, learners gain understanding about a behavior; through the psychomotor domain, learners perform the skills associated with the behavior; and through the affective domain, learners develop a willingness to perform the behavior correctly and safely. The cognitive, psychomotor, and affective domains are the *what, how,* and *why* of learning (Figure 4.2).

Each of the domains of learning are graduated into several levels of learning (Figure 4.3). The graduations start from the simple on the bottom to the complex at the top. The learner must pass through each step on

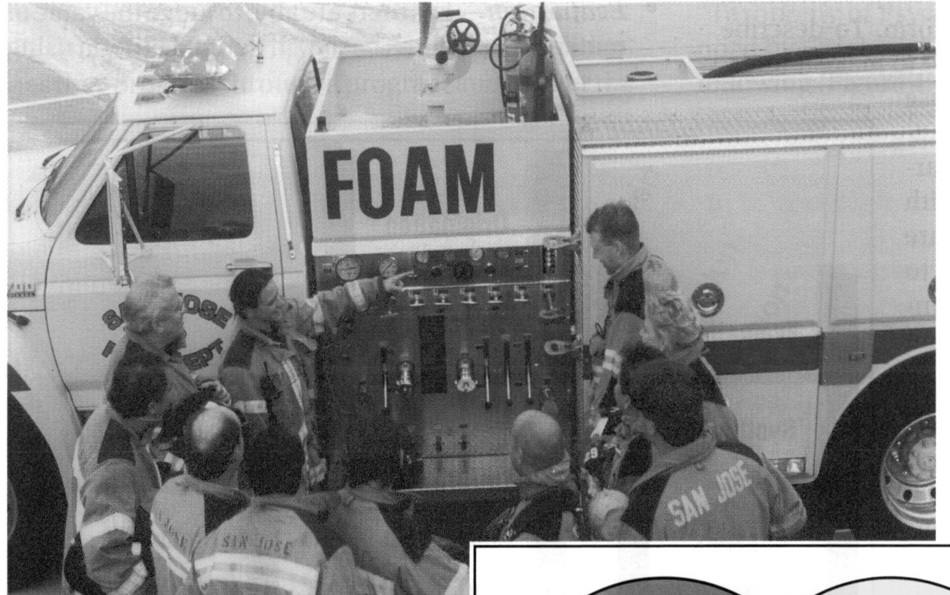

Figure 4.1 Individuals learn more effectively when their attention is focused on a goal. *Courtesy of Dana Reed, San Jose (CA) Fire Department.*

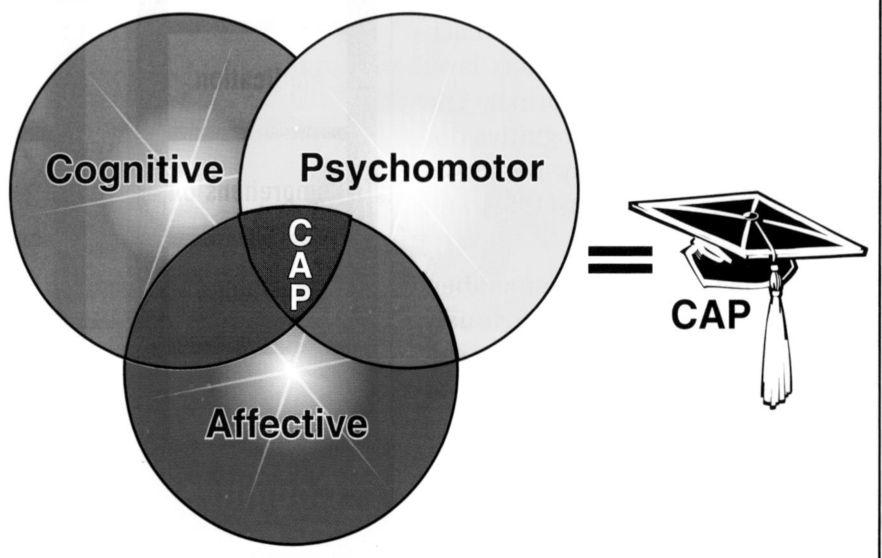

Figure 4.2 The three domains of learning: cognitive, psychomotor, and affective. For effective learning, the areas must overlap or interrelate with one another.

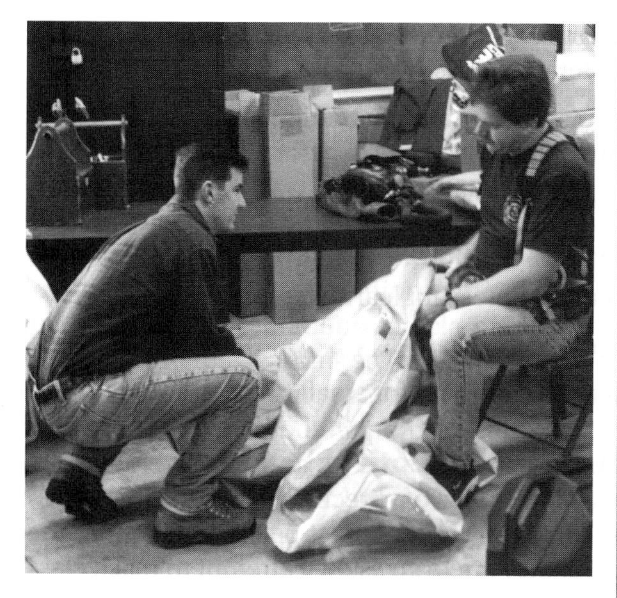

Figure 4.6 Learning will take place much easier when instruction is relevant to the job. *Courtesy of Larry Ansted, Maryland Fire and Rescue Institute.*

Figure 4.7 Learners will organize information in order to help themselves reach goals.

tend to organize it in their own way, but the effort may be time-consuming or incorrect.

Provide Opportunity for Participation

Learning is an active process. Learners who are actively involved in learning retain more information and recall it more quickly because they have discussed it, thought about it, and applied it while participating throughout the class. Those learners who must sit passively and take notes during a long lecture are less likely to remember the information or know how to apply it.

Provide Opportunity for Practice and Repetition

Learners retain learned material through practice. A person who sets out to do a task and performs it successfully retains the knowledge needed for performing the task in the future. The more often knowledge is used, the better it is retained and the quicker it becomes automatic. Learners should practice new skills as soon as possible after they are introduced (Figure 4.8). If there is a large gap of time between learning and application, other learning during the interim may interfere, and learners may forget what they learned previously. What individuals learned in previous situations (other classes or on the job) may interfere with new learning.

When individuals learn to perform a task one way, they develop a mental diagram (schema) that they follow as habit for each performance of that task. A *schema* is a framework of organized ideas or a knowledge structure in memory; it is how we organize information that enters our brains. When individuals must learn new ways, some may have problems adjusting because the new ways are different from the mental schemas they originally developed. If this new information does not fit an individual's preexisting schema, it may be rejected. However, with a lot of practice, a new schema can be created. Someone who has no previous experience with a new skill may be able to learn that skill easily because there is no previous learning, or schema, to interfere with new learning.

Bring Experiences to the Lesson

Individuals have different experiences and knowledge and may vary in how they receive and perceive new information. Existing knowledge has a tremendous influence on our memory when we are exposed to meaningful new information. Instructors must make new information meaningful in some way so that learners can link it to old information, store it in some logical place in memory, and recall it for later use. Instructors need to create relationships between new and old information. Explanations, discussions, and activities that are vivid and job-relevant help learners see relationships or links and transfer known or old information to new information.

There can be some resistance to linking information because of the individual's mental map (schema). As we learn, we develop a foundation on which we base all other new information, which we try to relate to something we know. Those with more knowledge and experience may have a greater mental schema, and they will discover that it is easier to find a link for new information. Those with less knowledge and limited experience will need to first create a foundation for that new information so they can later add new information or build on their mental network. Instructors can help link new information to old by relying on their experiences and those of other experienced individuals in class to illustrate relationships. Illustrations or examples from experienced individuals, if carefully guided by the instructor, can aid learners in gaining insight or solving problems. Instructors can further link information by encouraging learners to question the methods of the experienced

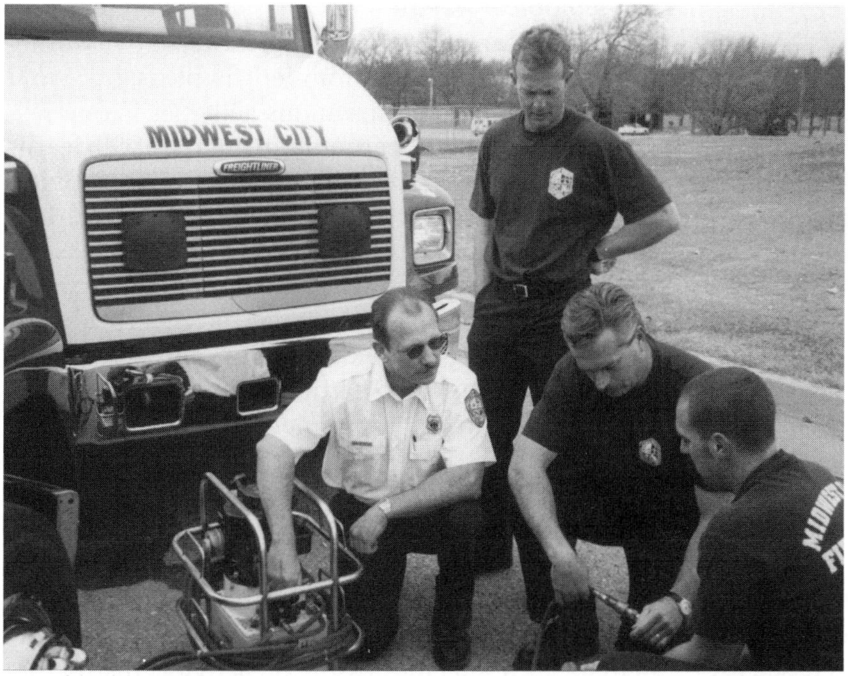

Figure 4.8 Practice helps the learner retain knowledge. *Courtesy of Jarett Metheny, Midwest City (OK) Fire Department.*

individuals and to discuss alternative ways to apply the new knowledge and skill.

Provide Feedback and Reinforcement

In all their tasks, learners must receive immediate and frequent feedback and reinforcement on their progress. Instructors must correct mistakes before learning is set, and they must reward successes to encourage continued learning.

Provide Appropriate Periods of Instruction

Trying to teach too much in a long, continuous session may interfere with learning capacity. Instructors who try to teach for more than an hour without a break will find that learners may not completely understand what was taught in the latter part of the lesson, especially after lunch or on a hot afternoon. Learners learn best in short, intensive sessions. Be sensitive to learner attention and give breaks as necessary. Especially when teaching complicated material, instructors should teach for short periods. Doing so allows learners to have time to absorb the material, discuss it among themselves, think of questions, and return ready to participate and learn more. Traditional sessions are 50 minutes with a 10-minute break, but instructors should not feel the need to stick with a rigid teach-and-break time frame if material, environment, or attention requires adjustment.

Motivation

[NFPA 1041: 2-4.2, 2-4.2.1, 2-4.2.2, 2-4.3, 2-4.3.1, 2-4.3.2, 2-4.4, 2-4.4.1, 2-4.5, 2-4.5.1, 2-4.5.2, 3-2.6, 3-2.6.1, 3-2.6.2, 3-3.2, 3-3.2.1, 3-3.2.2, 3-3.3, 3-3.3.1, 3-3.3.2, 3-4.2, 3-4.2.1, 3-4.2.2, 4-3.3, 4-3.3.1]

Motivating learners to achieve their best should be a priority but can be a challenge for every instructor. Instructors also need to be motivated. Certain job characteristics motivate instructors to participate as expected, and certain internal desires cause them to perform beyond the expected. Through their studies and research, psychologist Abraham Maslow (1943) and researcher Frederick Herzberg (1966) made assumptions that people have many needs. These needs originate from two human desires: (1) the desire to avoid pain, hardship, and difficulty and (2) the desire to grow and develop. Maslow's "Hierarchy of Needs" theory is the most familiar motivational scheme. These two theories, various motivation techniques, and motivation's relationship to learning are discussed in the following sections.

Maslow's Hierarchy of Needs

We all have certain wants or needs that we strive to satisfy and that motivate us into taking some kind of action to eliminate or fulfill a need. Psychologist Abraham Maslow identified needs as deficiency needs and growth needs and placed them in a hierarchy. *Deficiency needs* are those that individuals try to alleviate or rid themselves of such as hunger and physical discomfort, fear, loneliness, and lack of confidence. Listed in ascending order, the deficiency needs are *physiological, safety or security, social belonging,* and *self-esteem.* Though most people usually attempt to satisfy one level of need at a time, they may attempt to satisfy multiple deficiency needs at once. *Growth needs* are those that individuals usually attempt to satisfy after deficiency needs are met. Individuals are self-directed in seeking growth needs and motivated to seek self-fulfillment by realizing dreams or goals and by indulging in pleasant activities or surroundings. The growth needs are *self-actualization, desire to know and understand,* and *pursuit of the aesthetic* (the beautiful rather than the utilitarian) (Figure 4.9).

Deficiency-motivated individuals often depend on others for help; growth-motivated individuals are more able to help themselves. Instructors need to be aware of how these need types may affect learner motivation, interest, participation, and attention in the learning environment. The following sections give guidelines for helping individuals satisfy these needs and aid the learning process.

Physiological Needs

Instructors can satisfy physiological needs by assuring that the classroom environment is pleasant. Avoid scheduling too many strenuous activities in one session that will fatigue learners and interfere with learning. Eliminate irritating distractions. Provide comfortable seating, and arrange the room so it is suitable to activity-based learning and discussion. Adjust room temperature and ensure adequate ventilation so the classroom is comfortable. Provide adequate lighting so learners can see well. Give appropriate breaks so learners can take care of personal needs and get refreshments.

Security Needs

Fulfill the need for security by making sure all safety precautions have been taken. In addition, it may be the responsibility of the instructor to determine whether the participants are physically suited to participate in a specific lesson or activity. Instructors must guide and coach learners through tasks and en-

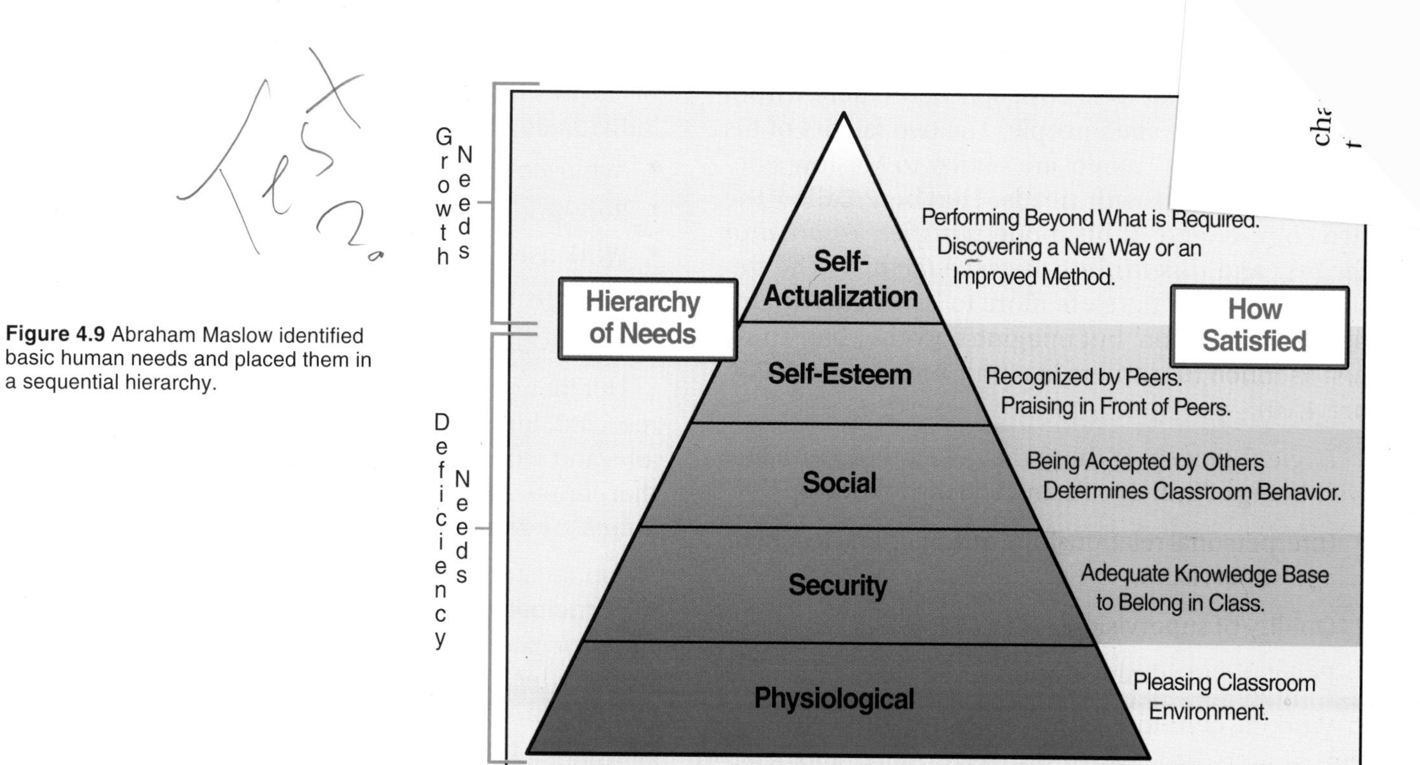

Figure 4.9 Abraham Maslow identified basic human needs and placed them in a sequential hierarchy.

sure that each individual is performing properly and according to safety guidelines. Be sure everyone has an adequate knowledge base and can keep up with other learners. Let learners know that instructors will work with them until they are able to perform appropriately and safely.

Social Needs

People normally have a high interest in satisfying social needs. There is a natural human tendency to have a feeling of belonging, and learners strive to associate with their peers and be accepted by them. This need to have peer approval governs learner activity and behavior in class. Instructors can help program participants feel comfortable in class by having ice-breaker introductory exercises. People in a group of strangers begin to feel comfortable and a part of the group if they at least know each other's names and where each person lives. Issuing name tags can help the process. Reassign individuals in group activities so everyone has an opportunity to work with and get to know everyone else in class.

Self-Esteem Needs

Learners' self-esteems can be satisfied by recognizing their competence or knowledge in front of the class. A learner who has the opportunity to demonstrate satisfactory performance of a difficult task will be highly thought of by other learners. Even those individuals who are having difficulties mastering skills need to be reassured that they are performing well as they accomplish each little step. A learner who is praised by the instructor, rather than ridiculed, works harder to earn more praise and recognition.

Self-Actualization and Other Growth Needs

Usually, it is the individual who pursues the growth needs, but instructors can enable and encourage that pursuit. Often, the effort and time needed to satisfy deficiency needs diverts a person's energies, and the desire to satisfy growth needs may remain unfulfilled for years. When people feel that their deficiency needs are met or under control, they feel ready to satisfy those dreams, goals, and pleasures they had to postpone. Some people are able to satisfy deficiency needs to a point where they begin to simultaneously fulfill growth needs such as going back to college while working two jobs to make the house payment. Some individuals regress to a lower level because they have become overwhelmed by the goals they have set for themselves. They may arrive at class unprepared or too tired to pay full attention and participate completely, or they may drop out. But they still have a desire to learn — a growth need — and they will try again. By providing encouragement, positive feedback, and relevant learning experiences, instructors can assist learners or encourage them to fulfill growth needs.

Herzberg's Theory

After extensive study of employees on the job, researcher Frederick Herzberg proposed that the job acteristics that satisfied people were different from those that dissatisfied people. The two factors of his "job-enrichment" model are similar to Maslow's deficiencies and growth needs. Herzberg called the two components of his model *satisfiers (motivator factors)* and *dissatisfiers (hygiene factors).* The descriptions seem to relate more to instructor or employee motivation, but imagine how they can apply just as much to learners in a training situation (Figure 4.10).

Hygiene factors, or those that affect the health and well-being of individuals, include the following:

- Interpersonal relationships with learners, teachers, and supervisors
- Quality of supervision
- Policies and administration
- Working conditions
- Personal life

Motivator factors, or those that stimulate action in individuals, include the following:

- Achievement
- Recognition
- Work itself
- Responsibility
- Advancement

Herzberg theorized that if hygiene factors were not resolved, individuals would be dissatisfied with their jobs and would perform poorly. Yet Herzberg found that eliminating job dissatisfaction did not necessarily increase commitment or job performance.

Job satisfaction and motivation to work are related to participation and performance. Individuals decide to *participate* based on the concept of a "fair day's work," which means that they do all that is necessary to meet minimum commitments in return for "fair pay" in the form of salary, benefits, acceptance, courteous treatment, and reasonable supervision. Herzberg

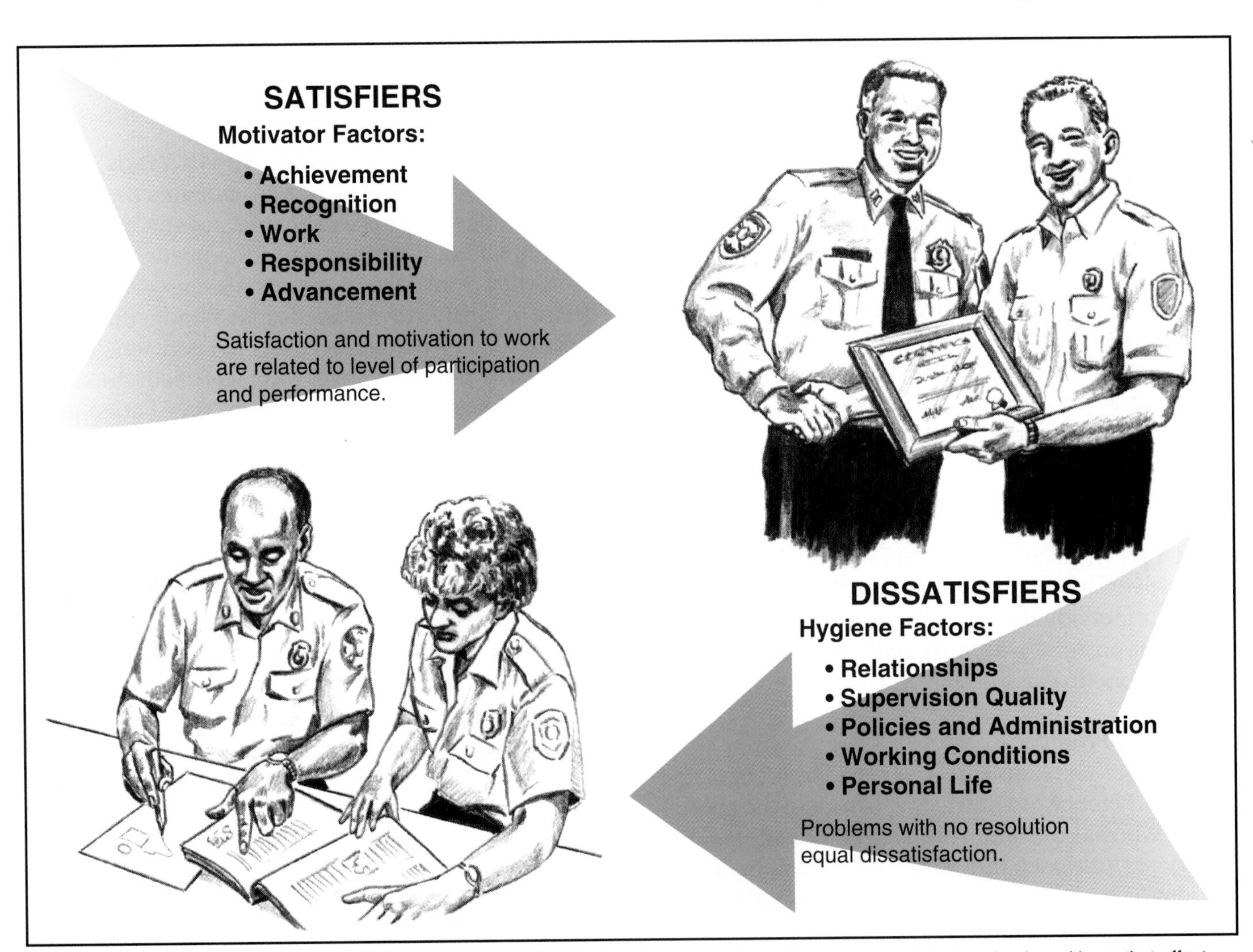

Figure 4.10 Frederick Herzberg determined that an employee's job satisfaction was affected by two types of factors: hygiene (those that affect health and well-being) and motivator (those that stimulate action).

found that though these hygiene factors did not increase job satisfaction, the absence of them could decrease job motivation. Hygiene factors are *expected* as part of fair pay and do not tend to motivate an individual to go beyond minimum expected participation. If an individual decides to *perform* beyond the "fair day's work for a fair day's pay" contract, that decision is voluntary. According to Herzberg, rewards for *participating* in a fair day's work are typically based on the *external* types of motivators that focus on work conditions. Rewards associated with *performance* that an individual invests into the work are typically based on *internal* motivators.

Although individuals are usually motivated from internal desires to fulfill certain needs, there are also a number of external motivational factors that stimulate people to pursue goals. *Internal motivation* is stimulated by personal growth needs and requires no outside stimulus to create a desire to comprehend, retain, and use the information received. *External motivation* is based on some reward or recognition, yet these types of motivational factors can be *internalized* or converted to internal motivational desires.

Techniques to Motivate Adult Learners

If there is a block or deficiency that prevents a learner from attaining course goals, learner reaction is either to "fight and drive through" the block or to "withdraw and give up." Some individuals need the attention and positive reinforcement from instructors who, by modeling certain behaviors and attitudes, provide the stimulus that influences and motivates reluctant learners (Figure 4.11). By using various instructional methods and relevant learning activities, instructors can stimulate individuals to fight and drive through learning blocks.

The needs theories of Maslow and Herzberg discuss factors that motivate individuals. Over time and exposure to their class members, instructors will find that they will develop an ability to recognize "motivational triggers" and can use them to encourage and enhance learning. Instructors play key roles in motivating adult learners and can use the following techniques:

- Provide opportunities for learners to be creative and to develop thinking skills.
- Share ideas and receive positive comments or participate in reasonable debates.
- Promote working together in peer groups to share and learn other methods.
- Show that classroom knowledge and skill can be applied to real-life situations.
- Show videos or provide demonstrations that have meaning relative to job requirements.

Motivation — Relationship to Learning

In their text *Learning in Adulthood,* Merriam and Caffarella (1991) estimate that at least half of the adult learners who take some form of training or education do so for job-related reasons. Adult learners see training and education as a direct benefit to their life situation and apply their new knowledge to solve work- or home-related problem situations. The relationship between motivation and learning is illustrated in the following series of progressive steps:

- Gain interest and ensure success by using a variety of teaching styles that match learning styles, abilities, and needs.
- Use activities that include discussions and expressing opinions to develop thinking skills and generate interest.

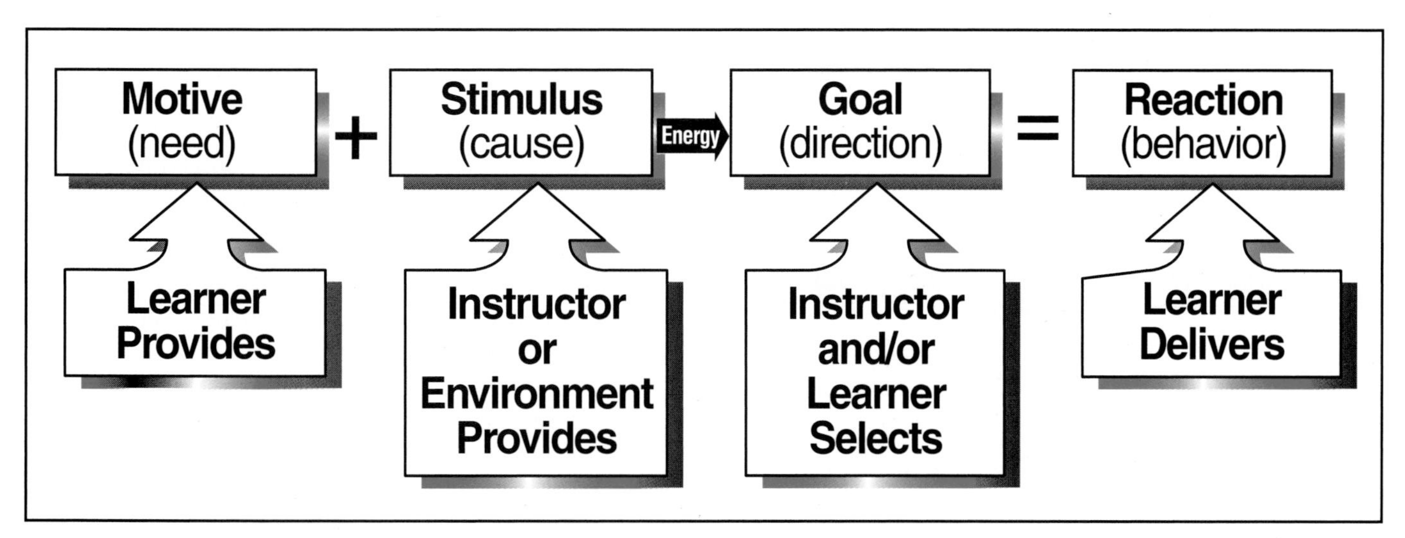

Figure 4.11 An instructor can help stimulate learners' interest so they become motivated.

- Generate interest and confidence by coaching and guiding skills and by offering encouragement and positive reinforcement. Increased confidence stimulates desire to learn and participate.
- Provide opportunities for participation in activities that keep attention and interest, stimulate thinking, develop thinking skills, and develop relationships with others. Interests, thinking skills, and relationships are carried over into work and life situations.
- Provide external motivations, such as rewards, recognition, and certificates, that cause learners to feel successful. Recognizing successful completion of tasks, lessons, and programs with external motivators stimulates internal growth desires and positive changes in values, attitudes, and beliefs.

Instructors can stimulate motivation in learners through certain attitudes and ideals, which demonstrate their own internalized growth needs. Consider the following actions and how they can motivate learners who are exposed to positive attitudes and ideals. By following these few simple procedures, instructors will see their actions aid learners in progressing through learning domains and mastering skills.

- ***Demonstrate enthusiasm*** — Show as much interest in the program as expected of class participants.
- ***Expect success*** — Convince learners that they are capable of mastering program goals. Instructor expectations are powerful learner motivators.
- ***Require outstanding performance*** — Encourage outstanding performance by guiding and coaching them to that level. Instructors who expect good performance will have learners who perform well.
- ***Encourage achievement*** — Set goals and motivate learners to achieve their best.
- ***Stimulate motivation*** — Make a conscious effort to determine learner motivators.
- ***Provide relevancy*** — Give learners the reasons for what they are learning; learners need to understand them.

Learning and Remembering

[NFPA 1041: 2-4.2, 2-4.2.1, 2-4.2.2, 2-4.3, 2-4.3.1, 2-4.3.2, 2-4.4, 2-4.4.1, 2-4.5, 2-4.5.1, 2-4.5.2, 3-3.2, 3-3.2.1, 3-3.3, 3-3.3.1, 3-4.2, 3-4.2.1, 3-4.2.2]

Many educational psychologists have done extensive research on how we learn and remember information. The theories, research, and conclusions fill many textbooks. This section will discuss a few areas of research that discuss how instructional methods affect or influence learning.

An educational theory promoted by the Greek philosopher Aristotle and later popularized by English philosopher John Locke (Harpham, 1992) is that an infant's mind is like a blank slate without content until exposed to experiences. Though other educational theories debate this, we know that learning begins the moment new life responds to the stimuli of the outside world. Many messages stimulate the senses, and some of those messages are important enough to work their way through the memory system to be stored. The mind looks at the world through the five senses, and each sense records a certain amount of information (Figure 4.12).

Dugan Laird, an author and consultant in the training and development field, describes *sensory-stimulation theory* in his book *Approaches to Training and Development* (1985). This theory says simply that *"for people to change, they must invest their senses in the process."* Instructors manage this process by stimulating what learners see, hear, smell, touch, and taste during a learning session. Laird states that learners pay *"more attention to sensory experiences than to mental processes or emotional involvement."*

Those who promote the sensory-stimulus approach to learning emphasize that the sense of sight takes in the most information with hearing next. People learn very little through the remaining three senses, though those senses often stimulate memories. Figures vary among researchers and educational theorists, but they believe the following information is accurate:

- Between 75 and 83 percent of what people learn is acquired through what they see. This means that instructors should use strong, colorful, and vivid visuals in their presentations so learners can see the information.
- About 11 to 13 percent of what people learn is acquired through hearing, which means that learners get very little information from lectures with no other sensory stimulus.
- About 1.5 percent of what people learn is through touch. Though touch is an important sense, especially when it gives information on hot, cold, texture, or shape, it is described as a popular rather than scientific concept in J. P. Chaplin's *Dictionary of Psychology* (1985). Does the smooth texture of a desktop stimulate memories? Does the rough texture of a rock remind you of other rocks? Touch

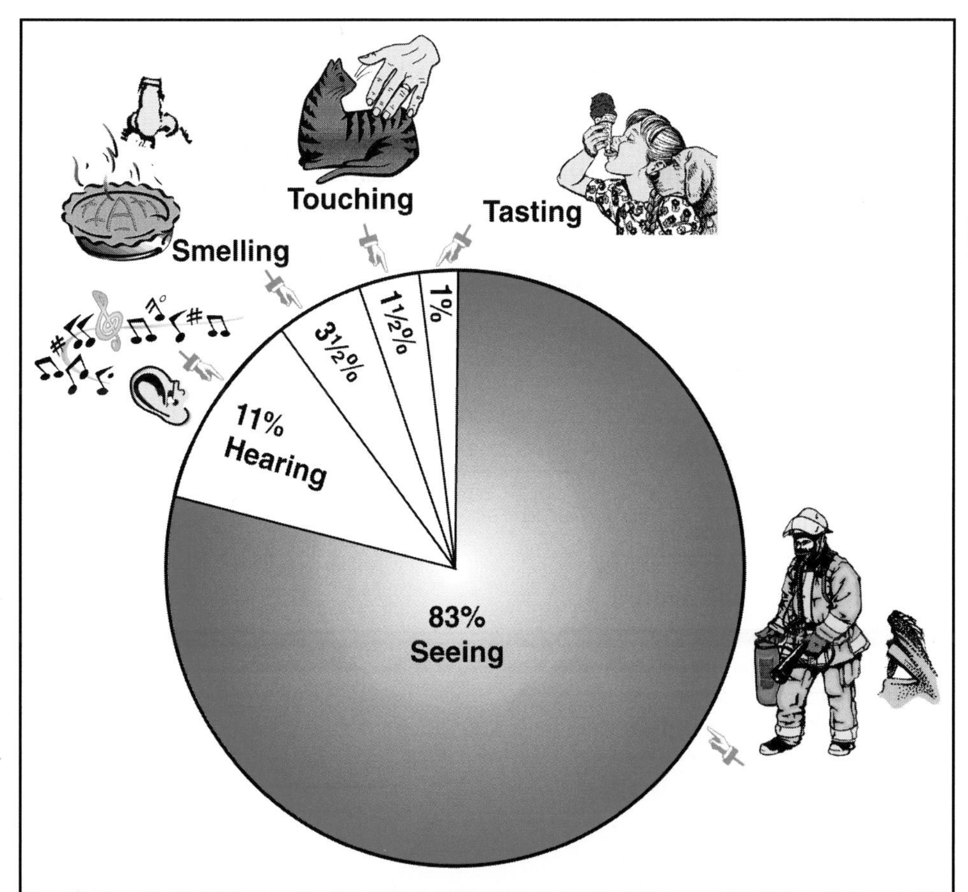

Figure 4.12 The five senses (seeing, hearing, smelling, touching, and tasting) are important to learning.

provides us with some information and some memories. You will remember certain touch experiences, such as when you touched a hot stove, you got burned. Based on that memory, you will not touch hot stoves again and will be cautious around other hot items because of that first vivid "hot-touch" experience.

- Learning by smell (about 3.5 percent) and by taste (about 1 percent) are often hazardous ways to learn especially in fire and emergency services training. Researchers are finding that the sense of smell has a very strong memory trigger. For example, visit your child's school on parent-teacher conference day, and it is likely that the familiar classroom odors will cause you to flash back to a particular day and time in your elementary school days. Visit someone in the hospital, and again the odors will trigger a memory of another visit or your own hospital stay from previous years.

Though people learn through the senses, what amount of that learning do they retain? Over the years, extensive research has been done to determine what methods have the most effect on learning and remembering. The research of educator Edgar Dale (1969) and psychologist Jerome Bruner (1956) overlapped in many areas. The philosophy of Dale, an audiovisual expert, was that learners must operate from a basis of experience, which helps them as they progress to more abstract concepts. Bruner, one of the great thinkers in education, held that there are three levels of learning: *enactive* (learning by doing), *iconic* (learning by using visual symbols), and *symbolic* (learning by using verbal symbols). Dale developed a "Cone of Experience," and in an apparent coincidence, Bruner developed a parallel scheme using his three levels of learning.

A "cone of learning" evolved that has been used in fire and emergency services training for years. This cone illustrates that individuals retain about 10 percent of what they read, 20 percent of what they hear, 30 percent of what they see, 50 percent of what they see and hear together, 70 percent of what they say or repeat, and 90 percent of what they say while doing what they are talking about (Figure 4.13). In addition, since we learn more as an active participant than as a passive participant, we can conclude from the figure that the most effective mode of learning is a method that includes receiving or learning a new idea by a combination of methods that causes individuals to be active or to participate while learning. As illustrated on the cone, the highest level of remembering occurs when an individual performs a task while saying or describing that task. On the other hand, the cone illustrates that individuals recall very little from passive methods such as reading an assignment or listening to a lecture. The conclusion appears to be that the more senses used in the learning process, the more information is remembered for later recall.

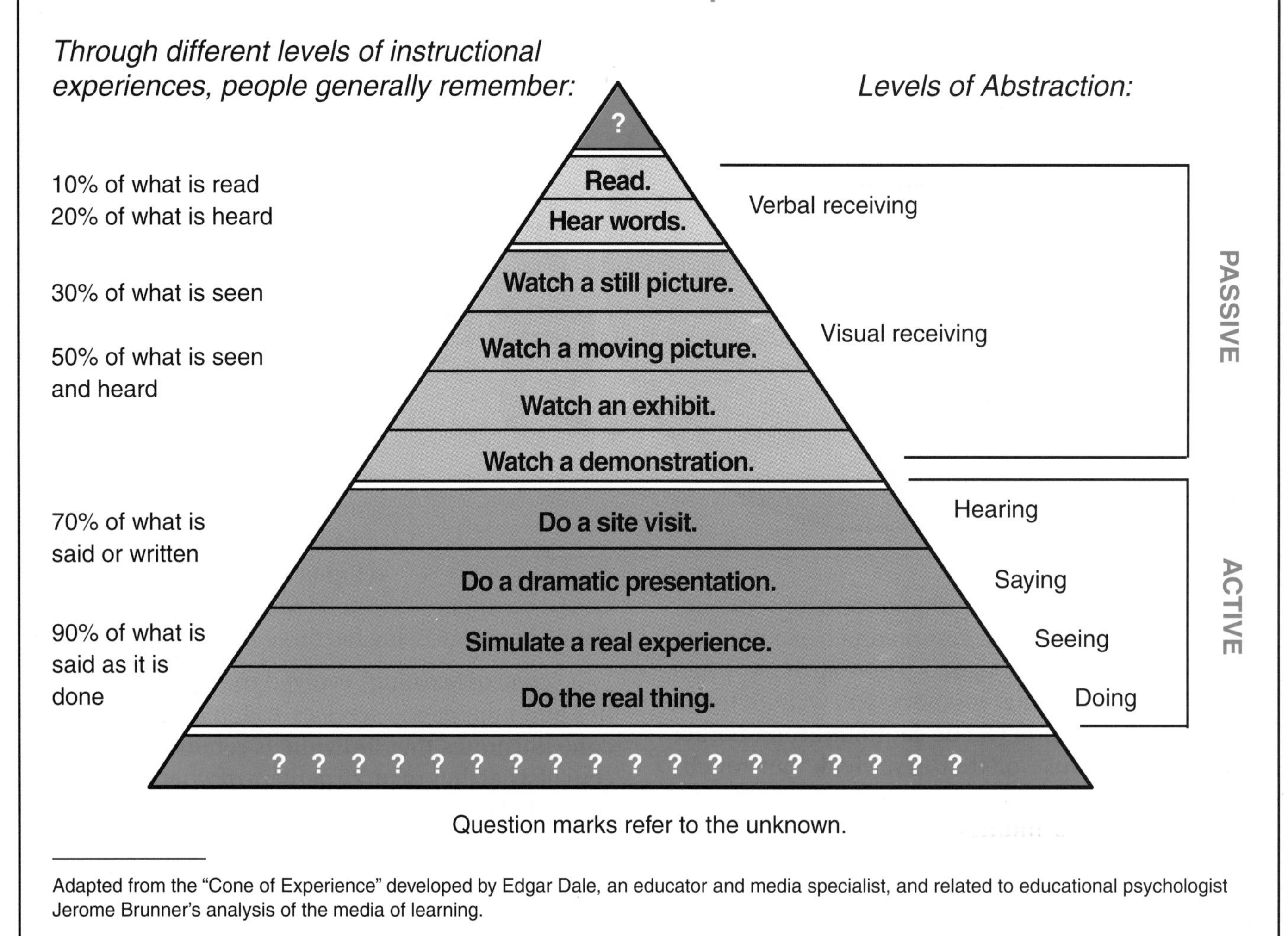

Figure 4.13 The application of a basic learning principle to Dale's Cone of Experience.

Memory

[NFPA 1041: 2-4.3, 2-4.3.1, 2-4.3.2, 2-4.5, 2-4.5.1, 2-4.5.2, 3-3.2, 3-3.2.1, 3-3.2.2, 3-3.3, 3-3.3.1, 3-3.3.2, 3-4.2, 3-4.2.1, 3-4.2.2, 4-3.3, 4-3.3.1, 4-3.4, 4-3.4.1]

What learners store in memory depends on how instructors gain attention, stimulate the senses, and ensure attention to stimuli. To make learning memorable, instructors must make learning vivid and interactive. Action and color that grab attention and embed themselves in the memory are easy to recall and relate to the time, place, and context in which they were learned. Learners must relate new information to what they know so it fits into their mental schema — the mind map that organizes knowledge.

Each individual organizes information differently in a schema that is based on the knowledge stored in different memory areas. Some learners have an extensive schema of knowledge, while the schema of other learners is limited. When instructors provide new information, learners take that information and try to link or relate it to what they already know. If there is some similar concept in their schema or knowledge network, they can remember the information more easily. If the concept is entirely new or unfamiliar, they must take more time to rehearse and store that memory so they can recall it later.

Remembering, or placing information in memory for later recall, is an amazing process. Knowing the process of remembering and recalling helps guide instructors in planning and presenting information. The following sections describe the memory components: sensory, short-term or working, and long-term. These components are major factors in learning, and they affect how we learn.

Sensory Memory

The mental storage system for attention-getting sensory stimuli or input (such as smells, sights, sounds, and sensations) is *sensory memory.* A sensory stimulus is either important enough to remember, so commonplace it is disregarded, or unimportant enough that it is forgotten. It is difficult to attend to more than one stimulus at a time and remember it well. Of the many stimuli around us, we notice and record only a fraction of it in sensory memory. In order to remember information, learners must give an appropriate amount of attention time to the sensory stimuli they are receiving on the current information before they can attend to other stimuli on new information. To aid memory, new information must relate to some other known and understood information. Most people have developed a preference for learning through a particular sense, and everyone tends to learn and remember more easily if information is presented to them in their preferred learning sense. Instructors cannot always present instruction through five different senses, but they can stimulate multiple senses through sound (thought-provoking discussion), visuals (colorful slides, pictures, graphs, and charts), activities (demonstrations and practice), and combinations of sensory stimuli (videos and role-playing). Think of how you as an instructor could present vivid information and use all five senses in a lesson on ladders, patient assessment, report writing, ventilation, oxygen therapy, fire behavior, or cardiopulmonary resuscitation (CPR). Even the most ordinary or routine lessons can be presented in ways that stimulate the senses with a little thought and planning. Instructors must give learners time to "take in" all the sensory information so that their memories can attend to it or decide where to store it for later recall.

Instructors who try to introduce or cover too much new information in a short time span will have learners who may not be successful in learning or remembering all the information or skills. It is helpful to the memory process if new information and skills can be linked to already known or learned information. The relationship helps learners grasp meaning, understand new concepts, and apply the new information more quickly. For example, explain new terms so that they actually link to something already familiar. The term *hypoxia* is not difficult but may be confusing to new emergency care learners. The *hypo* part means low — opposite of *hyper.* The term *hyperactive* should be a familiar one. The *oxia* part is familiar enough — it comes from oxygen, the substance we breathe. Once explained, a seemingly difficult or confusing new term suddenly makes sense. Instructors can illustrate the word parts by using different color marking pens as they write on an easel pad (also known as flipchart). Have members of the group hold their breaths as long as possible, tell them as they take that first gasp that they were *hypoxic,* and ask how they felt. The learners were actively participating in hearing, seeing, and doing an activity that involved at least three senses. The example was vivid and memorable and was likely sent to other memory components and through numerous memory links so that information will be easy to recall. The following section will show the link between sensory and short-term memory.

Short-Term or Working Memory

The memory component that holds information for about 20 seconds or so and is limited to about seven items or "chunks" of information is called *short-term memory.* Short-term memory deals with a tiny slice of several sensory events occurring in the present and therefore limits what we receive, process, and remember at the moment. Remembering requires "work" to process information if it is to be stored in long-term memory for later recall. In the earlier example, the learners "worked" or participated in remembering the term *hypoxia.* They were actively involved and consciously thinking while they learned the term. Short-term memory is the center for conscious thinking. It allows a person to choose whether to forget something or store it for later use. It connects sensory and long-term memory functions as it works to process sensory information to store and to recall from long-term memory.

Any information an individual processes and stores is easily bumped by other information. To remember or store something, an individual must rehearse it and find a relationship between the new information and familiar images, meanings, or other complex information. Then, short-term memory processes the information and sends it to long-term memory to store for later recall.

Because memory can only take in about seven items at a time, instructors must plan to provide simple but vivid sensory stimuli. Visual aids are sensory stimuli, and looking at them is a learning method that uses short-term memory. Short-term memory registers a limited number of items seen and tries to relate them to something already known. If there is a link and the information is presented vividly, the learner actively participates in the learning and sends the information to long-term memory.

To have a memorable impact, visuals must be simple and show key points that instructors spend time discussing and demonstrating. Visuals should not be cluttered with long descriptive paragraphs or busy pictures. Crowding visuals such as transparencies or slides to get all the information on one screen causes a memory overload; too many stimuli — words, facts, and figures — cannot be remembered. Some learners will remember only the first few items and forget the rest; some will remember only the last few and forget the rest (see Thorndike's "Laws of Learning" later in this chapter). Memory needs about 20 seconds per item for the short-term memory process to decide whether to save or forget the information. Remembering requires some rehearsal time (repetition) and time to find a link or relationship with a similar experience or piece of information in memory. Also, when introducing new information, consider that each new item that is introduced can be bumped by the next item if too many items are presented in too short a time period. Instructors must give some time for conscious thinking, rehearsal, and linking of each piece of information. This teaching method helps maintain information in short-term memory and transfer it to long-term memory.

Long-Term Memory

The memory component that holds information for a long time and is considered permanent storage is *long-term memory.* This memory component uses past information to understand events in the present. Researchers tend to agree that its capacity is limitless. Psychologist Endel Tulving (1985) describes three types of long-term memory: episodic, semantic, and procedural:

- ***Episodic memory***— Contains personal experiences and remembered life events. It is an autobiographical warehouse. Life episodes or experiences add to general knowledge.
- ***Semantic memory***— Contains general world knowledge including language. It stores word meanings and language rules and enables us to relate ideas or concepts and events.
- ***Procedural memory***— Knows how things are done and provides a blueprint for future action. It deals with motor skills and "remembers" actions and takes care of acquiring, retaining, and using skills such as writing, typing, riding a bicycle, and the many skill steps used in fire and emergency services.

The first two types of memories can be illustrated with the following example: The concept of *dinner* is common to all of us (semantic memory); what we ate for dinner last night is an episodic memory specific to each of us and our experiences. To illustrate procedural memory, imagine turning a corner while riding a bicycle. The speed and angle required for turning the corner on a bike without falling can be computed mathematically. But when you are turning a corner on a bike, you do not figure the math, you just perform the action — you just know how to gauge the speed and angle. Procedural memory is considered a little more advanced than simple instinct; it is the feeling of knowing an action is right or the knowing that an action feels right.

Factors that Affect Learning

[NFPA 1041: 2-4.3.1, 2-4.3.2, 2-4.5, 2-4.5.1, 2-4.5.2, 3-4.2, 3-4.2.1, 3-4.2.2, 4-3.3, 4-3.3.1, 4-3.3.2, 4-3.4, 4-3.4.1, 4-3.4.2]

Individuals who are trying to learn and understand but are not having success often become frustrated. Instead of learning, they are busy trying to cope with frustration. Instructors must realize there may be underlying problems with individuals who are not having success in class and look for underlying causes. Once instructors discover the causes, they may be able to help learners resolve some of the frustrations either personally or by directing learners to appropriate assistance. Table 4.1 lists some areas of learner frustrations. Note how many areas correspond to Maslow's Hierarchy of Needs illustrated in Figure 4.9.

Frustrations that come from fear and worry include the fear of not knowing how to study appropriately, fear of ridicule by the instructor or classmates, or fear of failure if they cannot perform as expected. Many learners come to class with personal worries such as leaving someone at home who is sick or trying to resolve money problems.

Table 4.1
Areas of Learner Frustrations

Fear or Worry	Discomfort	Poor Instruction
• Fitting in, acceptance • The class situation • Failure • Ridicule • Keeping up with requirements • Personal problems • Family • Health • Money	• Personal strength and stamina • Eye strain • Difficulty hearing • Classroom too hot or too cold • Uncomfortable seats or poor seating arrangements • Dangerous training conditions	• Class too advanced • Class too simple • Instructor unprepared • No opportunity for participation • No variety in presentation • Class too large • No direction

Many frustrations arise from the discomfort of the physical environment or class setting. Learning situations where learners must stand or sit too long make it difficult to concentrate and learn. Poor lighting and ventilation also have a negative effect on learning. Learners are also distracted from learning if they must train in dangerous conditions on a poorly organized training ground. They will be more concerned about their safety than about learning.

Other frustrations stem from boredom, which may be the result of poor instruction. If the individual is not interested in the subject and the instructor does not gain that learner's attention through motivational tactics and relevance, then boredom will be high. Lectures that are too long and instruction that provides little, if any, opportunity to practice quickly lose learner interest. Lack of training aids and improper teaching methods quickly bore learners and reduce learning.

Success in any learning situation depends upon the individual and the learning material. Instructors must look for and remove negative learning influences so they may enhance instruction and promote learning. There are several factors that can have a negative influence on learning.

The emotional attitude of the learner is vital to the learning process. Individuals have difficulty learning if they are concerned with "real" problems such as a death in the family or with "imaginary" problems such as feeling they are not accepted by the group. Successful learners can overcome negative emotional attitudes by focusing on their needs or desires to learn. Some learners need assistance from the instructor to overcome or work out their problems. Until then, these learners may reach "leveling-off" points or plateaus in their learning processes.

A plateau can be compared to the landing in a flight of stairs; it is a break in upward progress. Some learners stay there briefly, others become stuck because they are discouraged, or they become discouraged because they feel they are stuck. Learners sometimes create their own learning plateaus from emotional responses such as fear of failure and boredom. These emotions occupy their minds and interfere with their concentration and progress. After learners master the procedural steps of a skill, they need to practice until they meet a desired skill level. Once they reach this level, they will be exposed to more information and skills and expected to progress to the next skill level. Individuals may become discouraged if they have not been able to practice a task enough to feel proficient at a certain level, or they may find it more difficult to reach a particular skill level. At this point, further progress seems impossible, and an individual may feel like quitting. Athletes often experience plateaus in developing skills. Learners, like athletes, must be coached and receive positive feedback as they practice.

Instructors must let learners know these learning characteristics are normal, help them recognize signs of frustration, and work with them to overcome problems. One solution to the problem is to continue practicing until the skill is thoroughly understood and the procedures become automatic. Another solution is to take a break, direct learners to review and think about the task for awhile, and then have the learners return to it after a period away from it. When learners cannot

, it may be that they have formed
or tried to learn something beyond
ructors must also consider that they
provide proper assistance. Instruc-
v their instructional methods to ensure that they are able to communicate and demonstrate effectively.

Laws of Learning

[NFPA 1041: 2-3.2.2, 2-3.3, 2-3.3.1, 2-4.3, 2-4.3.1, 2-4.3.2, 2-4.5, 2-4.5.1, 2-4.5.2, 3-4.2, 3-4.2.1, 3-4.2.2, 4-3.3, 4-3.3.1, 4-3.3.2, 4-3.4, 4-3.4.1, 4-3.4.2]

Learners who are able to understand and complete a task feel successful, have positive effects from learning, and are further motivated to achieve. Through these learning processes, they are working their way through Maslow's Hierarchy of Needs and, at the same time, exposing themselves to the laws of learning.

Learning is a basic process of life and is based upon certain recognized principles. Successful instructors understand and apply the laws that govern the learning process. Noted educational psychologist Edward L. Thorndike (1914), and his contemporaries (Edwin R. Guthrie, Clark L. Hull, George A. Miller, Ivan P. Pavlov, and Edward C. Tolman) developed certain learning laws that have been fundamental to educational theory (Figure 4.14).

The laws of learning (readiness, exercise, effect, disuse, association, recency, primacy, and intensity) may have different effects on each individual learner. Instructors need to understand how the laws affect learners and be aware of the many personal characteristics that learners, particularly adults, bring to the learning environment.

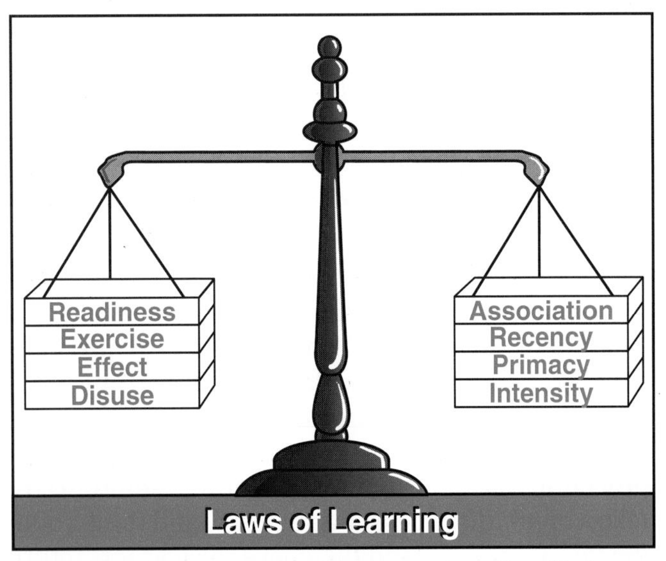

Figure 4.14 Laws of learning: readiness, exercise, effect, disuse, association, recency, primacy, and intensity.

Readiness

Readiness means a person is prepared to learn — not just ready and willing, but also mentally and physically able to learn new knowledge or skills. Prerequisite abilities may depend on prior learning as well as the level of maturity of the individual. This readiness to learn is evident in a class where learners have high interest in and anticipation of the activities in a lesson.

Exercise

The law of *exercise* stresses the idea that the more an act is practiced, the faster and surer the learning. Repetition is basic to developing adequate responses because no one ever becomes proficient at a skill without performing the operation over and over. The amount of repetition required varies from person to person. Learning is always based on activity, which requires some kind of exercise involving both mind and body. Keep in mind that practice does not always "make perfect." Mere repetition may be dull and meaningless if the learner cannot see and appreciate the reason for it. Repetition is next to useless without the essential elements of interest, meaning, and goal fulfillment. Practice must be accompanied by reward and feedback on results. Punishment or humiliation for poor or incorrect performance while practicing does not stimulate learning. The effects of reward, feedback, and punishment have ties with the law of effect.

Effect

Learning is always more *effective* when a satisfaction, pleasantness, or reward accompanies, or is a result of, the learning process. Learners will not learn or will only learn slowly when the learning situation is annoying or dissatisfying. This is not to say that learning can always take place in pleasant circumstances or ideal conditions. People often learn worthwhile lessons by "suffering the consequences" of their actions. However, if the goal is desirable and satisfies a need, individuals are usually willing to suffer setbacks while working toward success. Another factor influencing the effects of learning is the use of either praise or blame (punishment) as a learning tool. Research shows that praise is more effective than blame or punishment in motivating learners.

Disuse

This law assumes that habits and memories used repeatedly are strengthened and habits not used are weakened through *disuse*. This assumption may be true as learners are learning new skills and have not yet developed habitual responses, but once ingrained, the skills usually return with little prompting. Compare this idea with riding a bike, skating, typing, performing CPR, or using certain rescue tools infrequently. Once the skill is learned well, the process usually returns once the steps are begun because the knowledge was stored in long-term memory and the motor skill became a habit.

Association

When the mind compares a new idea with something already known, it is using *association.* This means that it is easier to learn by relating new information to similar information from past experiences. Making associations is a great aid to learning. Instructors should provide examples of associations between new (unknown) and known material, which helps learners make connections with previously learned materials.

Recency

The principle of *recency* simply means that the most recent items or experiences are remembered best. Reviews and warm-up exercises are based on the principle that the more recent the exercise, the more effective the performance. Practicing a skill just before using it ensures a more effective performance. The law of recency is closely associated with the law of exercise and with the idea that practice and repetition are aids to learning.

Primacy

The law of *primacy* states the principle that the first of a series of learned acts will be remembered better than others. This appears to be in conflict with the law of recency, but consider that each individual learns differently. Some learners remember what they heard and saw first — that initial exposure has the greatest impact; others remember what they heard and saw last because the most recent exposure is still fresh in their minds. The two laws of recency and primacy are the reasons why instructors include an overview and a strong introduction plus a summary and a strong review in their instruction.

Intensity

The principle of *intensity* says that if the stimulus (experience) is vivid and real, it will more likely change or have an effect on behavior (learning). For example, an instructor who demonstrates how to use a certain rescue tool and involves learners in the demonstration is providing an experience that is more likely to be remembered than a lecture on how to use that tool. In other words, an actual explosion is more effective than just the word *Bang!*

Approaches to Learning

[NFPA 1041: 2-4.3, 2-4.3.1, 2-4.3.2, 2-4.5, 2-4.5.1, 2-4.5.2, 3-3.2, 3-3.2.1, 3-3.2.2, 3-3.3, 3-3.3.1, 3-3.3.2, 3-4.2, 3-4.2.1, 3-4.2.2, 4-3.3, 4-3.3.1, 4-3.3.2, 4-3.4, 4-3.4.1, 4-3.4.2]

A goal of instruction is to provide opportunities for all learners to learn and become successful in meeting the program requirements or criteria. Opportunities to learn can be presented through three teaching approaches: traditional (norm-referenced), mastery, and criterion-referenced. Based on learner experience, instructors can manage learner performance through these approaches. They can help instructors organize lessons, set performance standards, and individualize instruction. By understanding and using a combination of approaches, instructors can effectively provide positive learning experiences. The three approaches are described in the following sections.

Traditional Learning

The *traditional* approach to learning promotes the concept that if a normal group of learners is exposed to a standard curriculum, their achievements after instruction will be normally distributed. In a normal distribution, there will be the usual number of high- (those scoring 90–100 percent), middle- (80–90 percent), and low-achieving (70–80 percent) learners and some failing (below 70 percent) learners. If a score of 70 percent is considered passing, then the majority of learners have mastered the lesson. If a 90–100 percent score is considered mastery, then only about a third of the group is considered competent. This last approach causes both teachers and learners to expect that only a third of all learners will adequately learn the criteria, and this leads to a disastrous self-fulfilling prophecy. The purpose of instruction is to set up learners for success, not failure. A better approach is the *mastery-learning* concept that promotes the idea that all learners can learn and starts the instructors and the learners in a positive position (see the Mastery Learning section that follows).

Traditional learning also measures learner performance against one another, which is called the *norm-referenced* type of measurement. The *norm* is the average learner in a standardized group. The group of learners who receive scores in the 80–90 percent range provides the measure, or label, of above average or below average for the other learners in the class. *Criterion-referenced* tests give instructors a better idea of who needs how much additional instruction and in what areas. This method of testing tends to reduce learner competition and emphasizes mastery of objectives (see Criterion-Referenced Learning section).

Mastery Learning

Benjamin Bloom (1984) promoted mastery learning with the belief that it is theoretically possible for the majority of learners to master knowledge and skill provided they are given the optimum teaching appropriate to their needs and are allowed to learn at their own pace. Bloom based his philosophy on the mastery-learning proposal of John B. Carroll (1971), an educational researcher. Carroll proposed that the focus of instruction should be on the time required for different learners to learn a given amount of material. He also suggested that instructors allow more time and provide more and better instruction for learners who learn less easily and rapidly than their peers.

Mastery is defined as a high-level or nearly complete degree of proficiency in the execution of a skill. It also means the skill or knowledge in a subject that makes one master of it. When testing for mastery, instructors base learner performance on criteria (standards) stated in objectives. Bloom states that the basic task of instructors is to determine what is meant by mastery and to search for methods and materials that enable the largest proportion of learners to attain such mastery.

The mastery-learning concept became widespread and effective as Robert Mager, an experimental psychologist, and Norman Gronlund (1985), an educational psychologist, began describing desirable ways to write instructional objectives. Carroll suggested that instructors follow simple steps to enable learners to master learning through carefully worded objectives that describe what learners will do to accomplish the tasks. The steps are described as follows:

- Specify what is to be learned in a clearly stated objective that includes performance, conditions, and criteria.
- Motivate learners to learn through interesting and relevant participatory activities.
- Provide instructional materials that enable and enhance learning.
- Present materials at a rate appropriate for a group of learners who have different learning styles and preferences.
- Monitor learners' progress and provide feedback.
- Diagnose difficulties and provide remediation in the form of retraining, coaching, or peer assistance.
- Give praise and encouragement (positive feedback) for all successful performances or increments of correct performance.
- Provide review and practice sessions.
- Provide activities, practice, and mentoring that maintain a high rate of learning over a given period of time.

The mastery approach uses criterion-referenced teaching, learning, and testing and focuses attention on objectives. Learners who have problems reaching the desired criteria level on their initial efforts are given additional instruction and more opportunities to perform to the acceptable level. Write the objectives to accomplish the following:

- Identify and describe the terminal behavior or expected end performance.
- Define the important conditions by which the learners will perform.
- Define the criterion of acceptable performance.

Though mastery learning has its successes, it is not without its problems, most of which stem from misconceptions and abuses of the techniques. Note the following criticisms but realize that they can also be overcome with effort and creativity:

- Mastery learning requires careful planning and preparation of instruction and evaluation and monitoring of learner progress, which requires more instructor time and effort than traditional instruction (see Traditional Learning section).
- Criteria for achieving mastery level may be difficult to determine or define.
- Learners who achieve mastery faster than others must be given alternative assignments (such as peer assistants) to keep them busy and motivated.
- Some learners initially put forth the least effort possible, assuming that they will receive help or acquire test-taking hints as they go through an evaluation, which causes instructors to administer and score several exams as the learners are reevaluated.

Criterion-Referenced Learning

The term *criterion* refers to a standard on which a decision or judgment is based. The criteria that learners must meet are stated in the course objectives, which guides learner performance. Learning by criteria is called *criterion-referenced learning.* Instructors involve learners in activities and practice that enable them to become competent in or to master the knowledge and skills criteria stated in the course objectives. Learning is a personal experience. Each learner has a different capacity and preference for learning and masters knowledge and skills in either more or less time than another learner. All learners must be given appropriate opportunity to master or become competent in the required criteria. Remember, Benjamin Bloom said that if instructors give learners appropriate time and opportunity, they can learn anything.

The knowledge and abilities expected of a learner are provided in a planned sequence of criterion-referenced instruction that is developed from some selected standard. The fire and emergency services profession uses standards from such organizations/agencies as the National Fire Protection Association (NFPA), the Occupational Safety and Health Administration (OSHA), and the Department of Transportation (DOT). At the end of instruction, tests are given that are based on the criteria selected for the program. A criterion test measures learner performance by comparing it to the standard or criterion stated in the program objectives. At the point of testing, learners should have mastered the criteria requirements. Criterion-referenced testing compares learner performance with stated criteria, not with the performance of other learners. Though each learner's performance varies somewhat from another's, the performance is acceptable if it meets the conditions stated in the criteria; it must not be compared to the performance of another learner.

What learners learn and how they learn it affects their performance. Criteria must be written clearly so that learners are able to discriminate or tell the difference between acceptable and unacceptable performance and so that instructors may observe and measure performance. Instructors must engage learners in the process of learning through purposeful activities. The activities must be designed to meet the criteria stated in the lesson objectives, which the instructor provides in a planned sequence of instruction based on a selected standard. Included in the objective is the expected end behavior or performance, the conditions of performance, and the criteria of performance that a learner is expected to master. The criterion refers to how well a learner must perform and describes acceptable performance. Instructors specify acceptable performance, and learners practice and master that performance. The criterion also provides the standard by which learner performance is measured or tested. A learner who demonstrates mastery of the performance criteria is judged or measured as competent. Sometimes individual characteristics affect how well or how quickly learners master criteria. These characteristic factors must also be considered and are discussed in the next section.

Learner Characteristics

[NFPA 1041: 2-4.3, 2-4.3.1, 2-4.3.2, 2-4.5, 2-4.5.1, 2-4.5.2, 3-3.2, 3-3.2.1, 3-3.2.2, 3-3.3, 3-3.3.1, 3-3.3.2, 3-4.2, 3-4.2.1, 3-4.2.2, 4-3.3, 4-3.3.1]

Personal characteristics affect how individuals learn; instructors can use these characteristics effectively to make learning successful. Characteristics include life experiences, motivation, time demands, confidence in ability, and variations in learning styles. Also, demographic factors (age, sex, culture and ethnicity, and educational background) affect learning.

Instructors have quite a challenge in meeting learner needs and in making diverse groups find advantages in various learning methods. Knowing how and why individuals learn provide a foundation and direction for planning interesting and motivating lessons.

Life Experiences

Adult learners have a variety of life experiences, many of which relate to the instruction in programs they attend. Instructors can use discussion techniques to establish relationships between adult learners' past experiences and the new materials being taught.

Motivation

Adults take classes for a variety of reasons. They sometimes want to be there and want to learn something. At other times, they are there because attendance is required by an employer or supervisor, and they are not motivated to learn. Instructors need to show how the learning is immediately useful because learners want to meet their needs of achievement, satisfaction, and self-fulfillment.

Adult Responsibilities

Adults have multiple responsibilities and obligations to self, family, friends, work, community, and church. In addition, they are motivated to attend classes. Instructors must satisfy learner needs in class or thoughts of other obligations will take the learner's attention. Instructors must involve learners in planning learning activities, setting goals, and agreeing on options for meeting requirements because adults need to believe they have control over their time and involvement.

Confidence in Ability

Many adults have been out of the school environment for years and have little confidence in their abilities to be successful learners. Their learning experiences must satisfy their needs, and adults must feel that they have been successful and have accomplished something at the end of each class. Instructors must give immediate feedback and encouragement throughout the program.

Variations in Learning Style

Because of varied experiences, successes, and failures, adults have learning needs that are much different from children just starting school. Instructors must provide adult learners with a variety of instructional methods, activities, and materials to stimulate and maintain interest and motivation.

Adults learn best when they actively participate in setting learning goals and setting their own pace for meeting the goals. Adults also learn best when learning is problem-centered, and they can work to resolve an issue. All learners want to receive feedback about progress and know where they stand in relation to meeting goals. Finally, adults learn best in an environment that is friendly, comfortable, and informal.

Demographic Factors

In addition to the characteristics given previously, there are a few demographic factors that affect adult learning. The factors of age, sex, culture and ethnicity, and educational background are discussed in the following sections.

Factors of Age

Everyone has heard someone say *"I'm too old to learn these new things,"* or *"I don't remember things as well as I used to,"* or *"I know all of this already."* Many adults make such statements and convince themselves that they are true, but research has proven these ideas wrong (Figure 4.15). The mind does not necessarily deteriorate with age. For example, in one research project, a group of 50-year-olds were given the same intelligence test they had taken 31 years before. They made high scores on every part of the test with the exception of the mathematical reasoning section. In

Figure 4.15 Research has shown that age does not hinder the ability to learn; people are capable of learning throughout life. A model for lifelong learning and goal achievement is John H. Glenn, Jr., astronaut and U.S. Senator, shown prepared for his first space flight 30 years ago and for his recent space flight in 1998. *Courtesy of Office of John H. Glenn, Jr., U.S. Senate, Washington, D.C., and the National Aeronautics and Space Administration.*

another study, men aged 20 to 83 took a course in world affairs at the University of Chicago. Follow-up studies showed the older learners were more successful and continued to study the subject longer than the younger learners.

Younger learners are more likely than older learners to view authority figures such as instructors with suspicion and distrust. They are more likely to ask *"Why?"* and to demand a definite answer. Instructors can effectively use this need to know why. Knowing why is a good way to stimulate motivation and make relationships between known and unknown material, as well as answer the question.

Factors of Sex

Recent brain research has shown that behavioral differences between men and women are caused by differences in brain functioning that are biological characteristics and are not likely to be modified by cultural factors alone. Richard Restak (1979, 1985), a Washington, D.C., neurologist, Georgetown University professor of neurology, and author of two books based on his extensive research, found many factors to support his theory. A few insights on these characteristics can help instructors plan lessons that enable both men and women learners to be successful.

Women are more likely to use their left brain (logical, analytical, and sequential processes) and interact with others when learning. Men use their right brain (intuitive, creative, and manipulative) for learning and react to an inanimate object as quickly as they do to a person. Women are more sensitive to lights and sounds than men. Women are more proficient with fine motor skills, but men are more coordinated with total body movements. Men can mentally visualize how to manipulate objects better than women. Women are more likely than men to respond to loss, failure, and stress by developing depression. Women learn by absorbing their environment, while men learn by manipulating their environment. Women who are active, independent, competitive, and anxiety-free do better in school than women who are not; men who are timid, anxious, and less active generally do better than men with opposite characteristics.

Not all men and women exhibit these characteristics; some cultural factors affect behavior and learning abilities. However, it is helpful for instructors to be aware of factors that affect how individuals learn. With these characteristics in mind, coupled with the factors of Maslow's Hierarchy of Needs, instructors can plan appropriate activities that enable all individuals to participate in methods that help them master knowledge and skills.

Culture and Ethnicity

There has been a shift in our social structure in recent years because the number and influence of ethnic groups are rising. Each cultural and ethnic individual or group brings to the classroom their customs, behaviors, attitudes, and values. Instructors need to recognize and understand that there are individual differences in a class. It is also important for instructors to encourage understanding and cooperation among all learners. When conflicts or disagreements arise among different individuals, instructors must intervene with positive emphasis on the "rightness" of each culture and the values or customs that can contribute to the learning situation.

Because of the diverse group of individuals that participate in training programs, it is not always possible for instructors to be familiar with the customs of every culture and ethnic group. Many organizations have diversity training programs that introduce methods of recognizing differences and working positively and successfully with diverse groups. These training programs also include suggestions of how to handle issues that arise when someone uses a phrase or performs an action natural to his or her culture but offensive to another culture, though the word or action was not meant to offend. Instructors should take advantage of opportunities to attend cultural diversity training programs. Though these types of training programs have been offered frequently over the past few years and many organizations make attendance mandatory, attending more than one program is often beneficial. Different speakers offer different viewpoints and guidance on how to manage classroom diversity. Once in the classroom, talking openly with the participants about differences helps individuals to understand each other and promotes cooperation in learning.

Educational Background

Educational background influences individual attitudes, confidence, and ability to handle new learning experiences. Educational experience includes education level (high school, college, graduate school), literacy level (reading and comprehension ability), and learning disabilities.

Instructors generally expect that training program participants will have a high school education or a general equivalency diploma. There may be individual exceptions to completing this minimum education

Case Study: Learner Characteristics and Learning Styles

A new class of recruits includes a variety of individuals who all seem willing to learn and participate though some are more obviously aggressive and outspoken than others. After reviewing their applications and having them contribute some background information in class in order to get to know one another, it is also apparent that these individuals have a wide variety of backgrounds and experiences, schooling and previous jobs, attitudes and values. But all of them want to be successful in this training program and start on a new career. You also find that some come with superior recommendations as well as excellent scores from other training programs and jobs, others have entered this program solely based on their scores on the entrance exam or prerequisite course, and for a very few there is no background information. Your challenge is to provide appropriate learning opportunities through a variety of media and activities so that these individuals can learn, participate, and apply their new knowledge and skills in their upcoming jobs. Place this group in your training situation at your training facility. This class will prepare them for a job in your organization.

Answer the following questions:

1. What responsibilities do you have as an instructor? What steps can you take to help these individuals be successful in class?
2. How would you determine the skills and abilities of all group members in such areas as reading comprehension, thinking, leadership, giving and following directions, and cooperation? Are there other critical areas that these individuals must display to be successful in the class and on subsequent jobs?
3. What group activities can you assign these individuals to check on their leadership abilities and thinking skills? How can you give each individual opportunities to practice and apply different learning characteristics and application abilities?
4. What resources does the organization have that will assist learners in weak areas such as reading ability and comprehension or in personal areas such as controlling aggressiveness, overcoming timidity or introversion, and focusing attention? At what point do you direct these individuals to organization resources? What steps do you take to do so?

level, but completing any education level does not guarantee the ability to read, write, or comprehend. In addition, any degree implies completion of a program, not necessarily quality of learning, nor does it indicate any learning disability.

Educational inabilities and disabilities are often discovered by the instructor as an individual tries to participate and progress through a training program. Though instructors may recognize that individuals are unable to read, write, or comprehend to the expected level of the program, they are not likely to be qualified to recognize or identify specific types of learning disabilities. In their text, *Psychology Applied to Teaching,* Robert Biehler and Jack Snowman (1990) describe points on determining learning disabilities: An individual who seems motivated and has apparent potential for learning but has a low achievement or performance level may have a *learning disability.* First eliminate the possibility of other causes such as visual or hearing impairments, motor handicaps, mental retardation, emotional disturbance, or disadvantages such as economic, environmental, or cultural factors. A learning disability is then determined by findings of disorders in the psychological processing ability in memory, auditory and visual perceptions, and oral language. Instructors who recognize that individuals are having difficulties in a program should consult the policies of their organization. Government agencies and large private organizations that provide training may also offer evaluation and assistance for employees with learning disabilities or refer them to appropriate agencies.

Learners as Individuals

[NFPA 1041: 2-3.3, 2-3.3.1, 2-4.3, 2-4.3.1, 2-4.3.2, 2-4.5, 2-4.5.1, 2-4.5.2, 3-3.2, 3-3.2.1, 3-3.2.2, 3-3.3, 3-3.3.1, 3-3.3.2, 3-4.2, 3-4.2.1, 3-4.2.2, 4-3.3, 4-3.3.1, 4-3.3.2, 4-3.4.1, 4-3.4.2]

Every class has a different group of individuals, some of whom require more instructor attention than others. Those who require more attention take instructor time and effort away from the total class. To maintain a reasonable balance among all learners, instructors must be able to recognize traits of time-consuming individuals and learn how to manage them effectively.

Different ability levels require varying amounts of instructor time. Learners who are quick to learn may require as much instructor time as learners who take longer to grasp a concept or learn a skill. Instructors must set the learning pace to the learners' rate of understanding, or learners become discouraged. Some learners require individual instruction or tutoring to keep up with the class.

Individual personalities place demands on instructor time and can take away from actual instructional time. Everyone is different in his or her attitude and emotional reaction. Some are timid or reserved; others show off and try to impress the group. Learners who are quiet (or nondisruptive) do not interrupt the lesson, but they may not be participating to the expected level. Disruptive learners can jeopardize classroom management. Each type of learner demands extra effort and special attention from the instructor.

Some individuals are labeled "problem learners" when they are really only expressing different personalities. Previous discussion in this chapter described a range of individual differences and characteristics among learners, which resulted in different ways of learning. Different types of performance are indications of learner ability and disability. Instructors must be aware that different social and cultural backgrounds shape learners and affect learning styles and abilities.

Consider the educational and literacy levels of the learner. *Educational level* is the number of years spent in school; *literacy level* is the level at which learners can read and write. Many adults return to college to take courses or pursue degrees with the intent of getting better jobs. Instructors may find that the educational level of learners is higher now than in past years, but the literacy level is declining.

After teaching the same subject several times, instructors will come to realize that each group has different characteristics, which will cause them to change their presentation style. The following sections describe some common learner personalities that instructors encounter in every audience. Instructors must be aware of, able to manage, and prepared to teach each type.

Individuals with Low Literacy Levels

How can instructors handle low literacy levels? The first step is through careful development of lesson plans and presentation methods. Instructors should use visuals and other training aids rather than long lectures. If asked for recommendations, suggest and select reading matter in textbooks, tests, and handouts that is written in short sentences and paragraphs, is double-spaced, contains directional headings and wide margins, and is printed in type large enough to be read easily. Vocabulary should be simple, and all terms should be easily found in a glossary. When giving exams, supply directions that are simple and to the point. Pictures, tables, graphs, and charts are helpful aids in illustrating and breaking long runs of text and may be used in exams also.

Individuals with Learning Disabilities

Learners with any kind of learning disability, such as those who have reading and comprehension problems, are at a disadvantage. Training organizations know more now about recognizing learning disabilities than in the past because the educational system has developed methods of identifying them. Instructors can look for the following indicators of learning disabilities in individuals:

- Problems with memory
- Problems with auditory and visual perception
- Problems with oral language
- Difficulty in speaking, listening, and writing
- Problems in reading such as word recognition and comprehension
- Problems in math areas such as calculation and reasoning

It could be said that almost everyone has some degree of learning disability, and almost everyone finds a unique way to compensate for that disability. However, there are those whose disability is a major stumbling block to learning, and they are not able to find any unique ways to compensate for their disability. Instructors must be alert for these learners and provide patience, understanding, and appropriate assistance so they may be productive and experience success. Several methods of helping individuals who have learning disabilities include tutoring, developing individualized instruction, and providing feedback on progress.

The personalities and participatory methods of some learners may be the cause of their disability or inability to learn. It is important that instructors know their audience and tailor the course to that audience.

Gifted Learners

Gifted learners are usually able to accomplish more than is expected of average learners, and they may study and learn very well without much supervision. Gifted learners are often ahead of other class members and are assets to the class if the instructor makes proper use of their abilities. Methods of motivating and maintaining the interest of above-average learners include giving them assignments that challenge their levels of ability and keep them busy with creative and stimulating activities in class. In addition, recall that "saying while doing" aids in retaining the most information and can be a learning/teaching method for gifted learners.

Instructors can couple a gifted learner with one who needs tutoring. While assisting other individuals, gifted learners improve their own retention. The one receiving the help may feel more comfortable learning from a peer rather than being embarrassed by showing incompetence in front of an instructor.

Slow Learners

Instructors are usually able to identify "slow" learners soon after starting work with a group. Though indications that a learner is slow may be familiar and strong, do not jump to conclusions. A learner may be slow to comprehend because an instructor does not give clear instructions. For learners with special needs, it may be necessary to arrange private conferences, special assignments, different types of study assignments, or individual instruction. The instructor may need to reevaluate and revise the subject matter or instructional methods used so that the individual is able to meet the course objectives and requirements. A learner may have difficulty grasping information and skills in one lesson or category, yet may excel in other areas. Therefore, instructors must develop assignments and lesson plans with these types of learner abilities in mind. Instructors should always provide positive feedback as learners accomplish course objectives.

Nondisruptive Learners

These types of learners include timid or introverted individuals and those who appear to be daydreaming, distracted, or uninterested. In programs where learners are attending instruction for a long period of time, instructors may use several methods to convert the nondisruptive type of learner. First attempt to check on the health and habits of the learner. By asking simple personal questions on breaks, learners may readily provide answers because they appreciate the interest and attention. Remind learners that, as a member of the class, they have a responsibility to learn and participate. Instructors who show an interest in their learners often are able to motivate them to try. These learners will normally make every attempt to fulfill the class requirements and please the instructor who shows (and continues to show) interest in them.

Nondisruptive learners are often shy, bored, or uninterested. By showing an interest in individual learners, instructors can help motivate them.

Shy or Timid

The shy or timid individual may be hesitant or at a loss for words when expected to respond or participate and may actually be afraid to respond aloud or participate during class discussion. Though the learner may know the material and may have much to offer, fear, bashfulness, and possibly lack of confidence keep this individual silent. This individual is likely learning in a passive manner rather than in an active manner.

Considerate instructors avoid calling on timid individuals for discussion or response until they have developed some levels of comfort in the class. Encourage these individuals to participate when the discussion is informal. Let these types of learners feel some success in their informal responses and participation before expecting them to initiate discussions or give presentations. Talk with these learners during breaks so they begin to feel comfortable with you and course expectations. Instructors who use these simple methods of making learners feel comfortable will help them overcome shyness and encourage them to take a more active role in the class.

Quiet or Bored

Many quiet learners may be above average in ability but because of circumstances — such as uninteresting subject matter, unfamiliar terms, boredom, and drawn-out points — they may drift mentally. Instructors should be alert for the signs of daydreaming and boredom: glazed looks, gazing around the room, doodling, or thumbing through materials not related to the subject. Instructors can redirect attention by asking direct questions or beginning activities that require learner participation rather than just listening.

Uninterested

Uninterested learners display little energy and attention. Curiosity is inherent, and lack of interest is not natural. Instructors should be curious about the lack of interest in these individuals. Check with the training office to see whether records show personal or other problems. Records may show that counseling or tutoring was provided in other courses, that the learner was admitted to a class beyond the level of readiness, or that there are health, emotional, family, or learning problems that must be addressed before the learner can make progress. Instructors will want to know outcomes in order to plan strategies for working with these types of individuals.

Disruptive Learners

Disruptive learners include those who sidetrack, distract, or stall class progress by diverting attention and interests from the lesson. Instructors must manage these types of learners with a certain amount of discipline. Use class time to pursue the lesson rather than allowing a disruptive individual to control it through antics. If these individuals do not cooperate and initial, tactful methods of directing them to cooperate fail, ask them to leave the session. Let them know you expect cooperation and constructive participation upon their return. Redirect their energy by calling on these individuals regularly so they expect and must prepare for participation or be embarrassed by their lack of ability and knowledge.

Talkative and Aggressive

Talkative, aggressive, and extroverted individuals monopolize the discussion. They talk so much that no one else has an opportunity to participate. If a private personal appeal does not alter the actions of these individuals, give them a special assignment in class. While the offenders are occupied, other group members will have an opportunity to participate. In almost every class, there is a small group who prefer to talk among themselves rather than attend to the current activity or discussion. In these situations, instructors must recapture the attention of the group. Tell the offenders that special problems can be discussed after class rather than taking time now from the whole class. The group can be separated by requesting that they move to different seats. With adults, this is not usually necessary because in most cases they will cooperate with your requests to participate appropriately. It may be beneficial to talk with problem individuals before class to find out if they have any special issues they want to discuss. Let them know you will try to find time in the lesson to discuss them. Remind individuals of the lesson topic and goals, and suggest that their issues can be discussed once the lesson objectives have been met.

Disruptive learners are often talkative and like to "show off." Redirect their energy by calling on them regularly so they must prepare for class participation.

Show-Off

Individuals who like to "show off" use a group situation to perform and gain attention for themselves. Sometimes their performances help put other learners at ease and get others to participate, but show-offs must know when to stop and when to share the floor. Call the class to order and review main points of the discussion or skill to redirect attention from the show-off. Ask others in the class to respond. If the show-off tries to respond to all questions, simply say that you want to hear from others or you want to give others a chance to respond. If this technique fails, a very direct and effective solution to this problem is to tell the show-off that the classroom is not the place for this type of behavior and disruptions of any kind will not be tolerated. The timeliness and manner in which an instructor delivers such an ultimatum determines its effectiveness. When learners understand that cooperation is important and no alternatives are acceptable, instructors will have few if any problems. If problems persist, follow the organization's discipline policies.

Managing Individual Learners

[NFPA 1041: 2-4.4, 2-4.4.1, 2-4.5, 2-4.5.1, 2-4.5.2, 2-5.4, 2-5.4.2, 2-5.5, 2-5.5.2, 3-3.2, 3-3.2.1, 3-3.2.2, 3-3.3, 3-3.3.1, 3-3.3.2, 3-4.2, 3-4.2.1, 3-4.2.2, 4-3.3, 4-3.3.1, 4-3.3.2, 4-3.4.1, 4-3.4.2]

Instructors must manage a variety of personalities and learning styles in order to provide effective learning opportunities for all learners. Most organizations have policies for managing issues of learner problems and performance; however, the best action is to prevent problems before they occur. To prevent problems, give the program participants a clear outline of course expectations, which must include information such as classroom and training ground behaviors, dress codes, and equipment needs.

Experience, common sense, and ordinary acts of understanding and empathy aid an instructor in managing learners. First, refer to the organization's policies, and follow them carefully. In addition, instructors must get acquainted with learners in order to identify problems early. It is also helpful to draw upon the experience of other instructors and find out how they have dealt with similar problems or with the same individuals.

The organization's policies may include guidelines on how to formally manage learner problems, or it may have personnel on staff who are qualified to provide counseling, coaching, peer assistance, or mentoring. Often, instructors are expected to provide these services to their learners.

What is expected of instructors who must counsel, coach, assist, or mentor learners through learning or discipline problems? Most instructors are in the teaching profession because they have an innate ability to work with and guide others. Guidance may start with a counseling session where instructor and learner have a simple discussion or conference listing concerns and considering solutions. In a coaching session, an instructor guides learner activities, reinforces correct behaviors, and redirects incorrect behaviors. Providing peer assistance may begin with assigning another learner to work with one who is having difficulty understanding lesson material or performing skills. A mentor acts as a trusted and friendly advisor or guide to someone new to a particular role.

Regardless of which method is used to manage learner performance, instructors will provide varied learning opportunities so learners may master knowledge and skills to expected standards or criteria. The following sections give additional definitions and guidelines.

Counseling

Counseling is a broad term used for a variety of procedures designed to help individuals adjust to certain situations. Counseling includes such actions as giving advice, having discussions, giving tests that help identify problem areas, and providing vocational assistance. If it is the instructor's job to counsel learners, it is usually to discuss progress in class. Instructors may also have to counsel learners when they have personal or social anxieties or crises that affect class activities or completion of course requirements. Instructors/counselors must not assume the role of therapist if the learner appears to have a psychological/emotional problem. The instructor/counselor's responsibility is to refer the learner to a professional therapist. Most employers have some sort of professional counseling available through their employee assistance programs.

Counseling is a means of redirecting learners or eliminating learning interferences. Learner-instructor counseling sessions or conferences must be done in private. The individual must feel that the instructor is sincerely interested and considers the individual as a potential colleague. It should be obvious to the individual that the purpose of the conference is to help work out seemingly difficult problems that may interfere with individual ability (Figure 4.16).

The instructor/counselor must encourage the learner to explain and express feelings about any troublesome situation. Attempt to see the problem from the individual's point of view. Explain to a learner who has a different understanding or perspective of policies, standards, or expectations that it is often difficult to determine who is right in some situations but that it is necessary for an organization to operate on basic rules, regulations, assignments, and acceptance of responsibilities in order to function effectively. Take care throughout a session to demonstrate interest and listening techniques and to maintain a positive atmosphere and regard for the individual.

Attempts to force individuals into acceptable behavior generally fail. Force causes resistance, and people do not like to be disciplined. Instructors must learn and practice appropriate counseling and positive reinforcement methods that stimulate and motivate learners to perform properly.

Once every attempt has been made to assist the learner and every remedy has been exhausted, then the instructor may have to treat the problem as a disciplinary situation. Organizations have policies to follow, and instructors must be familiar with them. If discipline becomes necessary, instructors must perform it privately, in a calm manner, and with an atmosphere that shows a willingness to help. Begin the session with encouragement and praise for good work, suggest a constructive course of action, and criticize the mistake instead of the individual. In most situations, the instructor who shows a sincere interest in the learner does more toward solving the problem than any action that intensifies the individual's feelings of inferiority or inadequacy.

Coaching

The term *coaching* usually refers to an activity associated with sports and athletics, but it is also an intensive kind of tutoring given to learners to improve skills in a subject or activity or to prepare for an examination. Coaching may also be used to guide individuals through learning challenges, which may include problems such as class disruptions. Disruptive behavior often results from frustration or confusion, and instructors can help learners overcome these behaviors through coaching techniques. Coaching can take place in a group setting or individually (Figure 4.17). Identify the problem source through observation and questioning, and then guide the learner to a solution.

Peer Assistance

Peer assistance refers to learners who assist other learners in the learning process. In the learning environment, a *peer* is someone who is equal in status either socially or psychologically to another. Some learners are intimidated by or afraid to perform in front of an instructor until they feel confident in their abilities. These learners feel more comfortable practicing with a peer. Learners who make good peer assistants are those who have grasped the knowledge and skill and can explain it well to others. They

Figure 4.16 The purpose of an instructor-learner conference is to work out problems that are interfering with learning.

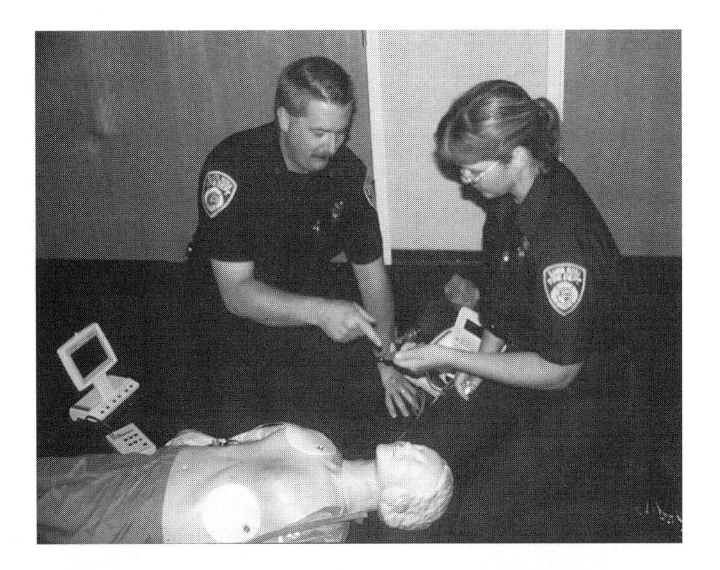

Figure 4.17 The instructor as coach guides individuals through difficult learning tasks.

may also have background experiences that enable them to describe to another how certain classroom activities can be applied on the job. Peer assistants can work with other learners who may have difficulty in grasping concepts or learning skills or who need more supervised practice time in a relaxed atmosphere.

Mentoring

Mentoring places a new learner under the guidance of a more experienced professional or another learner who acts as tutor, guide, and motivator. Mentoring situations occur outside the classroom, usually in the job environment. Many instructors have acted as mentors, either formally or informally. Instructors often guide the actions of new employees on the job, just as they guide the actions of new learners in the learning environment or training evolution. The typical mentor is usually someone other than the instructor who guides learner actions in real experiences on the job. Mentors must be chosen carefully and selected for their experience, interest, patience, and communication abilities.

References and Supplemental Readings

Apps, Jerold W. *Mastering the Teaching of Adults.* Malabar, FL: Krieger Publishing Co., 1991.

Armstrong, Thomas. *Multiple Intelligences in the Classroom.* Alexandria, VA: Association for Supervision and Curriculum Development, 1994.

Armstrong, Thomas. *Seven Kinds of Smart.* New York: Penguin Books, 1993.

Ashcraft, Mark H. *Human Memory and Cognition.* New York: Harper Collins College Publishers, 1994.

Biehler, Robert F. and Jack Snowman. *Psychology Applied to Teaching.* Boston: Houghton Mifflin Company, 1990.

Bloom, Benjamin S. *Human Characteristics and School Learning.* New York: McGraw-Hill Book Company, 1984.

Bloom, Benjamin S., M. B. Englebert, E. J. Furst, W. H. Hill, and D. R. Krathwohl, (eds.). *Taxonomy of Educational Objectives. The Classification of Educational Goals. Handbook I: Cognitive Domain.* New York: McKay, 1956.

Bruner, Jerome S., J. J. Goodnow, and G. A. Austin. *A Study of Thinking.* New York: Wiley, 1956.

Caine, Renate Nummela and Geoffrey Caine. *Making Connections: Teaching and the Human Brain.* Alexandria, VA: Association for Supervision and Curriculum Development, 1991.

Carroll, John B. "Problems of Measurement Related to the Concept of Learning for Mastery," in J. H. Block (ed.), *Mastery Learning Theory and Practice.* New York: Holt, Rinehart and Winston, 1971.

Chaplin, J. P. *Dictionary of Psychology.* New York: Laurel, 1985.

Cross, K. Patricia. *Adults as Learners.* San Francisco: Jossey-Bass Publishers, 1981.

Dale, Edgar. *Audiovisual Methods in Teaching* (3rd ed.). New York: Holt, Rinehart and Winston, Inc., 1969.

Galbraith, Michael W. (ed.). *Adult Learning Methods: A Guide for Effective Instruction.* Malabar, FL: Krieger Publishing Company, 1990.

Gardner, Howard. *The Unschooled Mind.* New York: Harper Collins Publishers, Inc., 1991.

Gronlund, Norman. *Stating Objectives for Classroom Instruction* (3rd ed.). New York: Macmillan Publishing Company, 1985.

Harpham, Edward J. (ed.). *John Locke's Two Treatises of Government: New Interpretations.* Lawrence, KS: University Press of Kansas, 1992.

Herzberg, Frederick. *Work and the Nature of Man.* Cleveland: World Publishing Company, 1966.

Knowles, Malcolm. *The Adult Learner: A Neglected Species* (4th ed.). Houston: Gulf Publishing Company, 1990.

Knowles, Malcolm. *Andragogy in Action.* San Francisco: Jossey-Bass, 1984.

Knowles, Malcolm. *The Modern Practice of Adult Education: Andragogy Versus Pedagogy.* New York: Association Press, 1970.

Kolb, David. *Experiential Learning: Experience as the Source of Learning and Development.* Englewood Cliffs, NJ: Prentice Hall, 1984.

Krathwohl, David R., B. S. Bloom, and B. B. Masia. *Taxonomy of Educational Objectives. Handbook II: Affective Domain.* New York: McKay, 1964.

Laird, Dugan. *Approaches to Training and Development.* Reading, MA: Addison-Wesley Publishing Company, 1985.

Lee, Chris. "The Adult Learner: Neglected No More." *Training,* March, 1998, pp. 47–52.

Mager, Robert F. *Developing Attitude Toward Learning.* Belmont, CA: Lake Publishing Company, 1984.

Maslow, Abraham. *Third Force Psychology.* New York: Grossman, 1971, p. 427.

Maslow, Abraham. *Motivation and Personality* (2nd ed.). New York: Harper & Row, 1970.

Maslow, Abraham. "A Theory of Human Motivation." *Psychological Review,* 50, 1943, pp. 370–396.

Merriam, Sharan B. and Rosemary S. Caffarella. *Learning in Adulthood.* San Francisco: Jossey-Bass Publishers, 1991.

Restak, Richard M. *The Brain.* New York: Bantam Books. 1985.

Restak, Richard M. *The Brain: The Last Frontier.* New York: Doubleday and Co., Inc., 1979.

Sergiovanni, Thomas J. *The Principalship: A Reflective Practice Perspective.* Boston: Allyn and Bacon, 1991.

Simpson, Elizabeth. J. *The Classification of Educational Objectives: Psychomotor Domain.* Urbana, IL: University of Illinois Press, 1972.

Solso, Robert L. *Cognitive Psychology.* Boston: Allyn and Bacon, 1995.

Thorndike, Edward L. *The Psychology of Learning.* New York: Teachers College, 1914.

Tulving, Endel. "How Many Memory Systems Are There?" *American Psychologist,* 40, 1985, pp. 385–398.

West, Laurie, Elizabeth Farris, and Bernie Greene. *Remedial Education at Higher Education Institutions in Fall 1995.* National Center for Education Statistics, Washington, D.C., October, 1996.

Wlodkowski, Raymond J. *Enhancing Adult Motivation to Learn.* San Francisco: Jossey-Bass Publishers, 1993.

Job Performance Requirements

This chapter provides information that addresses the following job performance requirements of NFPA 1041, *Standard for Fire Service Instructor Professional Qualifications* (1996 edition). Colored portions of the standard are specifically covered in this chapter.

Chapter 2 Instructor I

2-3.2 Review instructional materials, given the materials for a specific topic, target audience and learning environment, so that elements of the lesson plan, learning environment, and resources that need adaptation are identified.

2-3.2.1 *Prerequisite Knowledge:* Recognition of student limitations, methods of instruction, types of resource materials; organizing the learning environment; policies and procedures.

2-3.2.2 *Prerequisite Skills:* Analysis of resources, facilities, and materials.

2-3.3 Adapt a prepared lesson plan, given course materials and an assignment, so that the needs of the student and the objectives of the lesson plan are achieved.

2-3.3.1 *Prerequisite Knowledge:* Elements of a lesson plan, selection of instructional aids and methods, organization of learning environment.

2-4.2 Organize the classroom, laboratory or outdoor learning environment, given a facility and an assignment, so that lighting, distractions, climate control or weather, noise control, seating, audiovisual equipment, teaching aids, and safety, are considered.

2-4.2.1 *Prerequisite Knowledge:* Classroom management and safety, advantages and limitations of audiovisual equipment and teaching aids, classroom arrangement, and methods and techniques of instruction.

2-4.2.2 *Prerequisite Skills:* Use of instructional media and materials.

2-4.3 Present prepared lessons, given a prepared lesson plan that specifies the presentation method(s), so that the method(s) indicated in the plan are used and the stated objectives or learning outcomes are achieved.

2-4.3.1 *Prerequisite Knowledge:* The laws and principles of learning, teaching methods and techniques, lesson plan components and elements of the communication process, and lesson plan terminology and definitions.

2-4.3.2 *Prerequisite Skills:* Oral communication techniques, teaching methods and techniques, utilization of lesson plans in the instructional setting.

2-4.4 Adjust presentation, given a lesson plan and changing circumstances in the class environment, so that class continuity and the objectives or learning outcomes are achieved.

2-4.4.1 *Prerequisite Knowledge:* Methods of dealing with changing circumstances.

2-4.5 Adjust to differences in learning styles, abilities and behaviors, given the instructional environment, so that lesson objectives are accomplished, disruptive behavior is addressed, and a safe learning environment is maintained.

2-4.5.1 *Prerequisite Knowledge:* Motivation techniques, learning styles, types of learning disabilities and methods for dealing with them, methods of dealing with disruptive and unsafe behavior.

2-4.5.2 *Prerequisite Skills:* Basic coaching and motivational techniques, adaptation of lesson plans or materials to specific instructional situations.

2-5.4 Report test results, given a set of test answer sheets or skills checklists, a report form and policies and procedures for reporting, so that the results are accurately recorded, the forms are forwarded according to procedure, and unusual circumstances are reported.

2-5.4.2 *Prerequisite Skills:* Communication skills, basic coaching.

2-5.5 Provide evaluation feedback to students, given evaluation data, so that the feedback is timely, specific enough for the student to make efforts to modify behavior, objective, clear and relevant; include suggestions based on the data.

2-5.5.2 *Prerequisite Skills:* Communications skills, basic coaching.

Chapter 3 Instructor II

3-2.6 Evaluate instructors, given an evaluation form, department policy, and job performance requirements, so that the evaluation identifies areas of strengths and weaknesses, recommends changes in instructional style and communication methods, and provides opportunity for instructor feedback to the evaluator.

3-2.6.1 *Prerequisite Knowledge:* Personnel evaluation methods, supervision techniques, department policy, effective instructional methods and techniques.

3-2.6.2 *Prerequisite Skills:* Coaching, observation techniques, completion of evaluation forms.

3-3.2 Create a lesson plan, given a topic, audience characteristics, and a standard lesson plan format, so that the job performance requirements for the topic are achieved, and the plan includes learning objectives, a lesson outline, course materials, instructional aids, and an evaluation plan.

3-3.2.1 *Prerequisite Knowledge:* Elements of a lesson plan, components of learning objectives, instructional methods and techniques, characteristics of adult learners, types and application of instructional media, evaluation techniques, and sources of references and materials.

3-3.2.2 *Prerequisite Skills:* Basic research, using job performance requirements to develop behavioral objectives, student needs assessment, development of instructional media, outlining techniques, evaluation techniques, and resource needs analysis.

3-3.3 Modify an existing lesson plan, given a topic, audience characteristics, and a lesson plan, so that the job performance requirements for the topic are achieved, and the plan includes learning objectives, a lesson outline, course materials, instructional aids, and an evaluation plan.

3-3.3.1 *Prerequisite Knowledge:* Elements of a lesson plan, components of learning objectives, instructional methods and techniques, characteristics of adult learners, types and application of instructional media, evaluation techniques, and sources of references and materials.

3-3.3.2 *Prerequisite Skills:* Basic research, using job performance requirements to develop behavioral objectives, student needs assessment, development of instructional media, outlining techniques, evaluation techniques, and resource needs analysis.

3-4.2 Conduct a class using a lesson plan that the instructor has prepared and that involves the utilization of multiple teaching methods and techniques, given a topic and a target audience, so that the lesson objectives are achieved.

3-4.2.1 *Prerequisite Knowledge:* Use and limitations of teaching methods and techniques.

3-4.2.2 *Prerequisite Skills:* Transition between different teaching methods, conference, and discussion leadership.

Chapter 4 Instructor III

4-3.3 Design programs or curriculums, given needs analysis and agency goals, so that the agency goals are supported, the knowledge and skills are job related, the design is performance based, adult learning principles are utilized, and the program meets time and budget constraints.

4-3.3.1 *Prerequisite Knowledge:* Instructional design, adult learning principles, principles of performance based education, research, and fire service terminology.

4-3.3.2 *Prerequisite Skills:* Technical writing, selecting course reference materials.

4-3.4 Modify an existing curriculum, given the curriculum, audience characteristics, learning objectives, instructional resources and agency training requirements, so that the curriculum meets the requirements of the agency, and the learning objectives are achieved.

4-3.4.1 *Prerequisite Knowledge:* Instructional design, adult learning principles, principles of performance based education, research, and fire service terminology.

4-3.4.2 *Prerequisite Skills:* Technical writing, selecting course reference materials.

Chapter 5

Planning Instruction

Planning instruction is a critical part of developing fire and emergency services education and training programs. Essential to the planning process is having some idea of where learners are and where they need to be in terms of knowledge, skills, and attitude when the training effort is complete. How to get learners from where they are (or what they know) to the desired destination (or to what they need to know) is what planning instruction is all about. When planning a lesson or course, instructors create a *road map* that includes routes or methods that enable them to guide learners to the desired destinations or course goals.

There is an old saying that if someone has no destination in mind, any road will do. Having no destination is like driving down a road, racking up a lot of miles (kilometers) on the odometer, and getting great gas mileage, but never knowing where you are heading or why. When instructors and program planners do not identify a destination or program goal, they spend a lot of time, effort, and resources on going somewhere without determining where they are going or why, without knowing whether they have arrived, and without knowing how to measure results or effectiveness of the trip while en route.

Travelers who begin a journey with a specific destination can work through challenges of the trip while moving toward the desired end efficiently and safely. Similarly, instructors who plan classes to meet certain goals that can be reached logically by safe and efficient means can work through challenges that arise during the program. Whether planning a program of instruction or a single lesson, a helpful device called a *decision* or *planning model* can be used for selecting appropriate methods of instruction, the most direct routes or appropriate methods to reach course goals, intermediate points that measure instruction and learning, budgets to cover costs, and numerous other course-planning processes that direct successful programs. When using a planning model to work toward a well-defined goal, it is possible to "map out" and measure progress during the planning process and along the route to reach selected goals and determine when the desired destination has been met.

Program destinations are determined by identifying needs, by setting training goals to meet those needs, and by designing training and planning instruction to achieve those goals. These first steps are typical processes used to develop new courses or programs of instruction. The steps are also performed or reviewed each time instructors prepare a lesson. Just as program developers plan ways to meet goals within a program of instruction, instructors do the same as they prepare to teach. Developers and instructors ensure that instructional methods, course materials, training aids, evaluation instruments, exercises, and drills work to meet the program goals and performance objectives that are identified in the plan of instruction. At the same time, developers and instructors ensure that the program of instruction is presented safely and efficiently.

This chapter discusses a program planning model, the components of which have been developed by educators and experts to help program developers and instructors in the planning task. The chapter explains the planning processes as individual instructors, course developers or development teams, and training managers use them. A section also describes the *instructional model* for planning and developing instruction programs and lesson plans. This type of model includes and describes the steps of performing a needs analysis, planning and developing a program, developing course objectives, completing a task analysis, designing a lesson plan, and creating evaluation instruments. References and supplemental readings and job performance requirements are given at the end of the chapter.

Planning Models

[NFPA 1041: 2-2.2, 2-3.2, 2-3.2.1, 2-3.2.2, 2-3.3, 3-2.4, 3-5.2, 3-5.2.1, 3-5.2.2, 4-2.7, 4-2.7.1, 4-2.7.2, 4-3.2, 4-3.2.1, 4-3.2.2, 4-3.3, 4-3.3.1, 4-3.3.2, 4-3.4, 4-3.5]

Where do program planners and instructors start when they want to develop a training course? They usually plan a series of steps that lead to a chosen goal. How do they know what steps to take? They usually select a model designed to guide program developers through a planning process. And what planning model works best? There is such a variety of planning models that one is sure to meet any planner's needs. This section briefly discusses a sample model to give instructional planners an idea of its components and how to use them in planning instruction. A common model includes five steps that guide instructional planners through the processes of *identification, selection, design, implementation,* and *evaluation.* These types of models are usually presented graphically to illustrate the relationships of their component parts, and the models typically use arrows to show how the informa-

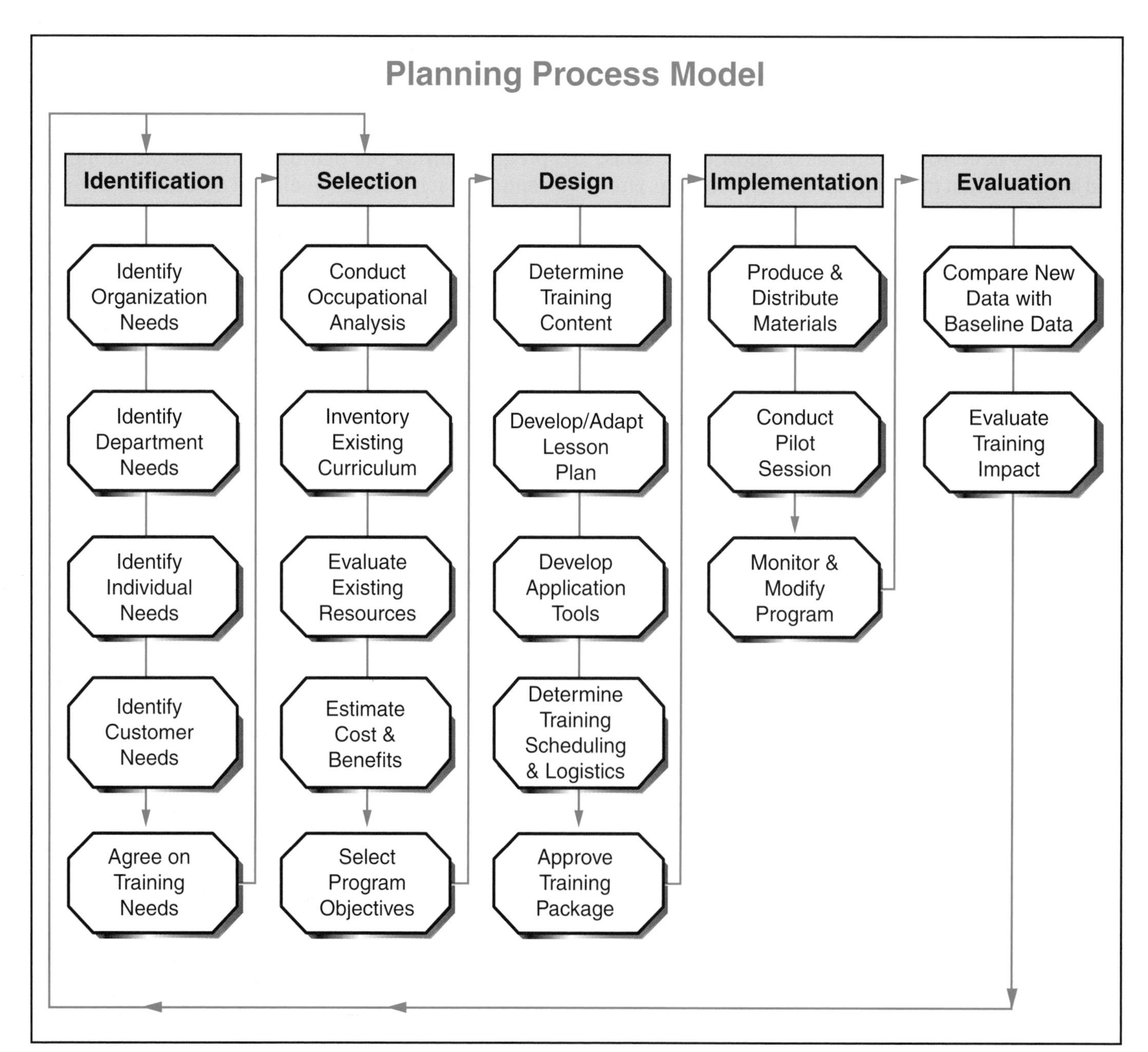

Figure 5.1a A five-step process model for program planning.

tion flows from one stage to another. The purpose of this five-step model or another similar program-planning model is to provide program planners with a systematic approach to design, production, evaluation, and use of a complete system of instruction (Gustafson and Branch, 1997).

A five-step model that has proven to be very effective in planning fire and emergency service instruction is shown in Figure 5.1a. This model is adapted from the IFSTA **Fire and Life Safety Educator** text and originated in the publication *Public Fire Education Planning: A Five Step Process* by the U.S. Fire Administration. The model gives the detail for the processes involved in each of the steps. Many other multistep models are similar, and Figure 5.1b shows a circular model that emphasizes the continuous processes of the five steps (Gustafson and Branch, 1997). The following sections describe how instructors, course planners and developers, and training managers, can use the five-step model for instructional planning.

Five-Step Planning Process

[NFPA 1041: 2-2.2, 2-2.2.1, 2-3.2, 2-3.2.1, 2-3.2.2, 2-3.3, 2-3.3.1, 2-3.3.2, 2-5.2, 2-5.2.1, 2-5.2.2, 2-5.4, 2-5.4.1, 2-5.5, 3-2.3, 3-2.3.1, 3-2.3.2, 3-2.4, 3-2.4.1, 3-2.4.2, 4-2.7, 4-2.7.1, 4-2.7.2, 4-3.2, 4-3.2.1, 4-3.2.2, 4-3.3, 4-3.3.1, 4-3.3.2, 4-3.4, 4-3.4.1, 4-3.4.2, 4-3.5, 4-3.5.1, 4-3.5.2, 4-3.6, 4-3.6.1, 4-3.6.2, 4-3.7, 4-3.7.1]

Following a planning model is helpful in developing programs and planning lessons because it provides an outline or guide that assists instructors and course developers in reaching the program destination (or goal) and lesson objectives. The steps of a typical five-step model are discussed in the following sections (Figure 5.2):

- Identify training needs.
- Select performance objectives.
- Design training.
- Implement the program.
- Evaluate the program.

Identify Training Needs

Instructors and program developers determine training needs through a process called needs analysis (see Needs Analysis section). With thoughtful questioning, carefully designed surveys, and in-depth research, a needs analysis accomplishes several tasks such as the following:

- Defines where personnel or organizations are at a designated time and where they need to be in terms of knowledge, skills, and attitudes
- Determines whether a need exists and indicates whether the need is for training, equipment to perform a task, or a change in procedures
- Identifies specific individuals, work groups, or organizations who need training, equipment, or procedure change
- Identifies a method to achieve desired levels of knowledge, skills, or attitudes

To determine the needs in the areas of knowledge, skills, or attitudes, program planners identify them by analyzing operations, injury records,

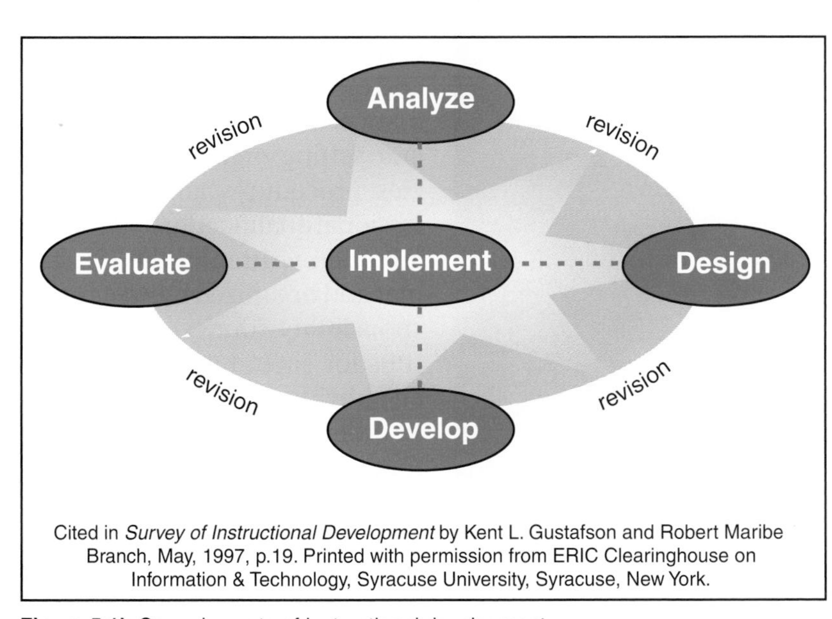

Figure 5.1b Core elements of instructional development.

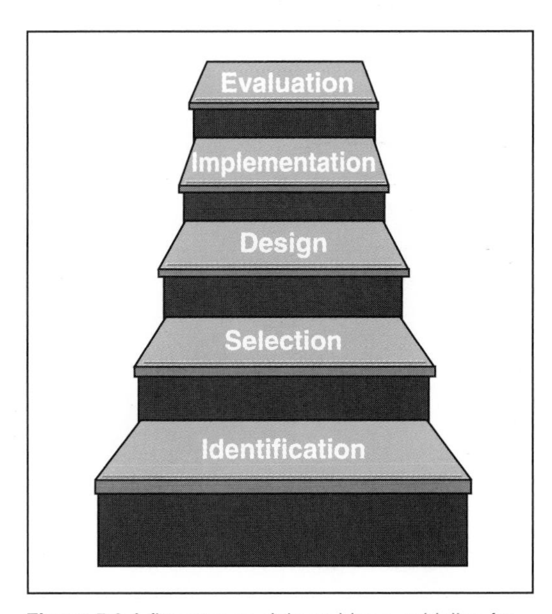

Figure 5.2 A five-step model provides a guideline for instructor planning.

accident reports, results of promotional exams, and a variety of other personnel, operation, and training records and resources. One of the ways selected to fulfill the identified needs and achieve a desired outcome is often a training program. Various methods are used to determine individual training needs or readiness for a training program such as the following:

- ***Pretests*** — A *pretest* is a type of test used by organizations and trainers to determine the current level of training or skill of potential program participants. Pretests help planners in making decisions on what to include in program content that will enable participants to reach desired levels of competence.
- ***Prerequisite training or skill level requirements*** — Organizations often set minimum levels of knowledge or skill *(prerequisites)* for personnel in certain types of jobs and for participants before they enter certain training programs. Knowledge and skill pretests and placement tests are often used to determine participant levels and eligibility for hiring, promotion, or entry into training programs.

Regardless of the technique used to determine training needs or readiness, instructors need to know the current level of participant knowledge or skill before proceeding with an instructional program. Otherwise the level or complexity of the instruction may be inappropriate for some or even all of the participants. When evaluating existing training materials for presentation, instructors and planners determine knowledge, skill, or attitude levels that the training program expects learners to both begin and end with. To effectively and efficiently develop and present an instructional program, developers and instructors must have a defined and desired starting point and an ending point in the education and training process.

Starting and ending points are often determined by the tasks that personnel must learn and perform on their jobs. Program planners often start their needs analysis by identifying and reviewing national standards and determining minimum *job performance requirements (JPRs),* which are performances required for a specific job based on those standards. JPRs are grouped according to job *duties* (functional areas of responsibility within a job) (Figure 5.3). After determining job performance requirements, planners evaluate resources and select training objectives that will be used in designing and implementing the revision of existing programs and the development of new programs.

Figure 5.3 Instructors and program developers review standards when planning new programs or program revisions so that training satisfies job-requirement needs. *Courtesy of Terry L. Heyns.*

Professional qualifications standards can be used for designing and evaluating training, certifying personnel, measuring and critiquing on-the-job performance, defining hiring practices, and setting organizational policies, procedures, and goals. Professional qualifications standards are often written in the JPR format and organized by areas of responsibility (duties). The list of JPRs for each duty defines what an individual should be able to do in order to successfully perform that duty. The standard containing each duty and its JPRs defines and describes the job, giving instructors and planners an end point around which to design and teach a training program.

Select Performance Objectives

If the data collected in the needs analysis identifies training as a means of correcting a problem, then the process of planning instruction proceeds to the step of selecting objectives and resources. *Performance objectives* address specific knowledge to learn and apply, skills to perform and master, or standards to meet. Performance standards or skills may be dictated by outside entities such as the National Fire Protection Association (NFPA) or legislation, for example, Code of Federal Regulations (CFR) 29 1910.120, *Hazardous Waste Operations and Emergency Response.*

In order to guide instruction effectively, program planners select, revise, or develop performance objectives to meet the following criteria:

- ***Identify the learner clearly***—Who is to perform the objective? It may be the generic learner or the specific firefighter, police officer, chief, hazardous materials technician, emergency medical technician (EMT), pump operator, etc.
- ***State precisely what the learner is required to know or do at the end of the lesson/course*** — State the conditions for performing the task: (1) equipment and assistance to be provided, (2) scenarios or moulage to be arranged, and (3) standard of performance required. Learners must know what will be given to them to complete the task so they may prepare appropriately to perform. Instructors and planners must have those required resources available throughout the training program and in the learner evaluation process.
- ***Specify a degree of satisfactory accomplishment***—How well must the learner perform the task? To what standard? The objective should cite a standard, time limit, manufacturer guideline, minimum testing score, and/or government regulation or other requirement.

Objectives designate a level of instruction. The evaluation of learning must be consistent with the level of instruction defined in the objectives. If an objective requires the participants to learn at the cognitive domain level three (application), the instruction and evaluation must be to that level and not to a more advanced level (see Tables 5.1 and 5.2).

The various objectives in a course and the requirements of a course require instructors to teach and learners to perform at different levels in one or more domains. These levels are determined by what was found in the needs analysis. When decisions regarding the level of instruction and learning domain have been made, then the performance objectives can be written. Program planners must evaluate and select objectives for completeness and consistency with the level of instruction and learning domain. Planners and instructors further evaluate objectives during the course to ensure that they achieve the change in student performance required by the needs analysis.

Performance objectives must clearly support the educational goals — the destination or end point — of the training program. The objectives state what tasks will be completed, and a task analysis is developed. Subject matter experts who can clearly identify the knowledge, skills, and attitudes to be performed or exhibited and evaluated will develop a task analysis that corresponds to each task stated in the objectives. As tasks are developed, planners and instructors determine the resources needed to appropriately complete the tasks and ensure that these resources are acquired for the training program.

Design Training

Designing a training program is an undertaking that can be both exciting and intimidating at the same time. It is exciting because successful programs can have long-range, positive benefits for learners and the organization. It is intimidating because a multitude of factors must be considered when setting out to design a training program. But a planning model can help program developers and instructors follow a route for designing and implementing programs that meet training needs and use resources effectively. To ensure program effectiveness, instructors work with developers to design and try out the program in pilot classes. Through the development process, developers and instructors conduct *formative* or *process evaluations* (frequent checks for accuracy, compliance with standards and organizational policy, and compatibility with time and facility restraints). It is evident the training design has many intricate aspects.

This chapter has already discussed some of the aspects of training design. These aspects include selecting a model or development plan, completing a needs analysis, and identifying resources to support the training goal and lesson objectives. These steps lead to decisions about the scope and sequence of training and the appropriate delivery methods. By their instructional level and learning domain, the performance objectives guide delivery methods.

Performance objectives also guide the training design of lesson plans or learning modules. The training content must be portioned into workable units that

Table 5.1
Examples of Verbs and Their Corresponding Learning Levels (Cognitive Domain)

Learning Levels	Samples of Corresponding Verbs
Basic Levels	
Knowledge (Level 1) Ability to recall or recognize information	**Recall, List, Label, State, Recognize (communication or situation previously presented), Quote, Define, Name, Show, Identify** *Illustration:* The learner recalls that the formula for computing the area of a rectangle is $A = L \times W$. The learner labels the sides of a rectangle.
Comprehension (Level 2) Ability to understand or use information within a limited context	**Restate, Interpret, Explain in your own words, Illustrate, Discuss, Describe, Compare, Demonstrate, Give an example of your own** *Illustration:* The learner can explain that in the formula given, *A* stands for area, *L* stands for length, and *W* stands for width. The learner can give an example of a similar area that is also rectangular in shape.
Intermediate Levels	
Application (Level 3) Ability to use abstractions in particular concrete situations	**Apply a rule, guideline, or principle; Calculate; Measure; Use; Manipulate; Predict trends, outcomes, or results; Solve problems** *Illustration:* Given a rectangle 4 inches (100 mm) long and 3 inches (75 mm) wide, the learner can calculate that the area of the rectangle is 12 square inches (77 cm^2). Given other examples of different sized rectangles, the learner can solve the problems of finding the areas.
Analysis (Level 4) Ability to break information into its parts to clarify relationships	**Test; Check; Inspect; Analyze; Determine conditions, properties, aspects; Verify; Divide; Classify; Organize; Deduce; Choose** *Illustration:* For a given rectangle, the learner recognizes the properties of the rectangle and determines that if either the length or width is increased, area will also increase. The learner verifies this property by checking the calculations.
Advanced Levels	
Synthesis (Level 5) Ability to create a new communication or concept through the examination of other communications or situations	**Create; Design; Compare; Describe; Develop (new information, not recital); Produce; Plan; Formulate; Devise; Support; Write; Report** *Illustration:* Knowing the formula for the area of a rectangle, the learner can develop a formula for computing the area of a parallelogram. The learner can describe steps taken in these calculations.
Evaluation (Level 6) Ability to use standards and criteria to make judgements	**Rank, Evaluate, Judge, Defend, Critique, Choose (based on standards), Diagnose, Appraise, Recommend, Decide, Justify** *Illustration:* The learner chooses from a list which is the correct formula for determining the area of a right triangle. The learner performs a self-appraisal on ability to select formulas.

Table 5.2
Performance Terms for Behavioral Objectives

Term	Description	Example
Describe *Synonyms:* discuss, define, tell how	Reports essential properties or characteristics of objects or events.	**Describe** the eight-step problem-solving process.
Define *Synonyms:* describe, delineate	Provides a description that gives a precise meaning or basic traits.	**Define** *scalar.*
Identify *Synonyms:* mark, match, choose, recognize	Selects a named, described or pictured item orally or by pointing to it, picking it up, labeling it, or marking it.	**Identify** ventilation tools in the given illustration.
Name *Synonyms:* describe, delineate	Supplies a title for objects, people, processes, events, or principles.	**Name** the type of heat generated by splitting or combining atoms.
List *Synonyms:* write, arrange	Recalls similar objects or events and records in a methodical or systematic arrangement.	**List** all of the types of splints carried on the ambulance.
Order *Synonyms:* arrange in order, list in order, sequence	Arranges, rearranges, lists in sequence, or places in order.	**List in order** the phases of negotiation.
Differentiate *Synonyms:* distinguish, discriminate	Recognizes as different and separates into kinds, classes, or categories.	**Differentiate** among hazard classes the following chemicals.
Classify *Synonyms:* sort, arrange, group	Puts into groups having common attributes, uses, characteristics, or functions.	**Classify** forcible entry tools according to their uses.
Construct *Synonyms:* draw, make, build, design, create	Makes an object, verbal statement, or drawing.	**Construct** a floor plan for a pre-incident survey.
Apply *Synonym:* use	Uses a stated relationship or principle to perform a task or answer a problem.	**Apply** the progressive system to properly discipline an employee.
Demonstrate *Synonyms:* show, perform, (any of various appropriate action verbs)	Performs operations necessary to carry out a specified procedure.	**Demonstrate** the ability to properly place a roof ladder. **Show** how to properly dispose of personal protective equipment.

instructors can present in a reasonable length session, laboratory period, or drill exercise. It is often helpful to prepare individual lessons in a standard format that instructors can follow when preparing and presenting a lesson. Organizations often design or select a certain format for presenting lessons and require that program developers create all training materials in that format. For example, visuals are slides or transparencies rather than computer-generated types (based on available resources), handouts are printed (not handwritten) in a selected font (for consistency and appearance), or videos are restricted to a 20-minute length (to encourage more instructor-learner interaction).

The lesson plan leads the instructor through the following four-step method (see Chapter 6, Lesson Plan Development, for details):

- ***Preparation*** — Prepare the instructor to deliver the lesson *and* prepare learners to participate in the lesson.
- ***Presentation*** — Present the lesson in the appropriate format.
- ***Application*** — Guide learners to apply new knowledge and skills through exercises, practices, or drills.
- ***Testing or evaluation*** — Test or evaluate learners' accomplishments against the stated performance objectives.

During needs analysis and program development phases of planning instruction, developers review existing training materials to select the course components that meet the training goals and objectives. Current materials may exist in-house or in other training resources such as other local, state/provincial, and national training organizations (National Fire Academy, Police Training Commission, etc.). These organizations have likely developed courses that meet the program requirements in areas where state/provincial and national training standards have been established. Also, commercial sources exist for training courses and provide up-to-date and effective training materials that support instructional programs.

Selecting an existing training program that meets the needs of an organization can reduce the cost of the program and shorten the time delay in beginning instruction. In most cases, a committee or board of subject-matter experts assists instructors in selecting training programs and materials and in evaluating various course materials for completeness and applicability to the training objectives.

The final product that an instructor delivers is constrained by considerations of cost and resources. The program must be one that can be realistically achieved with the funding available or planned. For budget reasons, the scope of the training selected must be limited to those knowledge, skills, and attitudes that the needs analysis identified as absolutely necessary to meet the overall goals. Adding "nice-to-have" materials may add to time and material costs. The program must be achievable in the environment that is likely to be available. In most cases, the environment is a classroom, fire station, or training facility. Programs that depend on state-of-the-art facilities for presentation are desirable if the required facility is available. If such facilities are not going to be readily available, then it is better to develop or look for programs that can meet training objectives in a setting that an organization can reasonably expect to provide.

Implement the Program

Throughout the process of instructional design, program developers must be aware of the reality of implementing the training. This chapter has briefly discussed some of the resource issues to consider. When program developers have completed the initial design steps through the final writing of lesson plans and development of evaluation instruments, their next logical step is to implement the program.

Implementation requires a complete delivery system, including such steps as the following:

- Acquire funding and facilities.
- Determine instructor and learner time requirements.
- Find qualified instructors or train them.
- Determine and create training materials.

Program developers must confirm, for example, that the program can be presented in the existing facility. The location, design, and equipment of a facility directly affect the kinds of activities that instructors can conduct and learners can perform to achieve course goals.

Before implementation begins, program planners perform a detailed formative evaluation. Ideally, formative evaluations have been in progress throughout the design phase. These evaluations include peer or supervisor reviews, consultations with subject-matter experts, reviews by prospective instructors, and any other steps that help to ensure the quality of the program and continually monitor the program to ensure its focus on meeting the performance objectives.

Once everyone is satisfied that the formative evaluation and the ensuing corrections indicate that the course is ready to move forward, program planners can take the necessary steps to implement the training program. Implementation actions typically include the following steps:

- Obtain final course approval.
- Assemble training materials.
- Schedule facilities and equipment.
- Qualify instructors.
- Establish appropriate training records systems or databases.
- Schedule and announce the program.
- Select learners.

Each organization or jurisdiction may have different requirements for obtaining program approval for implementation. The program developers may conduct a detailed presentation showing the approval authorities how the course supports organizational learning goals and achieves the learning objectives (through instructional methods and learning activities and with appropriate training materials). The presentation gives information on the financial and time commitments that the organization will have to invest by providing details on facilities and equipment required and on instructor and learner hours necessary. When the course is required by defined standards, the information on how the course meets those standards and what records must be kept to satisfy certification processes assist the decision makers in reaching a favorable conclusion. With approval, the process of scheduling classes and selecting learners can easily follow.

Evaluate the Program

Program planners, instructors, and learners can perform a *summative evaluation* after the first presentation of the course to determine whether the course met the educational goals of the organization. This evaluation answers the following questions:

- Did learners meet performance objectives?
- Was training conducted as designed and within the resources allocated?

There are many ways to accomplish summative evaluations. Examples include the following:

- Participants' grades on written tests or performance tests
- Learners' behaviors observed in the field
- Feedback from learners and instructors
- Feedback from field supervisors

A summative evaluation is performed after every course presentation. The trends identified in learner performance help evaluators judge whether the course was successful (meets organizational training goals) and determine whether the course continues or needs further adjustments. In order to accurately measure the improvement in learner performance, program evaluators need to compare the postcourse performance to the performance level observed in the needs analysis and pretests.

In addition to evaluating learners' performances and whether they meet course objectives and job performance requirements, there must be an evaluation system for course materials and instructor performance. A system for evaluating instructor performance is an important aspect of evaluating training.

Evaluate Course Materials

Both instructors and program participants should have input into the assessment of teaching and learning materials. The results should answer the following questions:

- Did the materials support the objectives?
- Were the materials relevant or applicable to job requirements?

Evaluate Instructor Performance

An excellent course design with appropriate training materials can be jeopardized by one or more incompetent instructors. When designing instructor evaluation criteria, include the following:

- Review of educational credentials
- Assessment of prior and continued training in the subject matter
- Assurance of experience level as an instructor

During the presentation of a course, the course instructor should be periodically evaluated by experienced instructors with subject-matter expertise. Learner input is also an important component in instructor evaluation. Program participants can complete carefully designed course-feedback instruments that elicit objective responses. The results of instructor evaluations may show the need for either enhanced instructor development or only the need for appropriate instructor orientation, particularly when new courses are being introduced.

The Instructional Model

[NFPA 1041: 3-5.2, 3-5.2.1, 3-5.2.2, 3-5.3, 3-5.3.1, 3-5.3.2, 4-3.2, 4-3.2.1, 4-3.2.2, 4-3.3, 4-3.3.1, 4-3.3.2, 4-3.4, 4-3.4.1, 4-3.4.2, 4-3.5, 4-3.5.1, 4-3.5.2, 4-3.6, 4-3.6.1, 4-3.6.2, 4-3.7, 4-3.7.1]

The specific steps used for planning and developing instructional programs and lesson plans are just as useful as the five-step planning model to instructors and program designers. The *instructional* model includes the following steps described in the sections that follow:

- Perform a needs analysis.
- Plan a program.
- Develop course objectives.
- Complete a task analysis.
- Design a lesson plan.
- Create evaluation instruments.

The components of the instructional model are similar to those related steps in the five-step planning model. Not all instructors perform all components. Some may never participate in performing the needs analysis or completing the task analysis that created the course. Course planners and developers usually perform the steps of the instructional model. Program developers then write objectives and develop a lesson guide to give to instructors. These *lesson plans,* sometimes called *instructor guides,* are outlines that map out the information and skills to be taught and state the format or method to be used in delivering the instruction. With the instructor guides in hand, instructors then plan how they will deliver the lessons within the time frame listed for each lesson in the program.

The course developers may also be the ones who create the evaluation instruments for testing learners, evaluating the program, and evaluating the instructor. For some training programs, instructors may create their own tests. This option may be based on the requirements of the standard, the specifications of the program or certification process, or desires of the organization.

Needs Analysis

A *needs analysis* is an assessment of training needs that identifies the gap between what exists and what should exist. It is a process of analyzing members of a group to determine their desires, problems, or needs, often with the purpose of developing a plan to provide training or equipment that fulfills those needs. A needs analysis has several purposes, including the following:

- Analyzes a problem and reports findings to decision makers (thus used as a decision-making tool)
- Identifies community or group interests, functions, and characteristics
- Finds problems and identifies resources for change
- Lists steps to use in evaluating problems
- Prioritizes needs and provides the basis for subsequent research because not all needs can be addressed at once
- Identifies the difference between what exists now and what should exist in the future
- Identifies ineffective programs

A needs analysis analyzes training to find out whether it matches job requirements. Where training is often used to resolve on-the-job problems, a needs analysis may find that the problem is not a lack of knowledge or skill, but rather a lack of appropriate equipment to perform the skill accompanied by poor motivation because of the lack of appropriate equipment. Through a needs analysis, analysts look for knowledge and skill levels, which help determine if training is required and how it will be delivered.

What else do analysts look for? What do they ask when conducting a needs analysis? One of the areas to examine is the interests or characteristics of the group that identify common needs. The characteristics of the group may be related in one of the following ways:

- ***Function*** — It has the same occupation and interest.
- ***Structure***—It is organized by space (area, building, or neighborhood).
- ***Political boundary***— It has common political philosophies or affiliations.
- ***Geographical area*** — It is organized by area (state or province, county/parish, or jurisdiction).

How do analysts collect information? There are many methods, which can be divided into formal and informal methods. *Formal methods* include carefully designed and executed surveys, opinion polls, checklists, observations, psychological profiles, research analyses, and tests. *Informal methods* include conversations, casual observations of activities and habits, and other unobtrusive measures.

What areas or populations are included in a needs analysis? How are boundaries designated, and who decides the boundaries? Usually an organization designates the population, group, department, or subsection of an area to be analyzed. Time and money often constrain the extent of a needs analysis. A target population may be an organization or department or a group within it. The analysis may ask for the following information:

- Knowledge and skill level of group members such as level of education, years of experience, and ability to operate certain kinds of equipment
- Personal factors such as age, seniority, interests, and learning styles to determine group similarities and differences
- Group goals and motivations
- Effectiveness of current training to determine whether goals and objectives are clear, training personnel are qualified, equipment is adequate, and evaluation methods are effective

After collecting this information, analysts must then prioritize group needs. Through a prioritization process, analysts identify and determine the importance of each need based on standards or selected criteria and select a top five or ten to recommend to management. In selecting the top needs, analysts determine effects on other priorities and whether these needs can be met with current resources. If additional resources are needed to meet critical needs, they must be listed with potential costs. It is also important to set time frames for completing needs and get commitments from the group and from the organization. Part of the needs analysis includes a goal that will plan a time line but allow for some flexibility. Often the goal includes a budget for the program, but a budget may have been determined by the organization with the understanding that fulfilling needs would be restricted by budget figures.

Performing a needs analysis can be a long process. An effective one takes time, planning, and commitment. It requires a variety of skills including those of communicating, researching, analyzing, writing, drawing conclusions, and reporting.

Program Development and Planning

The needs analysis in the instructional model overlaps some of the steps in the five-step planning model. The next step (which coincides with the *design* phase of the five-step model) is to develop a training program using the results of the needs analysis and the decisions made from it. The needs analysis assists in answering the following questions:

- What budget is allowed or proposed for training instructors and personnel, and how and when will they be scheduled or trained?
- When will courses be offered, and how will they be evaluated?
- What policies are in place or needed for budgeting and purchasing materials and equipment?
- What resources are already available, including facilities, equipment, personnel, funding, and consultants?

Based on the information gathered in the needs analysis (learning styles, education, experience, and job requirements), program developers can make a *diagnosis* or form an opinion about the target population in order to focus a training program on particular needs. Before designing the program, a *front-end* analysis examines carefully what personnel need to know or do competently on the job. Developers list and prioritize these tasks or competencies and rewrite these competencies into objectives or key concepts (see Course Objectives section).

As developers focus the program towards target population needs, they identify and locate resources for the program. These resources include such factors as instructors, equipment, and instructor and learner materials. The time element is also important. Developers must work within designated time frames or set time limits for the following components:

- Developing a program
- Training or orienting instructors
- Piloting the program and making revisions
- Implementing the program

As they develop the program, planners must make decisions about the desired knowledge and skill level of participants, create the program to meet those levels, and avoid developing a program that falls below or extends beyond the desired or standard criteria level. When the knowledge and skill level is determined, developers then plan and create program goals and objectives, outline instructional plans, and determine instructor needs.

Developers must also determine how instructors will observe or measure learner success or mastery in each competency. Developers must include time in the program for participants to practice activities to be performed and measured. What measurement tools must be developed? Checklists and written exams are typical and appropriate types, but the measurement

So that . . .

- — **NFPA 1041 3-1.2** requirements are met.
- — A **minimum score of 70 percent** is received.
- — A **rating of *satisfactory*** is received **based on an objective checklist.**
- — SCBA is **operating in 1 minute.**
- — The saw is **checked per manufacturer's manual.**
- — The ladders are **raised as stated in standard operating procedures (SOPs).**

Task Analysis

A *task analysis* is the process of identifying tasks for doing a certain job and creating a task list for that job. The job of cleaning the station might include the tasks of mopping floors, dusting furniture, washing windows, and cleaning the kitchen; for servicing the pumper, tasks might include changing oil, checking fluid levels, replacing engine belts, and checking lights.

A task analysis, often also called *job analysis* and *occupational analysis,* contains several components that enable analyzers to break down and identify every aspect of a job. The components include the following (Figure 5.4):

- ***Occupation*** — Occupation is a career or professional category such as law enforcement or fire and emergency services career, teaching or medical profession, etc.
- ***Block*** — Block is an occupation division that includes related tasks with common factors such as police departments (which have street officers, detectives, and undercover officers), fire departments (which have firefighters, technicians, and hazardous materials specialists), and emergency medical service (EMS) departments (which have first responders, EMTs, and paramedics). Block can also be a division organized by operations such as public education, training, patient care, suppression, and equipment maintenance.
- ***Unit*** — Unit is a grouping of similar operations within a block, for example, suppression includes ventilation and overhaul; patient care includes assessment, performing cardiopulmonary resuscitation (CPR), using automated external defibrillators (AEDs), and performing patient immobilization.

Instructor Tip: *Using the task analysis worksheet, complete a worksheet for several tasks so that you can provide your class with samples. Have several groups in the class complete task analysis worksheets for several tasks and then allow them to compare their results. This activity gives the class opportunity to practice, discuss, and reason why the components fit into certain categories and what flexibility there may be in creating a task analysis.*

- ***Task*** — Task is a combination of jobs performed in a unit or identifies a component of knowledge and skills of an occupation such as the tasks of performing CPR, which require skills and knowledge in opening the airway, checking for a pulse, ventilating, and compressing the chest.
- ***Job*** — A task component that focuses on a step in a series and leads to other task steps is a job. An example is *opening the airway,* which includes head-tilt, chin-lift maneuvers. A job also includes a performance objective of the behavior that the learner will be able to do on the job, given certain conditions and based on a standard.
- ***Performance objective*** — The performance objective describes what the learner is expected to do or describes the product or result of doing something — learning information and applying that information to performing a task or skill. The performance objective must also include two other components: (1) the conditions, which state what material, information, or equipment is given or required to complete the performance and (2) the criteria, which state the standard to be met or how well (compared to the standard) the information must be known or the skill must be performed. (Review the previous section on Course Objectives for examples.)
- ***Enabling objectives/operations*** — Operations are the steps required to perform each job. For example, the operation explains how hands are placed to perform head-tilt, chin-lift maneuvers. Knowledge and skill requirements are also listed. Operations become enabling objectives, which are written as action statements and listed in steps that *enable*

Task Analysis Worksheet

Occupation:

Block: **Unit:**

Task: **Job:**

Performance Objective:* Given (conditions) . . . , the student will be able to (behavior/measurable verb) . . . so that (criteria/standard) . . . .

Enabling Objectives/Operations (Actions to enable learner to perform objective above)	Key Points/Outline Facts (Information/steps to meet enabling objectives)
1.	I. A. 1. 2. B. 1. 2. Etc. . . .
2.	II. A. 1. 2. B. 1. 2. Etc. . . .
3.	III. A. 1. 2. B. 1. 2. Etc. . . .

* The performance objective is written with cognitive verbs if information is to be taught and with psychomotor verbs if skills are to be taught. Do not use "constructs" or vague verbs such as *know* or *understand* that cannot be measured. Be sure to write objectives for the learner to perform, not for the instructor to teach.

Figure 5.4 A task analysis worksheet and its components.

the learner to *perform* (operate) the job or behavior stated in the performance objective.

- *Key points/outline facts* — Key points are main ideas taken from the list of operations and developed into outline form so cognitive information and/or motor behavior for skills development is/are detailed.

Lesson Plan

Once the task analysis is finished, the most difficult part of developing a program is also finished. At this point, the outline for a lesson plan is basically complete. Every bit of knowledge or skill is listed in this outline and is the same information that must be given to a group of learners. By adding a few component parts (given in the following list), the lesson plan or instructional guide is ready (see Chapter 6, Lesson Plan Development, for examples):

- *Preparation section* — This part of the lesson plan is a *front-end* piece that prepares the participant for learning. In this section, the instructor provides a motivational statement that gets the learners' attention and interest, states the objective so the learners know what they are to do in this lesson, and lists the overview points so the learners know the key points to notice.
- *Presentation section* — This part of the lesson plan is actually the outline taken from the key points section of the task analysis. In this section, the instructor introduces each topic (taken from the list of overview points), uses appropriate instructional methods to discuss or demonstrate, uses instructional materials effectively, and relates known or familiar information and tasks to unknown or new and unfamiliar information and tasks.
- *Application section* — The application section of the lesson plan works together with the presentation section. An instructor who presents effectively not only *tells* information but *shows* how it is applied. The application section provides opportunities for learners to practice applying the information or skill.
- *Testing or evaluation section* — In this part of the lesson plan, instructors review main points with the learners and check for their understanding of the topics discussed, demonstrated, applied, and practiced. Various types of evaluation instruments can also be used at points during the lessons or course to measure knowledge and skill. Tests and evaluations can be one or more types such as written or practical, diagnostic, and comprehensive.

Evaluation Instruments

Evaluation instruments serve many roles and are more than just tests for learners at the end of a program. Organizations are interested in evaluations on the programs themselves and on the instructors who teach them, as well as on the accomplishments of the learners who participate in these programs. Evaluations point out program assets and detriments in areas such as time allotted, equipment used or needed, what instructors and learners expected, instructional methods used, learning environments used, and testing methods used. For instructors, evaluations provide them with an assessment of themselves and their instructional methods. For example:

- Have they maintained their knowledge and skills?
- Have they used time effectively?
- Did their lesson plans and activities meet objectives?
- Did they interact appropriately with other instructors and learners?

For learners, evaluations provide feedback on knowledge and skill level and assess their needs for further study or practice. From learner evaluations, instructors and program managers receive feedback on test validity and reliability, as well as on the effectiveness of the instructor and the program. Low scores do not always mean that the learner did not learn; they could mean that the instructor did not teach to the objectives, the program was not suited to the level of the learner, or the evaluation instrument (test) was not valid. In those respects, evaluations provide information for examining and analyzing reasons for failure and for determining areas to adjust and improve.

Five-Step Planning Process for Training Managers

[NFPA 1041: 3-2.3, 3-2.3.1, 3-2.3.2, 3-2.4, 3-2.4.1, 3-2.4.2, 4-2.3, 4-2.3.1, 4-2.3.2, 4-2.6, 4-2.6.1, 4-2.6.2, 4-2.7, 4-2.7.1, 4-2.7.2, 4-3.2, 4-3.2.1, 4-3.2.2, 4-5.3, 4-5.3.1, 4-5.3.2, 4-5.4, 4-5.4.1, 4-5.4.2]

A planning process similar to that used by course developers and instructors can be used to ensure that those who have responsibility for long-range strategic training plans are moving the organization in the direction to accomplish its mission and goals. A five-step process would include *identification* of strategic goals, *selection* of strategic objectives, *design* of a strategic plan, *implementation* of the plan, and *evaluation* at appropriate stages in the plan (Figure 5.5).

Identify Strategic Needs

To define where an organization stands in meeting its training needs, managers can assemble a variety of indicators. These indicators include the current training plan in use, past and present participation rates, accumulated learner pass/fail data, promotion statistics, and the number and types of required certifications completed.

Managers who want to determine strategic needs must be willing to look beyond current success to identify where the organization should be going in the future. Typically, organizations select a five-, ten- or twenty-year planning cycle. For fire and emergency services providers, managers must consider many factors in long-range planning. A few of the factors that might dictate training requirements include growth of the jurisdiction, change in population demographics, and movement of new industry into the jurisdiction, which may introduce new threats or hazards. The evolution of performance standards by national, state/provincial, or local organizations can also result in new training requirements.

Long-range planners need to set a specific time period for their plan and then identify specific goals to be achieved by the end of that period. These goals would include personnel strength, numbers and locations of facilities, types and numbers of equipment operated, and special capabilities or special teams to address specific threats or hazards. Comparing current capabilities with long-term goals constitutes a strategic needs assessment. The result of this comparison dictates training capabilities and identifies

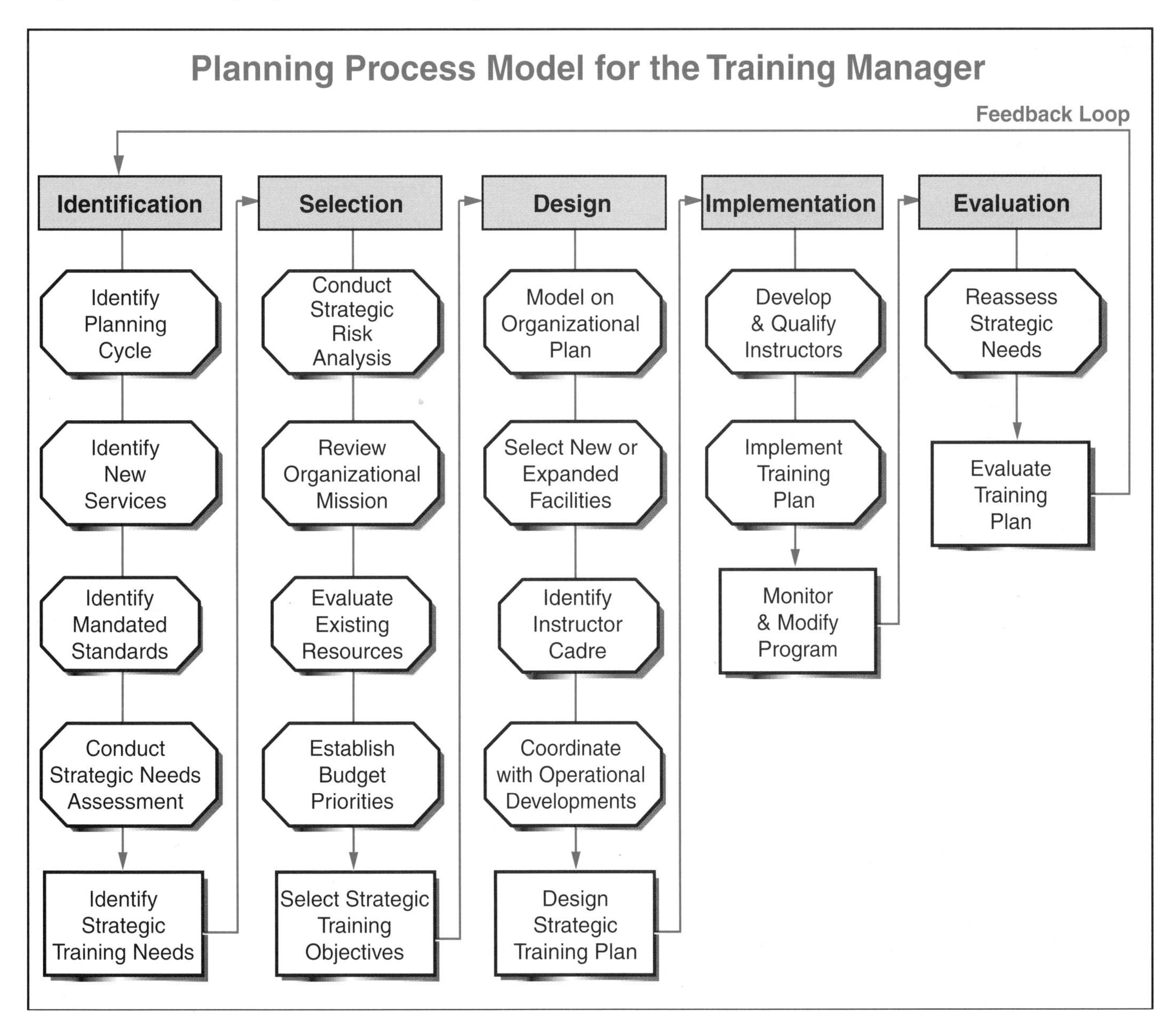

Figure 5.5 A planning process model for the training manager.

programs that need to be developed. It also dictates the direction toward which the organization must move in the following areas:

- Budgeting
- Planning personnel needs and meeting job requirements
- Procuring equipment
- Developing and scheduling appropriate training programs

Select Strategic Objectives

Based on the strategic needs assessment, the training manager must prepare to meet the organization's long-range goals with appropriate training programs that are introduced in a timely manner. Some of the strategic goals that require a specific training response include the following:

- Introducing new equipment
- Expanding service
- Recognizing or identifying new threats or hazards
- Developing new standards of performance

Appropriate and necessary strategic planning tactics for the training manager include making periodic reviews of the organization's risks. Analyzing risks that fire and emergency responders face is a necessary step to determine whether current training adequately prepares enough responders to reasonably deal with the level of threat. Recently, many organizations have adjusted their strategic training plans because they have recognized a needed change in fire and emergency response to address a heightened threat from terrorist weapons of mass destruction.

The goals that drive an organization's training needs also require that the training delivery system keep pace with growth. Funding for training facilities, instructor development, and training materials must grow as the training program evolves in accordance with the strategic plan.

Design Strategic Plan

The training manager should model the strategic training plan on the one developed by the parent organization (Figure 5.6). Time lines for introducing new or revised training programs must be coordinated with the organization's plans for growth in personnel and facilities, introduction of new equipment, and other anticipated developments. The strategic plan for training must show the following:

Figure 5.6 Training managers review and evaluate organizational plans for training and determine what changes are necessary to accomplish organizational goals.

- Start and stop times for course design and implementation
- Requirements for new or expanded facilities needed to meet increased capability goals
- Timetables to identify and train instructor staff to meet new training requirements

It is a challenge to develop a cadre of qualified instructors because of the variables that affect personnel actions. Activities such as promotion, retirement, sick leave, and a variety of other reasons cause instructors to become unavailable for instructional assignments. Qualifying new instructors may be an extended process because some training programs have long preparation times, specific qualification requirements, or special certification needs that must be completed. The strategic plan must address and include these variables in order to ensure that there are always sufficient qualified instructors available in order to prevent delays in starting a program or in continuing a critical training program already in progress.

Implement Strategic Objectives

The training manager should ensure that implementing strategic objectives tracks or coordinates with the organization's strategic plan. The implementation of strategic training plans must follow the lead and time lines of the organization's overall strategic plan. When an organization's strategic plan is either expedited or delayed, the status of training plans is also affected. A long-range plan document assists in developing items such as staffing requirements, annual budget plans, and capital improvement plans. The challenge for the training manager is to ensure that long lead items are ready when the strategic plan calls for implementation of new capabilities or equipment.

Human resources planning that is necessary for meeting strategic objectives should give priority to selection and training of instructor staff. The first step is to identify those individuals who are capable of becoming instructor staff. This step must be taken in time to allow these individuals to complete the appropriate training or certification requirements for becoming qualified to instruct the necessary offerings when these offerings are required by the strategic plan.

The long-range planning manager must plan and have in place a program for developing and maintaining a cadre of highly skilled and motivated instructors. Managers must consider the following traits and background information when selecting these individuals:

- Appropriate temperament and personality
- Experience in the desired area
- Necessary types of formal education and special qualifications or certifications
- Communication skills
- Motivation and interest in serving in an instructor role

Evaluate Strategic Plan

Usually during budget development but at least yearly, the training manager needs to review the progress on the strategic plan by answering the following questions:

- Have the actions completed in the past year achieved the desired results?
- Have the training programs met the needs of new personnel and those using new equipment?
- Were training facility expansions completed in time to train adequate numbers of personnel?

There are many other questions that managers may want to answer that indicate how well the plan is being implemented. Some changing factors may dictate a revision of the strategic plan. Some unexpected developments may require the addition of new objectives.

Managers should perform periodic evaluations, using both statistical and performance measures, to see whether the implementation of the training plan is supporting the organization's goals as anticipated. These and other similar measures tell the training manager what changes must be made in the plan in order to continue moving toward accomplishing the organization's training goals. Evaluate progress by answering the following questions:

- How has performance improved?

Case Study: Training Program Development

You have been assigned to the training division after working several years in the field. After completing all your instructor requirements and teaching your first class, you now know firsthand some of the problems of presenting a program for which you had no input. You know that not all activities fit every group, that all assignments cannot be completed in the time allotted, and that all learners do not learn the same way or need to learn the same things.

You have now been assigned to develop a training program to train station personnel to use a new piece of equipment. The program must be completed and ready for use in two weeks. Answer the following questions:

1. What is the first step you would take before you begin developing the training program?
2. What individuals or resources would you consult while planning and developing this program?
3. How would you plan a budget for this program and then convince your supervisors that this program requires the proposed dollar amount?
4. How would you test this program to see whether time frames and activities are appropriate?
5. What type of evaluation instruments would you develop?

- What new or improved capabilities exist?
- How many personnel have qualified in essential skills?

References and Supplemental Readings

Gagne, Robert, Leslie Briggs, and Walter Wager. *Principles of Instructional Design* (4th ed.). Fort Worth: Harcourt Brace Jovanovich College Publishers, 1992.

Gustafson, Kent L. and Robert Maribe Branch. *Survey of Instructional Development Models* (3rd ed.). Syracuse, NY: ERIC Clearinghouse on Information & Technology, Syracuse University, May, 1997.

International Fire Service Training Association. *Fire and Life Safety Educator* (2nd ed.). Stillwater, OK: Fire Protection Publications, Oklahoma State University, 1997.

Kidd, J. R. *How Adults Learn*. New York: Association Press, 1973.

Knowles, Malcolm. *The Adult Learner: A Neglected Species* (4th ed.). Houston: Gulf Publishing Company, 1990.

Knowles, Malcolm. *The Adult Learner* (2nd ed.). Houston: Gulf Publishing Company, 1978.

Maslow, Albert, "The Role of Testing in Training and Development," *Training and Development Handbook* (2nd ed.). R. L. Craig (ed.). New York: McGraw-Hill Book Company, 1976.

Morrison, James H. "Determining Training Needs," *Training and Development Handbook* (2nd ed.). R. L. Craig (ed.). New York: McGraw-Hill Book Company, 1976.

Pregent, Richard. *Charting Your Course: How to Prepare To Teach More Effectively.* Montreal, Canada: Magna Publications, Inc., 1994.

Strother, Richard, Pam Powell, and Laura Buchbinder (eds.). *Public Fire Education Planning: A Five Step Process.* U. S. Fire Administration, 1979.

Tough, Allen. *The Adults Learning Projects* (2nd ed.). Toronto, Canada: The Ontario Institute for Studies in Education, 1979.

Verduin, John R., Harry Miller, and Charles Greer. *The Lifelong Learning Experience*. Springfield, IL: C. C. Thomas, 1986.

Verduin, John R., Harry Miller, and Charles Greer. *Adults Teaching Adults*. Austin, TX: Learning Concepts, 1977.

Job Performance Requirements

This chapter provides information that addresses the following job performance requirements of NFPA 1041, *Standard for Fire Service Instructor Professional Qualifications* (1996 edition). Colored portions of the standard are specifically covered in this chapter.

Chapter 2 Instructor I

2-2.2 Assemble course materials, given a specific topic, so that the lesson plan, all materials, resources, and equipment needed to deliver the lesson are obtained.

2-2.2.1 *Prerequisite Knowledge:* Components of a lesson plan; policies and procedures for the procurement of materials and equipment and resource availability.

2-3.2 Review instructional materials, given the materials for a specific topic, target audience and learning environment, so that elements of the lesson plan, learning environment, and resources that need adaptation are identified.

2-3.2.1 *Prerequisite Knowledge:* Recognition of student limitations, methods of instruction, types of resource materials; organizing the learning environment; policies and procedures.

2-3.2.2 *Prerequisite Skills:* Analysis of resources facilities, and materials.

2-3.3 Adapt a prepared lesson plan, given course materials and an assignment, so that the needs of the student and the objectives of the lesson plan are achieved.

2-3.3.1 *Prerequisite Knowledge:* Elements of a lesson plan, selection of instructional aids and methods, organization of learning environment.

2-3.3.2 *Prerequisite Skills:* Instructor preparation and organizational skills.

2-5.2 Administer oral, written, and performance tests, given the lesson plan, evaluation instruments, and the evaluation procedures of the agency, so that the testing is conducted according to procedures and the security of the materials is maintained.

2-5.2.1 *Prerequisite Knowledge:* Test administration, agency policies, laws affecting records and disclosure of training information, purposes of evaluation and testing, and performance skills evaluation.

2-5.2.2 *Prerequisite Skills:* Use of skills checklists and oral questioning techniques.

2-5.4 Report test results, given a set of test answer sheets or skills checklists, a report form and policies and procedures for reporting, so that the results are accurately recorded, the forms are forwarded according to procedure, and unusual circumstances are reported.

2-5.4.1 *Prerequisite Knowledge:* Reporting procedures, the interpretation of test results.

2-5.5 Provide evaluation feedback to students, given evaluation data, so that the feedback is timely, specific enough for the student to make efforts to modify behavior, objective, clear, and relevant; include suggestions based on the data.

Chapter 3 Instructor II

3-2.3 Formulate budget needs, given training goals, agency budget policy, and current resources, so that the resources required to meet training goals are identified and documented.

3-2.3.1 *Prerequisite Knowledge:* Agency budget policy, resource management, needs analysis, sources of instructional materials, and equipment.

3-2.3.2 *Prerequisite Skills:* Resource analysis and forms completion.

3-2.4 Acquire training resources, given an identified need, so that the resources are obtained within established timelines, budget constraints, and according to agency policy.

3-2.4.1 *Prerequisite Knowledge:* Agency policies, purchasing procedures, budget management.

3-2.4.2 *Prerequisite Skills:* Forms completion.

3-5.2 Develop student evaluation instruments, given learning objectives, audience characteristics, and training goals, so that the evaluation instrument determines if the student has achieved the learning objectives, the instrument evaluates performance in an objective, reliable, and verifiable manner, and the evaluation instrument is bias-free to any audience or group.

3-5.2.1 *Prerequisite Knowledge:* Evaluation methods, development of forms, effective instructional methods, and techniques.

3-5.2.2 *Prerequisite Skills:* Evaluation items construction and assembly of evaluation instruments.

3-5.3 Develop a class evaluation instrument, given agency policy and evaluation goals, so that students have the ability to provide feedback to the instructor on instructional methods, communication techniques, learning environment, course content, and student materials.

3-5.3.1 *Prerequisite Knowledge:* Evaluation methods, test validity.

3-5.3.2 *Prerequisite Skills:* Development of evaluation forms.

Chapter 4 Instructor III

4-2.3 Develop recommendations for policies to support the training program, given agency policies and procedures and the training program goals, so that the training and agency goals are achieved.

4-2.3.1 *Prerequisite Knowledge:* Agency procedures and training program goals, format for agency policies.

4-2.3.2 *Prerequisite Skills:* Technical writing.

4-2.6 Write equipment purchasing specifications, given curriculum information, training goals, and agency guidelines, so that the equipment is appropriate and supports the curriculum.

4-2.6.1 *Prerequisite Knowledge:* Equipment purchasing procedures, available department resources and curriculum needs.

4-2.6.2 *Prerequisite Skills:* Evaluation methods to select the equipment that is most effective and preparation of procurement forms.

4-2.7 Present evaluation findings, conclusions, and recommendations to agency administrator, given data summaries and target audience, so that recommendations are unbiased, supported, and reflect agency goals, policies, and procedures.

4-2.7.1 *Prerequisite Knowledge:* Statistical evaluation procedures and agency goals.

4-2.7.2 *Prerequisite Skills:* Presentation skills and report preparation following agency guidelines.

4-3.2 Conduct an agency needs analysis, given agency goals, so that instructional needs are identified.

4-3.2.1 *Prerequisite Knowledge:* Needs analysis, task analysis, development of job performance requirements, lesson planning, instructional methods, characteristics of adult learners, instructional media, curriculum development, and development of evaluation instruments.

4-3.2.2 *Prerequisite Skills:* Conducting research, committee meetings, and needs and task analysis; organizing information into functional groupings; and interpreting data.

4-3.3 Design programs or curriculums, given needs analysis and agency goals, so that the agency goals are supported, the knowledge and skills are job related, the design is performance based, adult learning principles are utilized, and the program meets time and budget constraints.

4-3.3.1 *Prerequisite Knowledge:* Instructional design, adult learning principles, principles of performance based education, research, and fire service terminology.

4-3.3.2 *Prerequisite Skills:* Technical writing, selecting course reference materials.

4-3.4 Modify an existing curriculum, given the curriculum, audience characteristics, learning objectives, instructional resources and agency training requirements, so that the curriculum meets the requirements of the agency, and the learning objectives are achieved.

4-3.4.1 *Prerequisite Knowledge:* Instructional design, adult learning principles, principles of performance based education, research, and fire service terminology.

4-3.4.2 *Prerequisite Skills:* Technical writing, selecting course reference materials.

4-3.5 Write program and course goals, given job performance requirements (JPRs) and needs analysis information, so that the goals are clear, concise, measurable, and correlate to agency goals.

4-3.5.1 *Prerequisite Knowledge:* Components and characteristics of goals, and correlation of JPRs to program and course goals.

4-3.5.2 *Prerequisite Skills:* Writing goal statements.

4-3.6 Write course objectives, given JPRs, so that objectives are clear, concise, measurable, and reflect specific tasks.

4-3.6.1 *Prerequisite Knowledge:* Components of objectives and correlation between JPRs and objectives.

4-3.6.2 *Prerequisite Skills:* Writing course objectives and correlating them to JPRs.

4-3.7 Construct a course content outline, given course objectives, reference sources, functional groupings and the agency structure, so that the content supports the agency structure and reflects current acceptable practices.

4-3.7.1 *Prerequisite Knowledge:* Correlation between course goals, course outline, objectives, JPRs, instructor lesson plans, and instructional methods.

4-5.3 Develop course evaluation plan, given course objectives and agency policies, so that objectives are measured and agency policies are followed.

4-5.3.1 *Prerequisite Knowledge:* Evaluation techniques, agency constraints, and resources.

4-5.3.2 *Prerequisite Skills:* Decision-making.

4-5.4 Create a program evaluation plan, given agency policies and procedures, so that instructors, course components, and facilities are evaluated and student input is obtained for course improvement.

4-5.4.1 *Prerequisite Knowledge:* Evaluation methods, agency goals.

4-5.4.2 *Prerequisite Skills:* Construction of evaluation instruments.

Chapter 6

Lesson Plan Development

Once a program of instruction has been developed, the next step is up to the instructors who will be teaching the courses. Using the program instructor guide or outline and the resources list, instructors develop lesson plans from which they can teach the information required by the program. *In advance of instruction,* lesson plans help instructors identify and prepare instructional materials, review presentation topics, and gather or prepare visuals, handouts, and worksheets that aid the participants in the learning process. *During the instruction,* lesson plans help instructors focus on topics, meet objectives, and engage participants in learning activities. *At the conclusion of instruction,* lesson plans guide instructors in summarizing main points and guiding learners in drawing conclusions from which instructors can determine or assess learner understanding.

This chapter provides information on the importance, uses, and components of a lesson plan. The following sections describe the lesson plan, discuss its purposes, and explain the component parts and the general formats commonly used. Because instructors find they not only have to design their own lessons but sometimes use or adapt the plans of others, this chapter also gives information on creating a lesson plan as well as modifying one to meet specific needs. In addition, this chapter describes support and application components. References and supplemental readings and job performance requirements are given at the end of the chapter.

The Lesson Plan

[NFPA 1041: 3-3.2, 3-3.3, 3-3.3.2]

Planning what will be taught is a prelude to instruction. Planning a lesson helps instructors to carefully think out and write what is to be taught and to plan strategies for teaching. Instructors cannot just walk into a classroom and begin teaching without some plan as to what they will do, where they will go with the information, and how they will get there. Teaching without a plan gives no guarantee that course objectives will be met or that learners will actually learn what is intended in the course. Using a lesson plan does not ensure fulfillment of objectives either, but it increases the odds of success. Without planning the lesson ahead of time, instructors may find that they are lacking important support equipment or supplemental materials, which means that they cannot teach or demonstrate information. The result for both instructors and learners is that time is wasted because appropriate teaching and learning could not take place. What, exactly, is a lesson plan? What are its purposes, and what are its benefits? The following sections answer these questions.

Definition

A *lesson plan* is an instructional unit that maps out a plan to cover information and skills and makes effective use of time, space, and personnel. A lesson plan clearly states what an instructor will accomplish with the learners during a particular lesson. It is a step-by-step guide for any type of presentation. A lesson plan states all the steps and methods necessary for presenting the required knowledge or skills in the proper sequence. It also lists the appropriate support materials and indicates when they will be used. The lesson plan provides instructors with a teaching purpose and fulfills other important purposes and benefits as well.

Purposes

Purposeful planning means that there is thoughtful direction embedded in an undertaking or situation. Instructors can embed purpose in their program by planning actions and events that will occur in a class or training program. They must develop strong links or connections between what is new and what is now known. The lesson plan writer should put into the lesson plan such components as examples, questions, and activities that create these links and carry these components throughout the lesson to ensure that learners participate in applications and achieve successful evaluations. Instructors plan so that lessons proceed smoothly and can be repeated, if necessary, without starting "from scratch" and making the same plans all over again. Lesson plans provide uniformity; when instructors must teach the same lessons to subsequent groups, lesson plans standardize the instruction and enable instructors to provide the same information in a similar format each time they teach the lesson. Lesson plans give a clear path for both instructors and learners to follow. The more effective the lesson plan is at providing a clear route to achieving the learning objectives, the more effective the instructor.

When instructors carefully create and use well-planned lessons, learners can expect instruction that is organized and orderly. It makes sense through appropriate sequences, introduces material, takes learners through application exercises to meet objectives, and guides learners through a summary of key ideas to ensure that they focus on important points and understand them. The whole purpose of learning and learners' successes are dependent on effective instruction, which hinge on well-structured, purposeful lesson plans.

Instructors prepare lesson plans not only to guide instruction and learner activities but also to provide documentation to the organization's administration. Lesson plans indicate teaching-learning information, methods and activities, and time frames for lessons. A lesson plan provides documentation of the amount and type of materials, equipment, and other resources needed to teach the lesson. This type of information, in turn, provides justification for equipment purchases.

The lesson plan is also a document for testing. It acts as a document to show what was taught (based on the objectives) and verifies that the information presented is appropriate for testing.

Benefits

A *benefit* is something that guards, aids, or promotes well-being or provides an advantage. Considering the definition and purposes of a lesson plan, it is easy to see the benefits to learners, instructors, and administrators. Consider the following:

- ***Makes teaching easier*** — As a result of taking the time to create a step-by-step guide to teaching a lesson, instructors have a clear pathway that keeps them on track throughout the activities and distractions that occur in a lesson.
- ***Makes learning easier*** — Learners are provided with sequential, orderly instruction, which makes learning interesting and worthwhile, aids in learner success, and motivates them for further learning.
- ***Makes uniform instruction easy*** — Uniform lessons ensure that subsequent classes receive similar training. Former learners can act as resources for current learners. Lesson uniformity also assures the administration of a cadre of employees who perform consistently and to the requirements of the actual job.
- ***Makes documentation easy*** — The lesson plan itself acts as a document for qualifying and verifying instruction should questions arise about lesson topics, instructional methods, equipment needs, or test questions.

Instructor Tip: *Gather and review some lesson plans created and used by your organization or other instructors. Look for the points that The Lesson Plan section discussed. Do these lesson plans fulfill the definition, meet the purposes, and provide the benefits described? Do you think these plans provide a clear path for instruction and learning? Part of the instructor role is preparing a plan for teaching, which means that when using already-prepared lesson plans, you may have to add information so that the plan works for you and your class during instruction.*

Lesson Plan Components

[NFPA 1041: 2-4.3, 2-4.3.1, 3-3.2, 3-3.3, 3-3.3.2]

Program developers and lesson planners have typically separated lessons into *manipulative* (skill) and *cognitive* (information or technical) lessons. These types of designations are still often used, but regardless of type, the components are the same, and it is not necessary to follow a rigid format when creating or modifying a lesson plan. There are many contemporary formats such as the Hunter Model and those used by the National Fire Academy. The IFSTA Model also uses a variety of lesson plan formats; the latest example is in the **Essentials of Fire Fighting (4th edition) Curriculum.** See Appendix C, Lesson Plans, for sample lesson plan models. Instructors can use any format and add any component part they desire into their lesson plans. The only requirements for lesson plans and their components are that they serve useful purposes and provide clear pathways by which instructors can guide learning to meet objectives.

Basic steps that lead to developing a lesson plan that developers use routinely and which have a significant purpose in lesson planning include the following:

- *Identify* information and materials, teaching level, and learner needs (completed in a needs analysis).
- *Select* materials and instructional format.
- *Design* lesson and delivery methods.
- *Implement* instruction and learning activities.
- *Evaluate* the lesson, instructional methods, learning activities, and learner outcome.

Each basic step is incorporated into the lesson plan components. When creating or modifying lesson plans, instructors include these basic steps. The steps often overlap and are cyclic (repeat) throughout the lesson and appear throughout the instructional process. The components of a lesson plan are explained in the following sections by how they are used to prepare for or conduct a lesson (Figure 6.1). The uses of a lesson plan are divided into the following parts, and although a particular lesson plan component is described under a particular heading, it may be used somewhere else in the lesson plan.

- Instructor preparation
- Instructional process: preparation, presentation, application, and evaluation
- Learner reinforcement

Instructor Preparation

Before teaching any lesson or giving any presentation, an instructor reviews the material, resources, and requirements of the lesson. By reviewing the components of a lesson plan, an instructor knows ahead of time the lesson details and logistics and how to prepare appropriately for teaching.

Preparation provides the readied instructor with confidence and credibility. The keys to success as an instructor are preparation, *preparation*, and ***preparation!*** Time spent preparing to teach is obvious to learners and relays knowledge and a caring, interested attitude. The following lesson plan components prepare the instructor for presenting the lesson:

- ***Job or topic*** — The *job* or *topic* is a short descriptive title of the information covered. The title should briefly describe or give an indication of the lesson content. Topic titles are usually taken directly from the course outline.
- ***Time frame*** — The estimated *time* it takes to teach a lesson is the time frame. Time frames may be set for each objective so that the instructor has a better idea of how to set the pace of the lesson. Time estimates can allow for variations in class size, experience level of learners, etc.
- ***Level of instruction*** — Based on job requirements, the desired learning *level* is listed. There is more than one approach to establishing level of instruction. One approach is the taxonomy of learning domains: Bloom's (1956) cognitive domain,

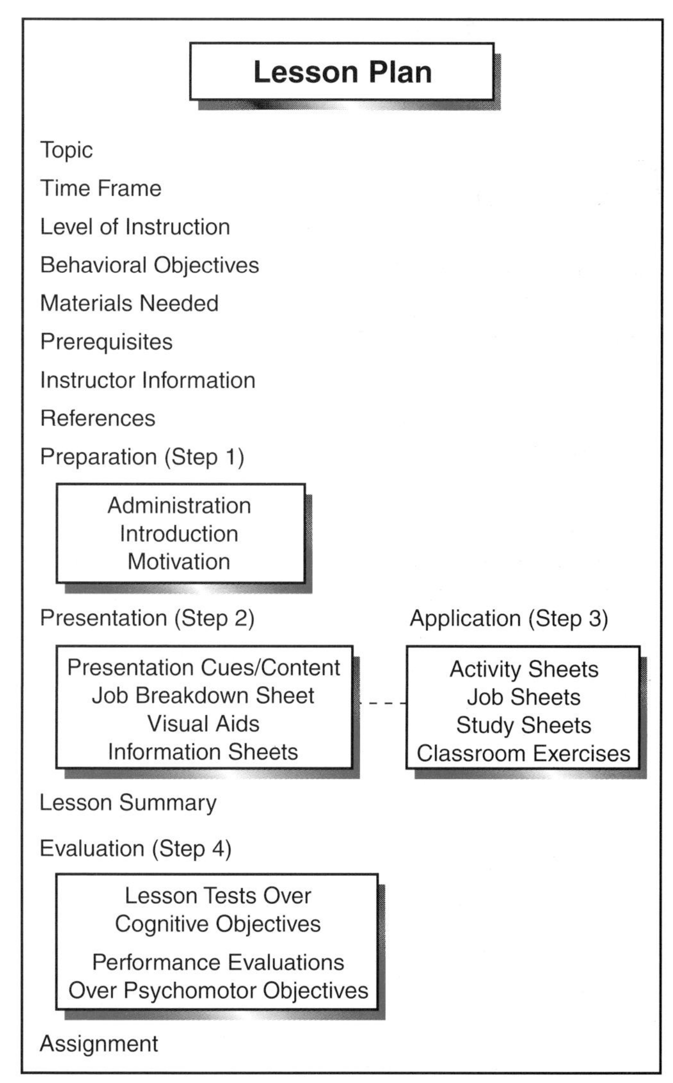

Figure 6.1 Components of a lesson plan.

Simpson's (1972) psychomotor domain, and Krathwohl's (1964) affective domain (see Chapter 4, The Psychology of Learning). The second approach is level of performance based on job performance requirements (JPRs). Performance represents the ability to apply knowledge and skill to any situation. The JPR approach is based on the belief that terminal performance is a product of all three domains and that one taxonomy can represent the five levels of performance (preoperational, basic skills application, superior skills, skills bridging, and creation and evaluation).

- ***Behavioral objectives*** — A description of the minimum acceptable *behaviors* that a learner must perform by the end of an instructional period is given in the behavioral objectives.
- ***Materials needed*** — A list of all *materials*, including quantity, needed to teach the number of participants in the program is included. This section of the lesson also includes any preparation, planning, or activities that the instructor needs to complete before delivering the lesson such as arranging to acquire and pick up equipment, copy handouts, reserve videotapes, etc.
- ***Prerequisites*** — A list of information, skills, or previous requirements that learners must have completed or mastered *before* entering this program or starting this lesson is given. Prerequisites for the current lesson may have been covered in the previous lesson or will be covered in this lesson before the expected outcome is reached. The instructor may give out-of-class assignments for the participants to complete in order to be ready for this lesson.
- ***Information to the instructor*** — A list of lesson *resources* can include the following information:
  - — Names of instructors or other personnel qualified to assist or provide logistical support
  - — Textbooks and other instructional materials
  - — Special equipment needs
  - — Specific instructional methods and learning activities required to meet objectives
  - — Unique training locations or change of training sites
- ***References*** — While preparing to teach a lesson, an instructor may refer to and review specific *references* and resources and list them on the lesson plan along with page numbers. These reference citations enable the instructor to qualify or verify information in the event that learners ask unique questions, question sources, or desire further information. When curriculum developers create a program, they may include a list of the references and resources used that instructors may also consult if it is different from their own lists when preparing for and planning lessons.

Instructional Process

The four-step instructional process contained in a lesson described in the following sections prepares the *participants* for learning, involves them in active participation in the instruction, and provides a way to measure their understanding. Adult learning theory indicates that instructional presentations should do the following:

- *Prepare* the learner for learning by motivating the person to answer the questions, *"Why am I here?"* and *"Why do I need to know this?"*
- *Present* information in segments, one segment at a time.
- Provide opportunity to *apply* the information.
- *Evaluate* learning through various checking and testing techniques. Repeat this process with each segment of information.

The four steps are performed in a continual process throughout the lesson and the program, and this repetitive process is referred to as the *present-apply strategy.* Also, instructors provide motivation at the beginning of the presentation, with each segment of information, during the presentation, and at the end of the presentation. This continuous motivation process is called the *Continuum Theory of Learner Motivation* by Raymond Wlodkowski (1993); it also continually answers the questions, *"Why am I here?"* and *"Why do I need to know this?"*

While actually teaching, instructors will find that these four steps overlap and are often performed together or repetitively throughout a lesson (Figure 6.2). With experience, instructors will notice that they begin to naturally and logically perform the steps — preparation, presentation, application, and evaluation — automatically, continuously, and with ease.

Preparation

Preparing the participants for learning (preparation) is the first of the four steps of instruction (Figure 6.3). The instructor establishes lesson relevancy to the job by performing the following actions:

- Introduce the topic.
- Gain attention.
- Arouse interest.
- State the objectives.
- Motivate participants by stating a *"need to know this information"* reason by relating it to an aspect of the job.
- Prepare participants to listen for key points by briefly stating the main topics that will be covered.

Presentation

In the second step (presentation), the instructor *presents* the information to be covered using an orderly, sequential outline (Figure 6.4). The outline lists with each key point the teaching methods, learning activities, demonstrations and practices, and instructional support materials such as audiovisuals, worksheets, and handouts that will be used to present the information to participants and involve them in learning. *Presentation* can be combined with the next and most important step, *application.*

How the Steps Work in Practice

In an actual teaching situation, the four steps in the instructional process tend to blend together. For example, demonstrations or explanations in the presentation step may be repeated during the application step. The same written test may be used during pretest and evaluation. The only real difference between the application and evaluation steps is supervision.

The steps of *Preparation, Presentation/Delivery, Application,* and *Evaluation* become logical and natural parts of the written lesson plan. With practice, the instructor simply does what is needed — without giving much thought to which step is involved.

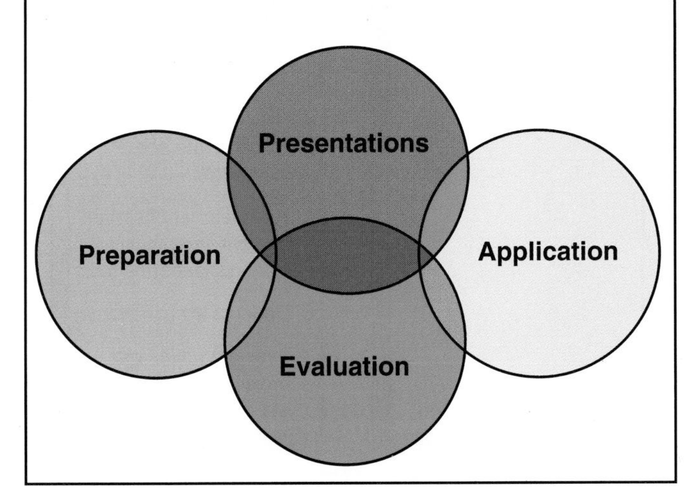

Figure 6.2 The four steps of the instructional process overlap in a lesson.

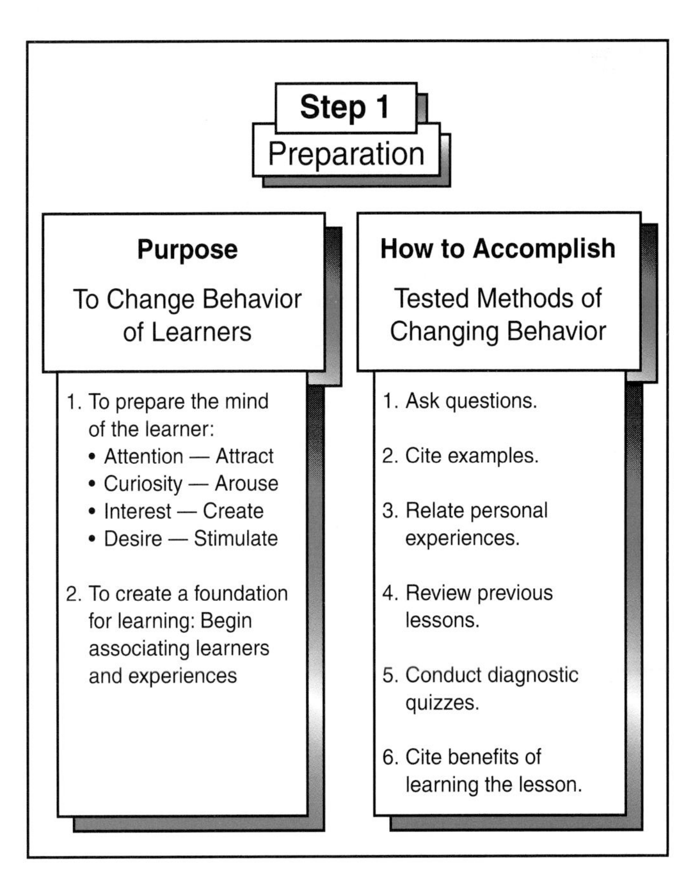

Figure 6.3 The preparation step: Prepare the learner for learning.

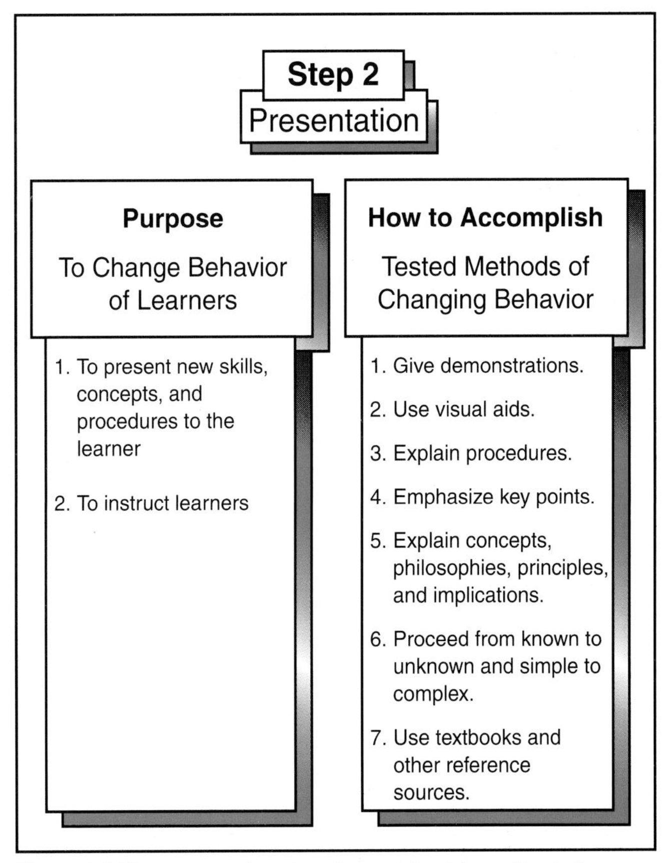

Figure 6.4 The presentation step: Present the information to be learned.

Application

The third of the four steps of instruction (application) is where the instructor provides opportunities for learning through activities, exercises, discussions, work groups, skill practice, and similar learning tasks (see Support and Application Components section for examples) (Figure 6.5). The purpose of the application step is learner reinforcement (see Learner Reinforcement section). Most of the learning takes place during the application step so it is critically important. Application can be combined with presentation so that the participants can *apply* the *presented* information through activities that require thinking (the *psycho* part of *psychomotor*) to manipulative or skill (the *motor* part of *psychomotor*) activities. Typically, application is related to performing the operations or steps of a task, but *skills* do not refer solely to activity steps of using equipment. A learner may demonstrate other skills such as the following:

- Give a presentation.
- Lead a group discussion or brainstorming session.
- Apply research methods.
- Demonstrate outlining and writing techniques.

Evaluation

The fourth step of instruction (evaluation) is where learners demonstrate how much they have learned through a written or practical test (Figure 6.6). In general, written tests are used to *evaluate cognitive* information; practical tests are used to *evaluate skill* ability. Other types of tests may be used to evaluate other learning areas (see Chapter 9, Testing and Evaluation). The purpose of evaluation is to determine whether learners achieved the program objectives.

Learner Reinforcement

It is not enough to prepare and present. Instructors must provide learners opportunities to *review, remember,* and *reinforce* the instruction. There are a number of ways throughout and at the end of a lesson that instructors use these three *R*'s to reinforce the lesson material. The following components reinforce lesson information:

- ***Summary***—In this part of the lesson, the instructor restates or reemphasizes the key points. During the

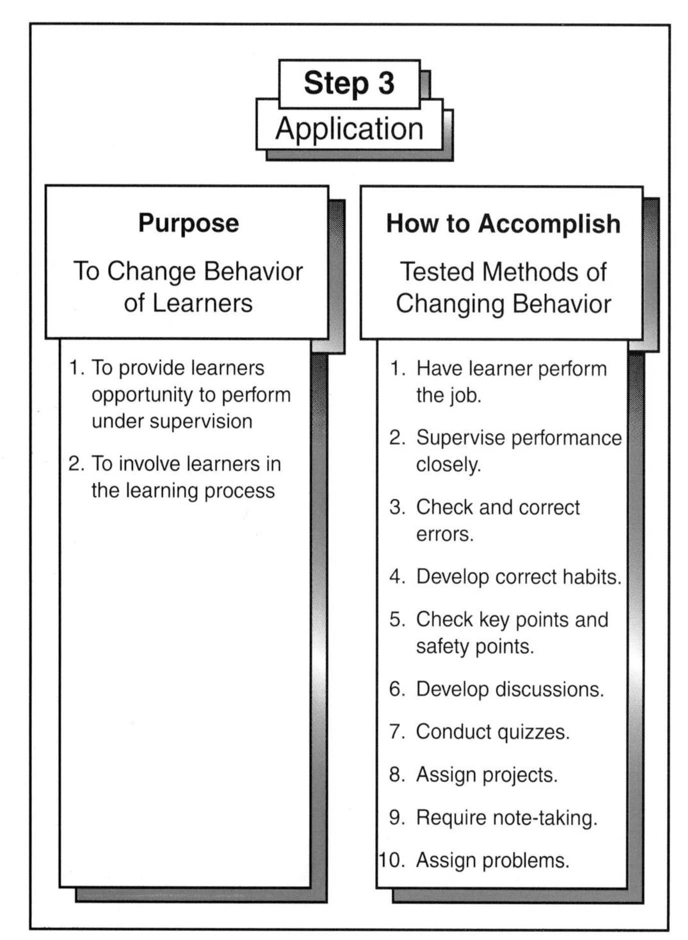

Figure 6.5 The application step: Provide opportunity for learners to apply information and skills.

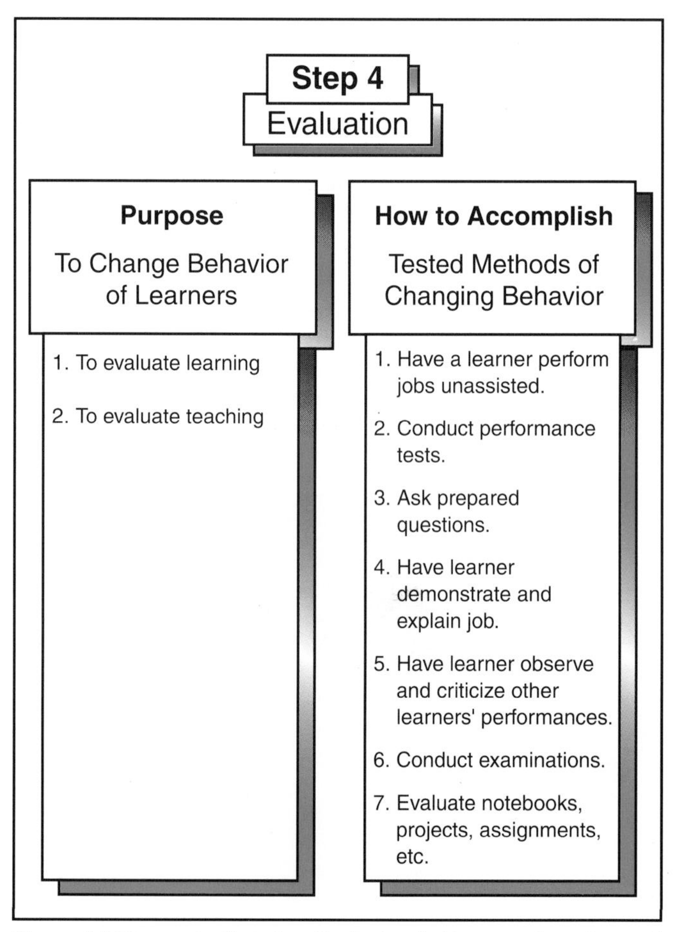

Figure 6.6 The evaluation step: Evaluate what learners have learned.

summary, instructors can also *evaluate* how much learners remember from the cognitive and skill parts of the lesson by having them participate. In addition to restating the key points, instructors can ask the learners to recall information or list steps, draw conclusions, and make comparisons. Summary activities such as these help reinforce the information covered and skills demonstrated or practiced and provide a platform from which to begin the next lesson.

- ***Assignment*** — An *assignment* is typically work that must be performed by learners outside class so they may reach a skill level, meet objectives, and/or prepare for the next lesson. There need not always be an assignment, and instructors may assign out-of-class activities or not as they desire or the program requires. Keep in mind that learners may forget new information and skills if they are not reinforced through relative and practical learning assignments that enable them to practice and apply what they learned.
- ***Job breakdown sheet*** — A form that lists a job and breaks it down into its parts by listing the operations (steps) and key points is a *job breakdown sheet* (or simply *job sheet*). Program developers and/or instructors can develop a job sheet for each of the psychomotor skills/jobs listed or discussed in the course outline. These sheets provide instructors and learners with sequences and details needed to teach and learn the knowledge and skill of a job (see Support and Application Components section for an example).

Instructor Tip: *When planning instruction, prepare a variety of learning/application activities. Having several activities preplanned gives the flexibility of using different methods of reinforcing learning. It has the added advantage of having plenty of activities planned for those learners who complete an assignment and are ready for another one and for those learners who are unable to perform or grasp the idea in one type of activity but could succeed in another type. A variety of activities can also provide alternatives that allow for weather conditions or equipment availability.*

Lesson Plan Format

[NFPA 1041: 2-3.2, 2-3.3, 3-3.2, 3-3.3, 3-3.3.2]

As the idea of the many lesson plan *components* is flexibility, so it is for lesson plan *format*. There are many ways to lay out a lesson plan, and training organizations may have a standard format that they use or require for certain programs. When planning lessons from program or instructor guides, instructors can adopt a method that works best for them. New instructors may wish to begin with a simple format, while experienced instructors may use a more detailed format or one that is unique to their own style and requirements (see Appendix C, Lesson Plans, for examples). Instructors will find the descriptions in the following sections helpful and will want to be familiar with their purposes.

Introductory Components

Introductory components do just that — introduce the lesson. They are important parts of the lesson, grab learners' attentions, and put learners on the right tracks. Introductory components include such information as what the instructor will cover and what the learners will be expected to do in how much time and to what level. These introductory areas are described as follows:

- ***Lesson title*** — The title hints at the topic and gives learners some idea of what to expect. It also reminds instructors of what they will teach and triggers reminders of equipment and other instructional materials they will need.
- ***Learning objectives*** — The purpose of objectives is *not* to tell instructors what to teach but to describe what learners will accomplish. They are *not* written for the instructors nor do they describe what instructors will accomplish; they are written for the learners. Learners must have a complete picture of what is expected of them. Properly written objectives tell learners exactly what they must do, what they will be given to do it with, and how well they must do it. Instructors use objectives as a basis for planning their instruction so that learners can meet the requirements of the objectives.
- ***Time frame*** — Having a time frame for accomplishing the tasks of the lesson helps instructors pace themselves while teaching. There may be a total time listed for an entire lesson; there may be increment times listed for each lesson section or skill. It is important that instructors plan their lessons to work within the time frame suggested by the

provide information to the administration and program developers that is important in making final assessments of the program. Is a high learner failure rate a result of the instructor's methods, knowledge, or abilities? The learners are the ones to provide insight because they were in attendance and have first-hand experience with the instructor. Many organizations provide forms at the end of a program so that learners may evaluate their instructors.

Assignments

At the end of a lesson, what must the participants do next? Do instructors just dismiss the class without having them think about or process the information further? Usually an instructor gives some kind of assignment. The most common assignment is to read the next chapter in a text in order to be ready for the next lesson. Some programs require out-of-class assignments, some of which may be part of an internship or an on-the-job experience that is related to the training program. Many instructors like to assign readings, research, or projects that reinforce the information introduced in class. Often, an out-of-class assignment is given because the program cannot or does not provide adequate time for thoroughly covering necessary information, much of which may be better learned or gained by on-the-job exposure and experience.

Summary

The purpose of a *summary* is to bring closure to a lesson or a course. It gives participants a sense of completion such as reaching the end of a journey or completing a goal. Often, participants may not remember all the key issues that were discussed by the end of a session. During a summary, perform the following activities:

- Review key issues.
- Remind participants of important ideas.
- Review important steps.
- Ask for conclusions.
- Involve participants in recapping the main ideas.

Creating a Lesson Plan

[NFPA 1041: 2-3.2, 3-3.2, 4-3.7]

A lesson may vary in length from a few minutes to several hours, depending on the objectives. When instructors create a lesson plan, determining the objectives should be the first step. There are a number of subsequent steps to creating a lesson plan, all of which guide the instructor through the preparation stage — a necessary component to complete before teaching a lesson or course. Consider the following steps when creating a lesson plan:

Step 1: Create or review the learning objectives.

A lesson must have objectives that guide the instructor in teaching and the participants in learning. If objectives are not given, create them. If objectives are given, review them. Be sure that the objectives are complete; that is, they contain behaviors or what the learner will do, conditions or information or items the learner will be given in order to complete the behaviors, and criteria or standards that state the degree to which the learner will complete the behaviors.

Step 2: Review the learner characteristics and needs.

Ask for information on the participants from the training organization, employer, or registration office. What are employer expectations, learner levels and learning styles based on previous courses, and test scores? Information such as this enables instructors to plan a wide range of activities that provide opportunities through which various types of participants can learn and be successful.

Step 3: Conduct research on the topic.

Usually, instructors have job experiences or are certified to teach certain programs. Even so, not every instructor is always aware of the newest equipment, the latest methods, or the most recent protocols for every type of training program. To appear credible and to conduct a credible program, instructors must research the lesson topic in order to present a lesson that provides current information and skills. See Appendix D, How to Research a Topic. The following activities can provide valuable information and resources:

- ✔ Check with the training organization and other employers.
- ✔ Look through recent national and local standards.
- ✔ Review jurisdictional operating procedures and equipment manuals.

- ✔ Look for resources on the Internet.
- ✔ Look for resources in journals and texts found at professional and university libraries.
- ✔ Meet with other instructors who have taught the same program.

Step 4: Construct course content outline.

When developing the course content outline, it is important to ensure that information supports the organization's policies and reflects current acceptable practices or standards as adopted by the organization. Outlines make it easy for instructors to teach and stay on track. For each objective, list a Roman numeral with an outline point that pulls out a key idea from the objective. Under each Roman numeral, outline subpoints or steps that include all the information needed to meet that one objective and incorporate all related organizational policies. Do the same for each objective.

Step 5: Identify instructional resources.

While researching the topic (Step 3), remember to list all resources. Add these resources to the lesson plan. If learners have in-depth content questions or if administrators or employers have questions about information sources, the lesson plan provides a quick referral for the information in that particular lesson. As a result, instructors also have recorded references for future research.

Step 6: Select delivery methods and strategies.

While planning a lesson, instructors may find themselves visualizing how they will teach each segment. They may recall how they learned this particular material or skill and plan to do something similar or create a different method that makes it easier for participants to become interested and to learn. Make notes about ideas and plan strategies that will accomplish instructional methods. Insert these plans next to each objective point in the lesson outline.

Step 7. Revise content outline and include instructor notes, audiovisuals, questions, performance steps, etc.

Review the outline after it is completed to be sure it includes all necessary content and meets objectives; make revisions where necessary. During the review, perform the following activities:

- ✔ Insert additional notes that provide guidance or reminders about instructional methods and activities.
- ✔ Note places where audiovisuals will be shown or where handouts and job sheets will be distributed.
- ✔ Write questions to ask participants at pertinent points.
- ✔ Check that all skill performance steps are included in the appropriate order (overall, the most important activity).

Step 8. Complete other sections of the lesson plan including performance job sheets, assignments, support components, etc.

- ✔ Create supplemental forms and application components that assist in guiding learners through lesson activities.
- ✔ Design assignments that reinforce information and skills.
- ✔ List instructional equipment or tools that must be available to accomplish the lesson objectives.

Step 9. Seek feedback from others on the lesson plan and modify as necessary.

Have experienced instructors review the outline. It may also be helpful to have both the training organization administrators and the participants' employers review the outline to ensure that the lesson meets training standards and job requirements.

Step 10. Create or obtain audiovisuals, props, models, etc.

Lessons should include visuals that appropriately illustrate the material or skill. Models that show equipment parts, props that allow practice on a smaller scale, and videotapes that show and describe procedures that are otherwise dangerous to practice are examples of appropriate visuals. Many instructors have the talent or creativity to develop their own teaching support materials. Sometimes, training organizations purchase instructional support materials to go with their programs. As instructors create or review these materials, they should ensure that they match (rather than contradict) the program objectives and will assist learners in reaching them.

Step 11. Practice presentation, and modify as necessary.

Planning the presentation is much different from actually delivering it. Practice the delivery to get an idea of how long it takes to cover the information. The difficulty in doing this is that trying to gauge how long questions, discussion, and activities will take is not always accurate. For the first presentation, keep activities simple and minimal, but always have additional activities planned in the event there is extra time. Teach the *need-to-know* information first; the supplemental or *nice-to-know* information can be added if time permits.

Modifying or Adapting a Lesson Plan

[NFPA 1041: 2-3.3, 3-3.3, 3-3.3.2]

Many training organizations and employers already have lesson plans from which contract instructors will teach. Receiving an "already-prepared" lesson plan does not mean that the instructor is ready to teach. Even though this lesson plan may include all the component parts and information listed in the preceding sections, an instructor must still prepare to teach. By reviewing the lesson plan, instructors may find they need to make modifications to the format for ease of use or to the content to ensure current information. Components may need modifications to ensure complete coverage and effective use of time, space, and personnel. Instructors may also have to make adaptations in the lesson to address certain groups or introduce information specific to a group.

When creating a lesson plan, recall that instructors started with the objectives as the first step; it is the same with modifying a lesson plan. Subsequent steps to modifying or adapting a lesson plan again guide instructors through the preparation stage necessary for teaching a lesson or course. Consider the following steps when modifying or adapting a lesson plan, and note that many of them are similar to the steps for creating a lesson plan:

Step 1. Review the learning objectives.

Be sure that objectives are complete; that is, they must contain behaviors, conditions, and criteria. If objectives are not complete, it may be necessary to review them with the training organization or employer and revise them as necessary.

Step 2. Review the learner characteristics and needs.

As done when creating a lesson plan, ask for information on the participants. Find out their expectations, learning levels, and styles in order to plan a wide range of activities that provide opportunities through which various types of participants can learn and be successful.

Step 3. Conduct research on the topic.

- ✔ Check with the training organization and employers.
- ✔ Look through recent national and local standards.
- ✔ Review jurisdictional operating procedures and equipment manuals.
- ✔ Look for resources on the Internet and in journals and texts found at professional and university libraries. (Refer to Appendix D, How to Research a Topic.)
- ✔ Meet with other instructors who have taught the program.

Step 4. Identify changes that need to be made.

Include any updates in the following areas:

✔ Standards, equipment, and materials

✔ Instructional methods and supplemental materials

✔ Learning activities and support materials

Step 5. Modify the outline.

Considering research and noted changes, indicate differences in the following areas:

✔ Delivery methods

✔ Audiovisuals

✔ Content

✔ Instructor notes

✔ Performance steps

✔ Instructional support

✔ Teaching and learning components and equipment

✔ Job sheets or handouts

Step 6. Seek feedback from others on the lesson plan and modify as necessary.

Have other instructors, the training organization, and/or the employer review the modified outline and provide comments and guidance on the adaptation.

Step 7. Create or obtain audiovisuals, props, models, etc.

Create or select appropriate visual materials that will support instruction and enhance and enable learning.

Step 8. Practice presentation, and modify as necessary.

Practice delivery to get an idea of how long it takes to cover the information and try to gauge how long questions, discussion, and activities will take. Always have additional activities planned in the event there is ext[illegible] the *need-to-know* information fi[illegible] the supplemental or *nice-to-know*[illegible] later in the lesson if time permits.

Support and Application Components

[NFPA 1041: 2-3.3, 3-3.2, 3-3.3, 3-3.3.2]

Those who teach tend to be multitalented, intelligent, and dedicated individuals, but they cannot perform the task of teaching alone. Teaching any program requires help and support from a variety of resources including personnel, equipment, and the inevitable paper materials in the form of informational handouts and study aids, job sheets, and worksheets or activity sheets. The purpose of these paper materials is to provide program participants with background or resource information that is not available in the text or is not easily copied from copyrighted standards or lengthy protocols. In addition, some application components list performance steps that learners can follow while practicing; some forms are designed to guide activities during group exercises. The overall purpose is to enable the learner to apply the lesson content (see Application section).

Support components are aids that an instructor can use as desired; however, they often serve as important and useful instructional adjuncts as well as helpful learning guides and reinforcers. The following sections describe several types of support and application components (samples are given in Appendix E, Support and Application Components).

Information Sheet

An *information sheet* is a fact sheet or type of handout that provides additional background information on a topic supplemental to what is provided in the text or other course resources (Figure 6.8). The information may be in the form of detailed text, or it may outline or summarize key ideas. The handout may also list the information references or include suggestions for further research. Information sheets should be designed to encourage program participants to learn. They are usually created for one of the following reasons:

- The information is unavailable to some learners because texts or other learning resources are limited.
- To get the information, learners would have to find and consult a number of texts, which may be difficult to obtain and would be time-consuming.
- The information is not available in any text.

eps for developing information sheets are as follows:

> ### Information Sheet
>
> **Step 1.** Create a title that indicates the subject area and relates the title to the lesson.
>
> **Step 2.** Introduce the information with a brief description that explains its importance and interests the learner in reading and studying it.
>
> **Step 3.** Present the information in the most appropriate form so that it is easy to read and follow. Include appropriate charts, tables, or illustrations on the form, or place them on separate pages; label them for easy referral.
>
> **Step 4.** Develop test questions based on the information sheet so learners can assess whether they achieved the lesson objectives. Test questions should stress important points in the information sheet and enable the instructor to check learner comprehension. Make the questions thought-provoking, and develop a sufficient number to cover the information.

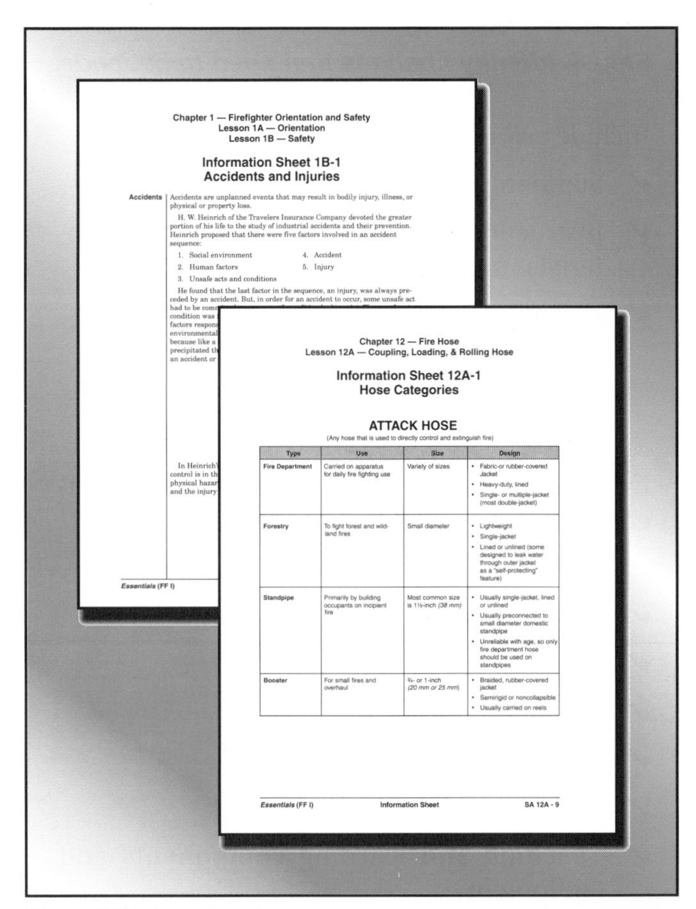

Chapter 1 — Firefighter Orientation and Safety
Lesson 1A — Orientation
Lesson 1B — Safety

Information Sheet 1B-1
Accidents and Injuries

Accidents Accidents are unplanned events that may result in bodily injury, illness, or physical or property loss.

H. W. Heinrich of the Travelers Insurance Company devoted the greater portion of his life to the study of industrial accidents and their prevention. Heinrich proposed that there were five factors involved in an accident sequence:

1. Social environment
2. Human factors
3. Unsafe acts and conditions
4. Accident
5. Injury

He found that the last factor in the sequence, an injury, was always preceded by an accident. But, in order for an accident to occur, some unsafe act

Essentials (FF I)

Chapter 12 — Fire Hose
Lesson 12A — Coupling, Loading, & Rolling Hose

Information Sheet 12A-1
Hose Categories

ATTACK HOSE
(Any hose that is used to directly control and extinguish fire)

Type	Use	Size	Design
Fire Department	Carried on apparatus for daily fire fighting use	Variety of sizes	• Fabric-or rubber-covered Jacket • Heavy-duty, lined • Single- or multiple-jacket (most double-jacket)
Forestry	To fight forest and wild-land fires	Small diameter	• Lightweight • Single-jacket • Lined or unlined (some designed to leak water through outer jacket as a "self-protecting" feature)
Standpipe	Primarily by building occupants on incipient fire	Most common size is 1½-inch *(38 mm)*	• Usually single-jacket, lined or unlined • Usually preconnected to small diameter domestic standpipe • Unreliable with age, so only fire department hose should be used on standpipes
Booster	For small fires and overhaul	¾- or 1-inch *(20 mm or 25 mm)*	• Braided, rubber-covered jacket • Semirigid or noncollapsible • Usually carried on reels

Essentials (FF I) Information Sheet SA 12A - 9

Figure 6.8 Information sheets.

Job Breakdown Sheet

A *job breakdown sheet* (or simply *job sheet*) breaks down a job into parts by listing the operational steps and their key points or steps for completing each operation (Figure 6.9). The purpose of these sheets is to provide the instructors and the learners with the sequences and details necessary to teach and learn a job that includes both motor skills and knowledge. The *operation (doing unit)* is a step or the smallest aspect in performing a job. A *key point (knowing unit)* is a step that is part of the process of completing the job and may include information that aids in knowing or understanding operations that enable the learner to perform the job correctly. Learners can use a job breakdown sheet to prepare for a competency profile or performance evaluation.

Key points that learners need to know in order to perform a job are listed on the right side of the page; skill operations that learners need to be able to perform based on knowledge are listed on the left side of the page. A completed job breakdown sheet lists the step-by-step procedures for doing a job in sequence and the key points that the instructor must stress and

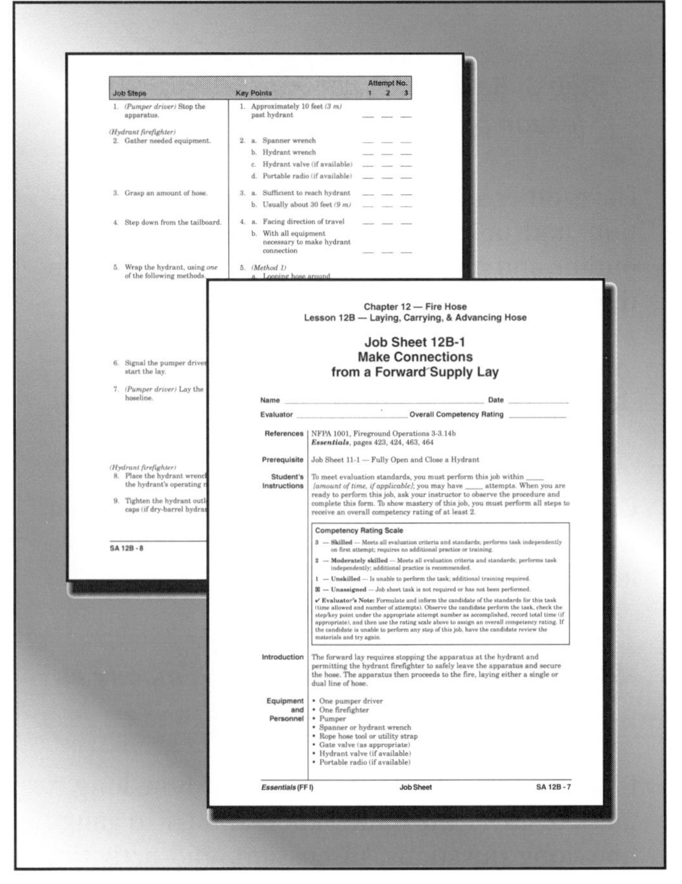

Job Steps	Key Points	Attempt No. 1	2	3
1. *(Pumper driver)* Stop the apparatus.	1. Approximately 10 feet *(3 m)* past hydrant	___	___	___
(Hydrant firefighter) 2. Gather needed equipment.	2. a. Spanner wrench	___	___	___
	b. Hydrant wrench	___	___	___
	c. Hydrant valve (if available)	___	___	___
	d. Portable radio (if available)	___	___	___
3. Grasp an amount of hose.	3. a. Sufficient to reach hydrant	___	___	___
	b. Usually about 30 feet *(9 m)*	___	___	___
4. Step down from the tailboard.	4. a. Facing direction of travel	___	___	___
	b. With all equipment necessary to make hydrant connection	___	___	___
5. Wrap the hydrant, using *one* of the following methods.	5. *(Method 1)*			

SA 12B - 8

Chapter 12 — Fire Hose
Lesson 12B — Laying, Carrying, & Advancing Hose

Job Sheet 12B-1
Make Connections
from a Forward Supply Lay

Name ______ Date ______

Evaluator ______ Overall Competency Rating ______

References NFPA 1001, Fireground Operations 3-3.14b
Essentials, pages 423, 424, 463, 464

Prerequisite Job Sheet 11-1 — Fully Open and Close a Hydrant

Student's Instructions To meet evaluation standards, you must perform this job within ____ *[amount of time, if applicable]*; you may have ____ attempts. When you are ready to perform this job, ask your instructor to observe the procedure and complete this form. To show mastery of this job, you must perform all steps to receive an overall competency rating of at least 2.

Competency Rating Scale

3 — **Skilled** — Meets all evaluation criteria and standards; performs task independently on first attempt; requires no additional practice or training.

2 — **Moderately skilled** — Meets all evaluation criteria and standards; performs task independently; additional practice is recommended.

1 — **Unskilled** — Is unable to perform the task; additional training required.

☒ — **Unassigned** — Job sheet task is not required or has not been performed.

✓ **Evaluator's Note:** Formulate and inform the candidate of the standards for this task (time allowed and number of attempts). Observe the candidate perform the task, check the step/key point under the appropriate attempt number as accomplished, record total time (if appropriate), and then use the rating scale above to assign an overall competency rating. If the candidate is unable to perform any step of this job, have the candidate review the materials and try again.

Introduction The forward lay requires stopping the apparatus at the hydrant and permitting the hydrant firefighter to safely leave the apparatus and secure the hose. The apparatus then proceeds to the fire, laying either a single or dual line of hose.

Equipment and Personnel
- One pumper driver
- One firefighter
- Pumper
- Spanner or hydrant wrench
- Rope hose tool or utility strap
- Gate valve (as appropriate)
- Hydrant valve (if available)
- Portable radio (if available)

Essentials (FF I) Job Sheet SA 12B - 7

Figure 6.9 Job breakdown sheet.

demonstrate while teaching the job. The content and format of a well-prepared job breakdown sheet are shown in the sample given in Appendix E, Support and Application Components.

Steps for developing a job breakdown sheet are as follows:

Job Sheet

Step 1. List the job to be done (this becomes the title).

Step 2. Divide the page into two columns.

Step 3. Head the left column *operations* or *doing units* where the instructor lists the actual motor skills.

Step 4. Head the right column *key points* or *knowing units* where the instructor lists pieces of knowledge without which the operations cannot be safely or accurately performed.

Step 5. List the steps of the job in sequence under *operations,* using action verbs (such as grasp, push, turn, lift, don, etc.).

Step 6. List cautions, warnings, safety factors, and conditions essential for performing the job operations under *key points.*

Learners typically practice skills with the guidance of instructors, but instructors cannot supervise every learner during every practice activity. Job sheets provide the skill steps learners need to know and practice and allow them to practice in groups on their own as they coach each other, discuss and think about the activities, and develop higher level (analytical and synthesis) cognitive skills. These self-practice exercises allow them to prepare for competency profiles or performance evaluations where they perform without instructor guidance, exercise thinking skills, and perform at mastery level for an evaluator (Figure 6.10).

Worksheet or Activity Sheet

A *worksheet* or *activity sheet* (also called *practical activity sheet*) provides learners opportunities to apply rules, analyze and evaluate objects and situations, or use multiple skills while completing activities. Instructors create learner worksheets or activity sheets from the information content of the lesson plan (Figure 6.11). Any worksheets that the instructor develops must support the objectives and provide activities that

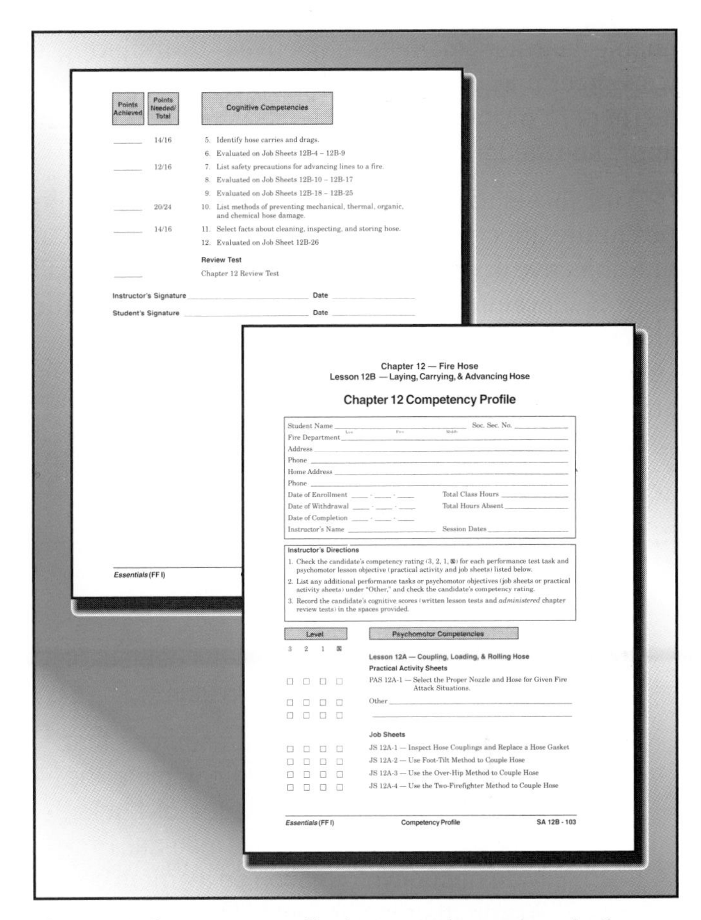

Points Achieved	Points Needed/ Total	Cognitive Competencies
______	14/16	5. Identify hose carries and drags.
		6. Evaluated on Job Sheets 12B-4 – 12B-9
______	12/16	7. List safety precautions for advancing lines to a fire.
		8. Evaluated on Job Sheets 12B-10 – 12B-17
		9. Evaluated on Job Sheets 12B-18 – 12B-25
______	20/24	10. List methods of preventing mechanical, thermal, organic, and chemical hose damage.
______	14/16	11. Select facts about cleaning, inspecting, and storing hose.
		12. Evaluated on Job Sheet 12B-26
		Review Test
______		Chapter 12 Review Test

Instructor's Signature ______________ Date ______________

Student's Signature ______________ Date ______________

Essentials (FF I)

Chapter 12 — Fire Hose
Lesson 12B — Laying, Carrying, & Advancing Hose

Chapter 12 Competency Profile

Student Name ______________ Soc. Sec. No. ______________
Fire Department ______________
Address ______________
Phone ______________
Home Address ______________
Phone ______________
Date of Enrollment ___-___-___ Total Class Hours ______________
Date of Withdrawal ___-___-___ Total Hours Absent ______________
Date of Completion ___-___-___
Instructor's Name ______________ Session Dates ______________

Instructor's Directions

1. Check the candidate's competency rating (3, 2, 1, ☒) for each performance test task and psychomotor lesson objective (practical activity and job sheets) listed below.
2. List any additional performance tasks or psychomotor objectives (job sheets or practical activity sheets) under "Other," and check the candidate's competency rating.
3. Record the candidate's cognitive scores (written lesson tests and *administered* chapter review tests) in the spaces provided.

Level 3	2	1	☒	Psychomotor Competencies
				Lesson 12A — Coupling, Loading, & Rolling Hose
				Practical Activity Sheets
☐	☐	☐	☐	PAS 12A-1 — Select the Proper Nozzle and Hose for Given Fire Attack Situations.
☐	☐	☐	☐	Other ______________
☐	☐	☐	☐	______________
				Job Sheets
☐	☐	☐	☐	JS 12A-1 — Inspect Hose Couplings and Replace a Hose Gasket
☐	☐	☐	☐	JS 12A-2 — Use Foot-Tilt Method to Couple Hose
☐	☐	☐	☐	JS 12A-3 — Use the Over-Hip Method to Couple Hose
☐	☐	☐	☐	JS 12A-4 — Use the Two-Firefighter Method to Couple Hose

Essentials (FF I) Competency Profile SA 12B - 103

Figure 6.10 Competency profile sheet or performance evaluation form.

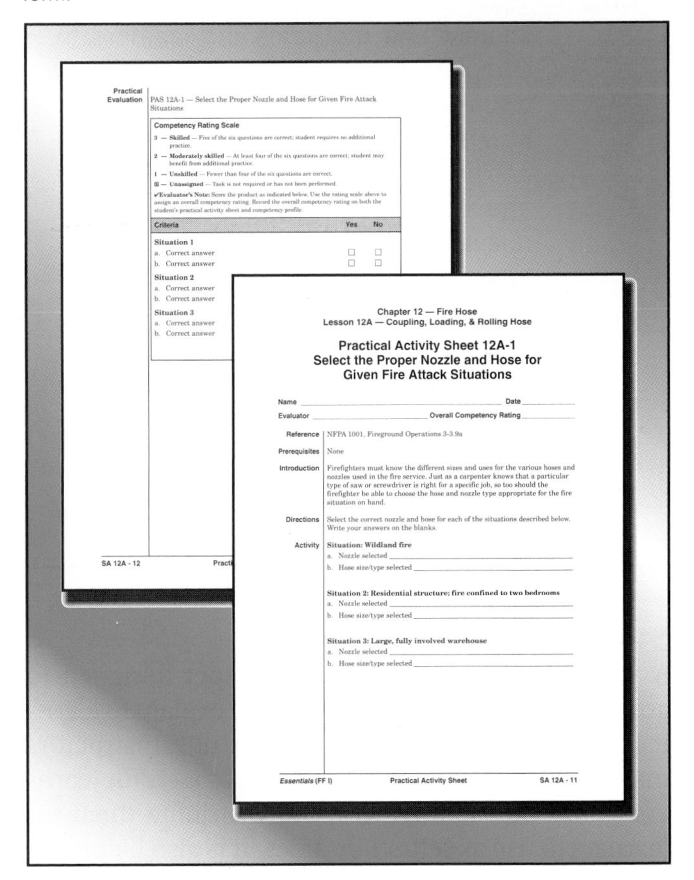

Practical Evaluation

PAS 12A-1 — Select the Proper Nozzle and Hose for Given Fire Attack Situations

Competency Rating Scale

3 — **Skilled** — Five of the six questions are correct; student requires no additional practice.
2 — **Moderately skilled** — At least four of the six questions are correct; student may benefit from additional practice.
1 — **Unskilled** — Fewer than four of the six questions are correct.
☒ — **Unassigned** — Task is not required or has not been performed.
✓**Evaluator's Note:** Score the product as indicated below. Use the rating scale above to assign an overall competency rating. Record the overall competency rating on both the student's practical activity sheet and competency profile.

Criteria	Yes	No
Situation 1		
a. Correct answer	☐	☐
b. Correct answer	☐	☐
Situation 2		
a. Correct answer		
b. Correct answer		
Situation 3		
a. Correct answer		
b. Correct answer		

SA 12A - 12 Practi

Chapter 12 — Fire Hose
Lesson 12A — Coupling, Loading, & Rolling Hose

Practical Activity Sheet 12A-1
Select the Proper Nozzle and Hose for Given Fire Attack Situations

Name ______________ Date ______________
Evaluator ______________ Overall Competency Rating ______________

Reference: NFPA 1001, Fireground Operations 3-3.9a

Prerequisites: None

Introduction: Firefighters must know the different sizes and uses for the various hoses and nozzles used in the fire service. Just as a carpenter knows that a particular type of saw or screwdriver is right for a specific job, so too should the firefighter be able to choose the hose and nozzle type appropriate for the fire situation on hand.

Directions: Select the correct nozzle and hose for each of the situations described below. Write your answers on the blanks.

Activity: **Situation: Wildland fire**
a. Nozzle selected ______________
b. Hose size/type selected ______________

Situation 2: Residential structure; fire confined to two bedrooms
a. Nozzle selected ______________
b. Hose size/type selected ______________

Situation 3: Large, fully involved warehouse
a. Nozzle selected ______________
b. Hose size/type selected ______________

Essentials (FF I) Practical Activity Sheet SA 12A - 11

Figure 6.11 Worksheet or activity sheet.

enable learners to meet objectives. Completing a worksheet may also be an objective, which requires learners to participate in activities that include and apply knowledge or skills in previous objectives.

Worksheets that require learners to exercise abilities in the affective domain may support more than one objective in which more than knowledge and skill must be developed or demonstrated. Recall that the affective domain has learners change or adjust, develop, practice, and adapt attitudes, values, beliefs, and appreciations.

Use the following steps to develop a worksheet or activity sheet:

Activity Sheet

Step 1. Create a title that reflects the subject or topic, and relate it to the lesson.

Step 2. List all the materials and resources that learners need in order to complete the activity. List titles and pages of books, journals, or other reference material. Provide enough information so learners can locate the resources easily.

Step 3. Write a brief introduction that arouses interest and motivates the learners to complete the activity. In the introduction, discuss the skill or activity, and explain how it relates to the topic or training area and to the lesson objectives. Explain how and why the activity is important and relevant to the job and how the activity helps master the skill.

Step 4. Provide clear directions that explain how to complete the activity sheet.

Step 5. Provide answers or solutions on a separate page. The answer sheet may be given either with the activity sheet or after the activity sheet is completed.

Study Sheet

A *study sheet* is an instructional paper designed to arouse learner interest in a topic and explain to learners what specific or unique areas to study. Instructors may want to distribute these types of sheets for learners to use during instruction or distribute them for learners to use as self-study aides. It is also helpful to include a self-study test with the study sheet, which enables the instructor to measure and provide feedback on how well learners understood the material (Figure 6.12).

Use the following steps to create a study sheet:

Study Sheet

Step 1. Create a title that reflects the subject or topic, and relate it to the lesson.

Step 2. List all the materials and resources that learners need to complete the study sheet. List titles and pages of books, journals, or other reference material. Provide enough information so the learners can locate the resources easily.

Step 3. Write a brief introduction that arouses interest and motivates the learners to complete the study sheet.

Step 4. Design the study sheet, and present the study information in a format that enables the learner to use and learn the material.

Step 5. Put a study-sheet test (if one is included) on a separate sheet of paper.

Step 6. Design study questions to make the learner think and to assess understanding of all aspects of the topic. Include enough questions to thoroughly cover the material.

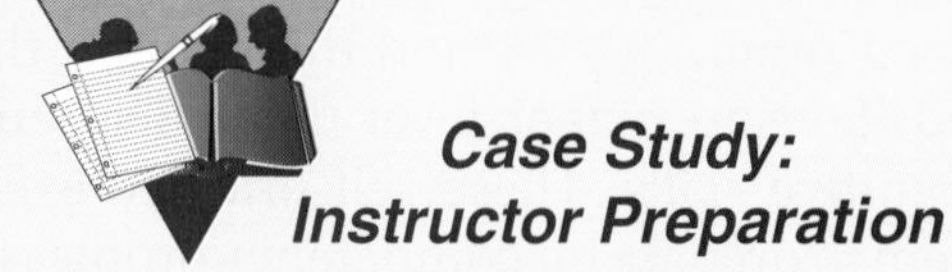

Case Study: Instructor Preparation

You are preparing to teach your first class as a new instructor. Much of your preparation time is spent in reviewing details of the topic so that you can answer questions the class may have. You also want to prepare some handout materials that will help learners study and learn the information and practice the skills.

Answer the following questions:

1. What are some of *your* resources? Who can you ask or talk to about ideas, where to get

materials, and what study aids are appropriate and effective for the type of class you are teaching?

2. What support tools, application components, and handouts would you develop for your first class? What information do you think would be appropriate for learners to have so they can master the particular skills or learn the information more easily?
3. What is your organization's policy on out-of-class assignments? If they are not mandatory, do you have time in the program to cover all areas adequately? How else can you have the class master skills? If out-of class assignments are required or expected, which ones would you assign to help learners master skills?

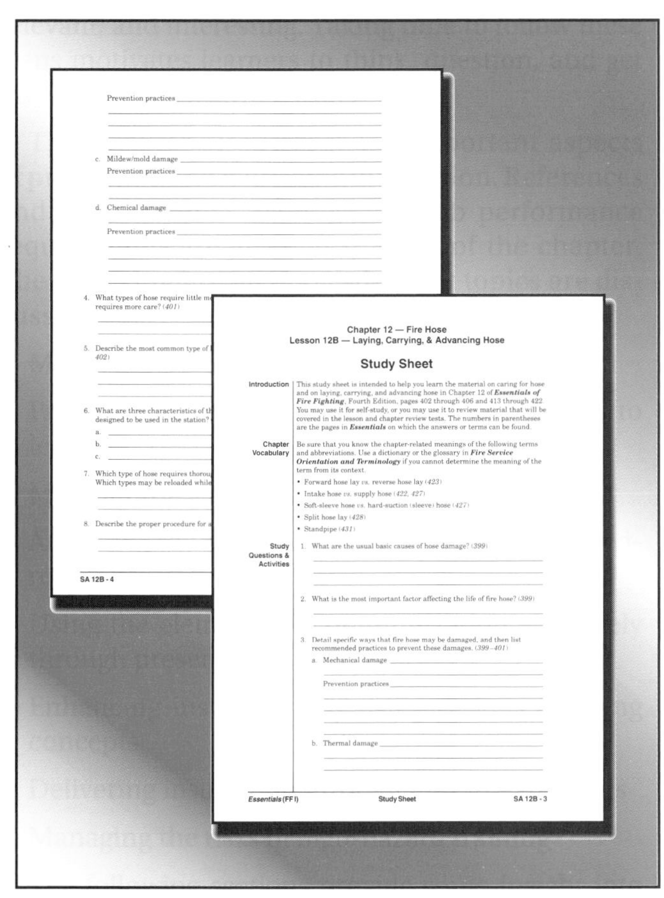

Chapter 12 — Fire Hose
Lesson 12B — Laying, Carrying, & Advancing Hose

Study Sheet

Introduction This study sheet is intended to help you learn the material on caring for hose and on laying, carrying, and advancing hose in Chapter 12 of ***Essentials of Fire Fighting***, Fourth Edition, pages 402 through 406 and 413 through 422. You may use it for self-study, or you may use it to review material that will be covered in the lesson and chapter review tests. The numbers in parentheses are the pages in ***Essentials*** on which the answers or terms can be found.

Chapter Vocabulary Be sure that you know the chapter-related meanings of the following terms and abbreviations. Use a dictionary or the glossary in ***Fire Service Orientation and Terminology*** if you cannot determine the meaning of the term from its context.

- Forward hose lay *vs.* reverse hose lay *(423)*
- Intake hose *vs.* supply hose *(422, 427)*
- Soft-sleeve hose *vs.* hard-suction (sleeve) hose *(427)*
- Split hose lay *(428)*
- Standpipe *(431)*

Study Questions & Activities

1. What are the usual basic causes of hose damage? *(399)*
2. What is the most important factor affecting the life of fire hose? *(399)*
3. Detail specific ways that fire hose may be damaged, and then list recommended practices to prevent these damages. *(399–401)*
   a. Mechanical damage ______
   Prevention practices ______
   b. Thermal damage ______

Essentials (FF I) Study Sheet SA 12B - 3

Figure 6.12 Study sheet.

References and Supplemental Readings

Bloom, Benjamin S., M. B. Englebert, E. J. Furst, W. H. Hill, and D. R. Krathwohl, (eds.). *Taxonomy of Educational Objectives. The Classification of Educational Goals. Handbook I: Cognitive Domain.* New York: McKay, 1956.

Hunter, Madeline. *Mastery Teaching.* London: Sage Publications, 1982.

Kizlik, Robert. "Five Common Mistakes in Writing Lesson Plans." Http://www.adprima.com/mistakes.htm, 1998.

Krathwohl, David R., B. S. Bloom, and B. B. Masia. *Taxonomy of Educational Objectives. Handbook II: Affective Domain.* New York: McKay, 1964.

Simpson, Elizabeth J. *The Classification of Educational Objectives: Psychomotor Domain.* Urbana, IL: University of Illinois Press, 1972.

Wlodkowski, Raymond J. *Enhancing Adult Motivation to Learn.* San Francisco: Jossey-Bass Publishers, 1993.

Job Performance Requirements

This chapter provides information that addresses the following job performance requirements of NFPA 1041, *Standard for Fire Service Instructor Professional Qualifications* (1996 edition). Colored portions of the standard are specifically covered in this chapter.

Chapter 2 Instructor I

2-3.2 Review instructional materials, given the materials for a specific topic, target audience and learning environment, so that elements of the lesson plan, learning environment, and resources that need adaptation are identified.

2-3.3 Adapt a prepared lesson plan, given course materials and an assignment, so that the needs of the student and the objectives of the lesson plan are achieved.

2-4.3 Present prepared lessons, given a prepared lesson plan that specifies the presentation method(s), so that the method(s) indicated in the plan are used and the stated objectives or learning outcomes are achieved.

2-4.3.1 *Prerequisite Knowledge:* the laws and principles of learning, teaching mentods and techniques, lesson plan components and elements of the communication process, and lesson plan terminology and definitions.

followed to ensure that everyone can participate with confidence and without injury.

An important aspect of teaching is how instructors communicate. An earlier chapter discussed the communications model and the importance of feedback. Besides instructors' actual interaction with learners, instructors must consider how they actually present themselves to their learners. A professional appearance demonstrated through prepared presentation methods and a positive attitude and demeanor add to instructor credibility in any *public* speaking situation.

Many theories promote ideas of how to teach in order to enhance learning. What is the best way to present new material, maintain learner interest, encourage success, promote participation, and reinforce learning? As potential instructors reading the paragraphs that introduce the topics of this chapter, recall and compare these topics with examples, exposures, and experiences gained through other teaching-learning situations.

Instructors also want to know what delivery methods effectively get the information to learners. How can instructors best use the methods of lecturing, demonstrating, discussing, reviewing case studies, role-playing, brainstorming, mentoring, individualized instruction, and other alternatives?

Along with managing many learners and their multiple learning styles, instructors must often manage several instructors and their different teaching styles during various types of simple-to-complex training procedures. How can instructors supervise other instructors during high-hazard training exercises?

Instructor Tip: *When teaching for the first time or before teaching a new class or topic, observe the methods of several other instructors. Watch how others manage learner groups, present information, and demonstrate skills. Look for advantages and disadvantages of their presentation methods; think about which methods accomplish objectives and which methods appear to create problems and cause disruptions in learning and lesson organization. Draw conclusions about what methods seem to work best, and adapt those until you have the experience and confidence to develop your own special methods.*

Behavior Management

[NFPA 1041: 2-3.2, 2-3.2.1, 2-4.5, 2-4.5.1, 2-4.5.2]

Behavior management can be a difficult skill for a new instructor to master. It is, however, one of the most important skills for a new instructor to acquire and maintain. In an adult learning environment, some instructors believe that they do not need to perform behavior management because *peer pressure* will control the classroom. It is a fact of human nature that the members of a group respond to peer pressure. But an instructor cannot shirk responsibility and depend on this fact to take care of a problem individual. In some situations, peer pressure works. Along with effective peer pressure, an instructor may still need to privately counsel an errant individual.

Instructors who fail to manage disruptive behavior neither make progress nor maintain the respect of other learners when an errant individual is allowed to constantly disrupt the class. Behavior management includes *prevention* of disruption in the learning environment. When prevention fails, instructors must manage behavior problems in a manner appropriate to the situation.

Through behavior management, instructors can create an effective environment in which all learners can participate with positive results. Some steps to take include the following:

- Follow policies.
- Establish a learning attitude.
- Organize and prepare.
- Use effective coaching techniques.
- Stimulate interest and learning through motivation methods.
- Manage disruptive behavior quickly, tactfully, and fairly.
- Facilitate learning.

Follow Policies

One of the first orders of business for a classroom instructor is to review with the program participants the rules, regulations, and policies of the training environment. Instructors who teach in and participants who return to the same familiar learning environment come to know and understand the organization's rules and regulations. But the instructor must consistently promote, support, and enforce rules and regulations so that participants respect and act on them accordingly. Instructors and program participants who travel to different learning sites must become familiar with and follow the policies or *ground rules* required of each location (Figure 7.1). These ground rules generally address the following concerns:

- Safety rules
- Facility layout
  — Using and cleaning
  — Safety plan
  — Emergency reporting procedures
  — Evacuation
  — Off-limits areas
- Attendance, absenteeism, and tardiness policies
- Expectations and regulations for responding to emergencies from class
- Class participation
- Methods of evaluation
- Assignment due dates
- Class cancellation (inclement weather if applicable)
- Participant parking
- Dress/grooming regulations
- Breaks/rest policies
- Smoking areas (if permitted)

Just as the ground rules in major league baseball stadiums seldom if ever change from season to season, the ground rules for established training facilities change little from course to course. Likewise, instructors will find that when delivering an established program, the ground rules do not change from offering to offering. For instructors and program participants who are new to or unfamiliar with an organization's programs and policies, frequently reviewing the requirements of these ground rules is of the utmost importance until everyone becomes familiar with and automatically adheres to them.

Figure 7.1 Instructors must review policies and set the ground rules.

It is important that instructors keep adult learners informed about the requirements and expectations of the program (see Chapter 4, The Psychology of Learning). The most effective way to do this is in writing. College-level programs provide a syllabus that describes the basic course requirements. Similarly, fire and emergency services training programs should provide each participant with a similar document that clearly defines course and instructor expectations. The document should include such items as schedule information, assignment due dates, grade percentages, information on exam types and dates, grading methods, and any other information that informs participants of what is expected of them for completing the course. This way, quite literally, both the instructors and the participants are "on the same page." The syllabus or a separate handout also should include any policies on such ordinary events as inclement weather and what to do or who to call if participants have an emergency that causes them to be late or miss a class.

Different policies may require different methods of introducing and enforcing them. Examples for using some ground rules effectively include the following:

- *Forms that require a signature* — Multiple-copy forms list procedures, rules, or other requirements that participants must agree to follow. Instructors read the procedures aloud as the participants read from the form. Participants then sign the form, keep one copy, and return the original to the instructor for the files.
- *Learning contracts* — These are formal agreements between learners and the instructor that establish a certain amount of work that must be finished in order to successfully complete the course. These contracts are actually a method of teaching and should contain the following:
  — What the individual should learn
  — How the individual can demonstrate that he or she has learned
  — Degree of proficiency the learner must demonstrate

— How the individual will proceed

- ***Organizational or program rules and regulations*** —Printed handouts give special policies and procedures that everyone participating in the program must know and follow.
- ***Course description and detailed syllabus***—Printed handouts describe the purpose and goals of the course and contain the following:
  - — Information on the course schedule
  - — Expectations for each meeting date and assignments
  - — Exam dates
  - — Requirements for earning grade points

These and similar documents can be helpful when dealing with either large numbers of learners with varying learning styles and motivations or programs containing multiple projects and assignments such as those required by advanced or officer training programs. See Appendix F, Organizational or Program Rules, Policies, and Regulations, for samples.

In preparing these documents, take care to ensure that personal requirements *do not conflict* with those of the sponsoring organization. For example, instructors may want to require that participants not miss any class sessions, but the sponsoring organization may allow a certain number of absences. In a case such as this, it is the instructor who must make adjustments to personal requirements in favor of organizational policies or negotiate a compromise with the organization.

Establish Learning Attitude

To be effective, the instructor can and must be the leader in establishing the classroom climate. A key factor in that climate is the *enthusiasm* of both the instructor and the learners. Instructor actions and the apparent interest in the presentation have a great effect on establishing a productive learning environment. The amount of enthusiasm that an instructor projects is directly related to the level of enthusiasm and interest returned by the learners. One of the challenges of teaching is for instructors to maintain a positive level of enthusiasm in both themselves and the learners.

Some instructors confuse enthusiasm with *theatrics.* The job of an instructor is to convey new ideas, facts, and procedures, not to entertain. In his book *Learning to Teach,* Richard Arends (1994) reported on a study by Nauftulin et al. in 1973 who found that a presentation that is very entertaining is not always very informative. Between being entertaining to maintain interest and being informative to impart knowledge, new instructors may ask, *"Where is the line between using humor, storytelling, and learner involvement to get the point across and using the same methods for entertaining learners?"* Teaching might be easier if there were a clear-cut formula or consistent theory to use.

Lacking specific formulas and theories, it may be best, then, for new instructors to carefully follow the lesson plans, take a *strictly business* approach at first, and then expand teaching techniques through practice and use of various teaching methods and learning activities. The primary instructor role is to teach with the result that learning takes place, not to perform with the result that amusement takes place.

Instructors must show enthusiasm in their teaching methods for the skills they demonstrate and the information they impart. Learners recognize enthusiasm, and because it is contagious, are affected by it and use it to learn. Figure 7.2 is a learning aid that can be used to measure the enthusiasm of an instructor. The evaluation sections of this form are taken from an evaluation form used by a training organization.

Organize and Prepare

Being organized is another attribute of behavior management that contributes to the success of the new fire and emergency services instructor. Being organized has certain implications; being *well-organized* can result in a variety of benefits. The process of being well-organized involves several concepts such as the following:

- ***Getting organized*** — *Organizing* means arranging items, units, or parts in such a way that they work together in a coordinated whole that shows a relationship among all the parts. Everyone has his or her own method of organizing lessons and their parts, but the method must enable an instructor to pull a unit from some storage place and see at a glance what preparation must be completed and what materials or equipment must be printed or gathered in order to be ready *before* teaching.
- ***Being prepared*** — *Pre* is a prefix meaning *earlier than, prior to,* or *before.* Being *pre*pared is being ready *before* class begins. Well-prepared instructors have planned, prepared, and organized handouts, audiovisuals, props, and other pertinent equipment and materials so that they are available and ready before the beginning of the appropriate class. Being prepared does not mean that the instructor gives learners "busy work" such as small

Evaluation Form

PART I: Points are earned for performance in the following areas:

Not done or not observed	1
Presented or completed at a basic or beginning level	2
Presented or completed with comfort; shows planning and makes some transition into topics	3
Presented or completed with confidence; shows planning, smooth transition of topics, and uses some applications	4
Presented or completed based on educational methods; well planned with transitions and applications to enable learning	5

Cognitive Development: (includes interactive discussion, demonstration, group activities, etc.)

Introduces new (unknown) knowledge and relates to old (known) knowledge ______
Interests all participants by using various teaching styles and learning activities ______
Encourages questions and discussion to aid comprehension and thinking ______
Begins with simple information or steps and guides learners into complex areas ______
Encourages learner ideas and discussion for use of new information ______
Relates known information and application to new areas ______
Has group draw conclusions in a review and summary of main points ______

Psychomotor Skills: (applies knowledge through discussion, demonstration, practice, labs, evolutions, etc.)

Provides opportunities to integrate learning through multisenses ______
Provides opportunities to enhance mental, emotional readiness ______
Models a skill, guides practice; encourages imitation from peers ______
Provides time for practice to develop mechanics, confidence ______

PART II: Points are earned for performance in the following areas:

Not presented to professional standards or educational expectations for the type of lesson/course	1
Demonstrated interest and concern with meeting professional standards or educational requirements expected	2
Presented or performed at the standard or expected level for the lesson/course	3
Sets a professional example and/or demonstrates sound educational methods in action and/or appearance	4
Demonstrates consistent professionalism and/or educational methodology throughout the lesson	5

Self Presentation:

Appears neat, clean; dresses appropriately for presentation ______
Engages learners in appropriate eye contact ______
Uses language appropriate to lesson and learner level ______
Projects voice; uses expression, voice inflection ______
Avoids/minimizes hesitation sounds *(um, uh)*, fillers *(ok)*, etc. ______
Uses appropriate gestures for emphasis ______

Program Facilitation:

Achieves lesson objectives through cognitive development and psychomotor skills by meeting all of the above evaluation categories ______
Adapts lesson as appropriate using interactive participation to meet standards, guidelines ______
Assures safe environment; provides written and oral safety guidelines for skills/hazards ______
Uses various teaching methods to address many learning styles to ensure that learners are exposed to multiple opportunities for meeting objectives and learning successfully ______
Uses a planned lesson to cover an appropriate amount of material in a given time ______
Attends to administrative details (attendance; forms; equipment acquisition; support instructors; class location change, date, time, etc.) ______

Figure 7.2 An evaluation form for measuring instructor enthusiasm. *Courtesy of Maryland Fire and Rescue Institute.*

group or desk assignments while running down the hall to the copying machine to print handouts. Being prepared is starting class on time (not late) and delivering a planned and prepared lesson complete with appropriate application activities (not calling a break 10 minutes into the lesson to check that there is enough training equipment for the class).

Being well-prepared and organized add to instructor credibility. Credibility grows for those who are well-organized. Credible instructors have success with maintaining class motivation and interest because they have gained a measure of respect through the appearance that they care. Instructors show they care about learner outcome when they have a prepared and meaningful, relevant lesson that shows planning and organization. Preparation takes work — the kind of work that pays off for both the instructor and learners. Learners do not see it when it is done at the proper time and place. But learners notice when instructors are going through the motions of looking prepared when, in fact, they are not.

Research has demonstrated that being well-organized and prepared are factors in the successful management of learners in regard to their behaviors. In their book, *Educational Psychology*, Good and Brophy (1977) report on the results of a series of studies done by Kounin and associates in 1970 on the traits of teachers who were successful in dealing with classroom problems and those who were not. The research showed that success in behavioral management was, in part, a result of success in instructional management. Successful instructors kept learners actively engaged in lessons and seatwork, minimizing the amount of time learners spent bored or restless. The research also showed that most disruptive outbursts occur when learners are bored or restless rather than when actively involved in the class. Kounin and associates concluded that organized teachers did not have to contend with disruptions nearly as often as less-organized ones.

Use Coaching Techniques

Coaching can be described as an intensive process of directing the skills performance of an individual. In psychomotor learning, the coaching process includes observation, evaluation, and suggestions for improvement (Figure 7.3). This phase of teaching requires instructor skill in two areas: (1) mastery of the topic being taught and (2) verbal interpersonal communications. To the fire and emergency services instructor then, coaching becomes two tasks: (1) demonstrating a correct example of a skill and (2) providing the verbal directions for a learner to use to achieve the same results.

The instructor/coach must be able to clearly recognize, constructively criticize, and carefully correct flaws in the learner-demonstrated skill. The instructor/coach must also effectively praise and positively reinforce those actions that are being performed properly. To generate learner enthusiasm about their performances, instructors must strike a balance between correcting what is performed wrong and praising what is performed right or to the selected standard. Too much weight on either side of the scale unbalances the learning process and limits or inhibits it. Feedback from the instructor/coach also must be objective and precise in its description of the elements that are good or bad. Instructors must not fall into the trap of just providing critical phrases such as *"That's wrong — do it again — but the right way this time."* That type of feedback does not inform the

Figure 7.3 Coaching includes observation, evaluation, and suggestions for improvement. *Courtesy of Jarett Metheny, Midwest City (OK) Fire Department.*

learners of the error or how to correct it. A better example would be *"Before you go further, do you remember what the rope needs to have on the running end before you tighten it?"* A questioning process allows errant learners to stop, think, and recall or build into their memory the information that they are attempting to apply. Some learners perform this thought-construction process quickly while some take more time, but they remember better because they recalled it themselves. If a learner does not recall the information, then the instructor can carefully review the steps and coach the learner through the steps again.

Often in delivering the verbal information followed by questioning — a technique to ensure understanding — instructors often give the pat reply of *"OK"* to all responses, even if they are off-track or clearly wrong. Instructors are reluctant to discourage responses, so they accept all answers while tactfully trying to steer the group to the right conclusion. But taking a wrong answer and asking why it is wrong or where it may fit appropriately in the lesson with what fact is a technique that allows learners the following opportunities:

- Think and analyze problems.
- Compare facts and ideas, and apply them to different situations.
- Critique and find solutions.
- Explore and discover new methods of application.

As instructors guide this type of discussion, they find that they have covered more lesson material than they would have otherwise and that the participants have truly learned the material so that they can apply it appropriately in a variety of situations rather than in just staged classroom situations. Again, this type of instructional method allows learners to build their knowledge from what they discover and associate it with the class material and skills.

Motivate Learners

Maintaining learner motivation is a challenge for instructors. Learners have a variety of reasons for attending training programs. Many actually come because they want to; a significant number come because they have to. In the "have-to-be-there" situations, fire and emergency or public safety services organizations often mandate and compel employees to attend training classes to maintain certification or comply with other continuing education requirements. Having to be there often causes a lack of motivation, which can present problems for an instructor. For example, learners who are bored, restless, or in attendance because they are directed to be there get little if anything out of the class, contribute nothing to the class, dampen the interest and participation levels of other learners, and thereby hamper instructor efforts to accomplish any learning. What can instructors do to turn around this situation?

One method that can be effective with adult learners is to address their need to understand the value of the topic(s) being covered. The use of phrases such as *"Because I say so"* or *"You're not getting your EMT/confined space/haz mat cards until we finish this lesson"* do not keep learners motivated. Adult learners generally need "to feel a connection" to the material they are covering. *To feel a connection* means that adults see a reason that validates the lesson for them; they see some relevance in the lesson that relates the information to what they actually do on their jobs. When adult learners can identify *what is in it for them,* they begin to be motivated, to become interested, to participate and contribute, and as a result, to actually want to be there.

Instructors who enable learners to see the benefits of the program have successful classes. How do instructors make learners *see* benefits? Ask them. Why must they recertify in automated external defibrillators (AEDs), cardiopulmonary resuscitation (CPR), hazardous materials, rescue tools, and so forth when they have not had the opportunity to perform these skills all year, when they do not have that equipment anyway, or when they are never called for that kind of emergency? Let them compare pros and cons, determine reasons why and why not, come up with ideas that may alter the situation, and draw their own conclusions. When reviewing and practicing a skill, why do they think they need to know this? Where else could it be used besides on the job? What other purpose can it serve? These types of activities get them off to a good start and generally work toward motivating a majority of those who don't want to be there. Why? Because they are building their own knowledge base and are not limited to adopting the constraints that someone else dictated.

Manage Disruptive Behavior

What does the term *disruptive* mean? To some, chaos is exciting; to others, it is disruptive. To some, a participant who shouts out an answer without being called on is normal; to others, it is disruptive. Many instructors believe that eliminating disruptions is a method of maintaining classroom control, which they believe is paramount to successful teaching and learning. What is *classroom control?* There are controlled discussions, behavior controls, controlled activities, thought controls, safety controls, control devices, and controlled intersections. *Control* means to exercise restraining influence over someone or something. The purpose of control, then, appears to be that some individual or group tries to ensure that something either happens in a designated, limited, and expected manner or does not happen at all. Should instructors attempt to control the adults they are teaching? In what manner? Through their thoughts, actions, creativity, inspirations? Why?

There are times when the most well-prepared lesson does not go well. And, unfortunately, there are always learners whose actions disrupt the class. When these events occur, are they indications that the instructor has lost control of the class? What is the typical reaction of instructors in such cases? Sometimes instructor reaction takes the form of punishment or stricter controls, neither of which always work.

Instructors list or define many events that they consider disruptions, usually events caused by learners. Some learners create disruptions in many forms such as the following:

- Arriving late
- Speaking out
- Talking with others off the subject
- Sleeping accompanied by snoring
- Intentionally showing off
- Interrupting
- Sidetracking
- Seeking attention
- Acting blatantly insubordinate and disrespectful

In the face of these types of events, instructors are challenged to maintain their composure and control the atmosphere of the classroom. The *thing* being controlled may not necessarily be the behavior of the learner(s) but *the atmosphere of the classroom.* Consider that adults usually resent any implication that someone else controls their behaviors, but instructors can *redirect* behaviors to beneficial outcomes if they handle situations appropriately.

Give some thought as to why some learners behave in ways that disrupt the class. There may be legitimate motivations for behaving as they do. For instance the sleeping and snoring individual may be doing so because he or she is working two jobs to support a family and a sick spouse. Many adults who attend training classes have true concerns with competing commitments such as juggling family and job while finishing training requirements, worrying about money problems, and dealing with other similar events in their lives. Before deciding to label, dislike, or punish the "perpetrators," instructors must learn something about the backgrounds of these individuals and attempt to understand and work with their needs. Learners appreciate any concern an instructor shows because instructors who show concern also show respect for the individual. Most learners return that respect by cooperating with instructors and fulfilling class requirements *to the best of their abilities.* The best of his or her ability may not be the instructor's ideal, but it is with an understanding of the circumstances for which an individual should not be punished. The result is that an instructor can manage disruptive behavior and maintain control of the class with only a little effort at being flexible and humanistic.

There are many reasons why individuals display disruptive behavior, and these may be the fault of the instructor. Instructors may (intentionally or unintentionally) act in certain ways that cause unfavorable learner reactions (see Table 7.1). Instructor actions stimulate learner reactions. Instructors must act appropriately to ensure that their actions do not stimulate unfavorable learner reactions.

Instructors can use various tactics to manage disruptive behavior in the classroom. If an individual does not improve, the next option is the counseling phase.

Tactics

Many individuals bring their disruptive behaviors with them. Some other experiences have created these re-

Table 7.1
Unfavorable Learner Reactions to Instructor Actions

Instructor Actions	Learner Reactions
Intimidates	Feels insecure
Gives dull lectures	Feels unstimulated, bored
Rambles without a goal	Shows no interest
Shows impatience	Shows fear
Runs overtime	Gets fidgety
Attempts to overcontrol	Rebels

actions or prepared these learners to react these ways in anticipation of what the instructor might do. One tactic instructors may use with bored, *"Why am I here?"* participants is to plan the lesson with activities that guide learners to discover exactly why they *are* there—activities that enable them to figure out just *what is in it* for them. Ask these individuals to express their motivations and see whether they can find meaning for themselves.

Some individuals inundate an instructor with questions on a subject (*sharpshooting*). Sometimes they are truly curious and interested; sometimes their intent is to embarrass and discredit the instructor or "shoot you down" as the "expert." The following list is adapted from Milt Badt's book, *Adult Learning in the Classroom* (1996), and offers suggestions for dealing with these classroom sharpshooters:

- Show confidence in your role as content expert. Any sharpshooters would not be in the class if they were the experts.
- Tell the class you will get the information and get back to them if you are unsure of an answer. Then do it! Being responsive is more important than being perfect.
- Respond to the learner, but continue to manage class time, and consider the interests of others when engaged in a dialogue with an individual.
- Always smile. A smile is disarming and indicates willingness to listen and discuss.
- Let sharpshooters off the hook easily (if they trap themselves — and they probably will). In doing so, you gain their respect and that of the rest of the class.
- Let participants know that their questions have merit, and then find some. Chances are that other learners have the same questions and are too bashful or embarrassed to ask, but they can contribute to finding the merit. Everyone gets involved, all benefit from the discussion, and everyone grows comfortable with participating from the positive reinforcement.

A learner who is disruptive may be seeking needed attention, but instructors must be careful not to reward this behavior by paying undue attention to that individual's actions. Some behavior extinguishes itself on its own; some does not. If it does not, simple initial steps often work: Look at the learner long enough to make eye contact; it can be very effective. Eye contact lets the individual know that *"I see what you are doing,"* and the implied *"Now stop it,"* goes right along with it. The nonverbal cue may not be enough. If that is the case, call on the individual and ask a question about the topic. In this subtle manner, instructors can communicate the knowledge that they are aware of what the individual is either doing or not doing.

To regain the attention of individuals who are talking among themselves on items other than the topic, instructors may try the tactic of calling on one of the individuals. Without trying to embarrass the person, make a statement that summarizes the topic and ask a simple opinion question that directs him or her back to the lesson. By this method, instructors are again letting the individual know they are aware of the disruptive activity and are requesting cooperation. Doing this more than once is not recommended. If the individual does not cooperate, instructors should take further steps to correct the problem (see Chapter 4, The Psychology of Learning). One of the advantages of being in quasi- or paramilitary organizations such as the police, fire, and rescue services is that the instructor generally has the authority of an officer in the management of the classroom. Hopefully, instructor authority does not need to be tested often.

Consider that there may be class situations where the initial nonverbal methods are not sufficient or appropriate. This is especially true in practice or skills sessions. If safety is an issue, there is zero tolerance of disruptive behavior. Injuries resulting from disruptive behavior are never acceptable.

Counseling

As with any situation, the best measures, methods, and tactics used for managing behavior may not succeed. Then what? How do instructors handle discipline in an adult setting? If an individual's behavior has not changed after one or two incidents with the measures mentioned in the Tactics section, then it is time to document the situation on paper. A simple notation on the class roster, schedule, or file is initially a sufficient reminder of the incidents. If disruptive behaviors continue, proceed past the nonverbal and verbal tactics that are easily met while conducting the class, and move to the documented counseling session. Documenting the disruptive behavior by memo or counseling form puts the issue in the *formal stage*.

Training organizations usually have a formalized process to follow when documenting behavior issues. One of the steps in the process may include completing a type of *counseling form*. It is important for a new instructor to be aware of the organization's process. In addition, realize that this is a major change from the simple and immediate types of measures that can be

Instructor Tip: *Instructors should remember that when they do all the talking, the participants are only learning from one perspective and only through one sense. The following old adages have shown merit over the years, and instructors should consider their meanings and intents. Essentially, they suggest that we learn when we participate in the learning and discover knowledge for ourselves.*

- *"I hear, I forget; I see, I remember; I do, I understand." (Old Chinese Proverb)*
- *"I want you to learn, but not to be taught." (Moshe Feldenkrais)*
- *"What you truly learn best will appear to you later as your own discovery." (Moshe Feldenkrais)*
- *"Teach people, but let the teaching be hidden; put forth new knowledge as something they had forgotten." (Alexander Pope)*
- *"Imagination is more important than knowledge." (Albert Einstein)*
- *"I can explain it for you, but I can't understand it for you." (Author Unknown)*

Preparing to Teach

[NFPA 1041: 2-3.2, 2-3.2.1, 2-3.2.2, 2-3.3, 2-3.3.1, 2-3.3.2, 2-4.2, 2-4.2.1, 2-4.2.2, 2-4.4, 2-4.4.1]

Where must an instructor start when preparing to teach an individual lesson or an entire program? What elements must instructors consider and plan that will start them in the right direction? There are three major elements that guide instructors in the right direction; each of them is an important step in preparing to teach: *session scheduling, session preparation,* and *session logistics.*

Session Scheduling

Organizations that have structured programs that meet on a regular schedule probably have someone assigned the job of scheduling class offerings. In these situations, session scheduling for instructors may be simply a matter of knowing where and when the class is meeting and setting aside the appropriate personal time to be there.

Instructors who do not teach scheduled programs must attend to those necessary items such as where, when, how often, and how long the class will be held. If a training facility is needed, instructors must find out if others have scheduled their commitments to that space and at what time. Someone designated as training supervisor usually schedules all course requests in advance and may require a one- to three-months' notice, if not longer, to reserve the use of certain equipment and facilities. Ideally, instructors check with the appropriate individuals on the process required to make reservations, including making request submissions in writing and getting approvals in writing (Figure 7.5). Because training equipment and space is at a premium, once a schedule is set, adhere to it. When the scheduled use of equipment or a room must change, provide the required type of notification (written or oral) as soon as possible. It may be that the training supervisor can make efficient use of available equipment and space by redirecting it to another program.

Session Preparation

To prepare for a given lesson, the instructor must not only know *what* topic is to be taught but must know *about* the topic to be taught. Knowing about the topic is a basic assumption that is often not the actual case, and circumstances may arise when instructors find themselves other than ideally prepared. Many program participants probably have had experiences where they have arrived at class and heard the instructor remark *"I didn't know I had to do this until 15 minutes ago"* and knew that he or she had done little to prepare for the class. Instructors should never let on that they are unprepared because all credibility is lost along with the opportunity to motivate learners.

It sometimes happens that a scheduled instructor is sick, injured, or suddenly transferred to replace another sick or injured individual. When it does happen, the "last-minute" replacement instructor must make every effort to quickly prepare and deliver a lesson that makes the time in class worthwhile. Instructors who are put in the position of walking into the classroom with no "lead time" should make the class session a *discovery zone* where everyone learns something. This may mean using each objective as an overview point and proceeding from there. It may mean that the group learns or reviews the basic steps on whatever equipment or materials are available such as getting sections of rope and practicing knots or getting some mannequins and practicing CPR.

Figure 7.5 Instructors may have to submit requests and make reservations for facilities and equipment. Some organizations have computer programs for reserving equipment. *(left) Courtesy of Jarett Metheny, Midwest City (OK) Fire Department. (right) Courtesy of Robert Wright, Maryland Fire and Rescue Institute.*

Ideally there is plenty of lead time to prepare for a class, but even at the last minute, the most important preparation items include the following:

- ***Read the lesson objectives*** — Become familiar with what the objectives require the learners to know and perform. The remaining items help the instructor prepare for meeting the lesson objectives.
- ***Review the lesson plan/lesson guide*** — Is there a standard lesson guide that maps out a suggested teaching format and the material to be covered? In planning a lesson, instructors use an organization's standard lesson guide to determine what material must be covered, in what time frame, and with what assistance, materials, and equipment. By reviewing the lesson guide, the instructor can plan a lesson that covers the lesson requirements. On short notice, the stand-in instructor may receive a lesson plan from the instructor who was originally scheduled and can review it to determine the format and activities planned.
- ***Check what equipment is needed*** — Is equipment ready and available, or does it need to be scheduled ahead of time? What arrangements must be made and in what time frame? Are there enough equipment and other lesson materials or handouts for each learner? Where are equipment and materials? Instructors must be familiar with how to operate equipment and with those materials used for learner activities (Figure 7.6).
- ***Determine what skills must be taught*** — Practice the skill steps, or at least review the steps mentally by looking at pictures, equipment, or handouts. Any instructor who lacks recent experience or practice in the skill must take time to review it or determine a method of presenting it so that the skill is clear to the learners. Ideally, the instructor presents a skill to a level of mastery. On short notice, instructors can introduce the skill to the class in steps that act as review and practice for them as they teach it.

Figure 7.6 Instructors must be familiar with how equipment operates.

- ***Review required lesson audiovisuals*** — Preview videotapes or slides to prepare for engaging learners in related discussions and applications. This means being familiar with the lesson plan to know when and how audiovisuals are used. Sometimes the instructor's guide for the course has a column that has pictures of the visuals on one side of the page and instructor's notes on the other side. If there is a videotape or film that goes with the session, watch it before the learners do. For last-minute situations when the instructor does not have time to review videotapes or films, have the learners watch for specific points related to the lesson such as looking for errors or differences in procedures. Learners can also list skill steps and write down questions about them.

- *Check paperwork requirements* — Are there forms for attendance or to check off skills completion? Do handouts have to be duplicated, and who is responsible for ensuring that they are? Be sure to check on, arrange for, and have on hand and ready all handouts, rosters, and other paperwork before class begins. For last-minute situations, there may be assistance available from a staff person or an individual in class who can get or copy needed forms as the class starts.

As instructors gain experience in a given subject in a familiar program, the time it takes to prepare decreases somewhat. In the beginning, it is not unusual to spend one to three hours of preparation time for every hour of delivery time. For some programs, preparation time can extend to more hours or even days of scheduling, coordinating, and confirming that class and teaching needs are met.

Session Logistics

The term *logistic* means rational calculation and reasoning. Organizations usually have a department or an individual who manages the logistics of scheduling by rationally juggling what are usually limited materials and equipment to meet the demands of multiple people. Logistics can consume much of the time that goes into class preparation. This is especially true with classes that have a lot of practical work like basic fire, rescue, and emergency medical technician (EMT) training programs along with police and highway safety training programs.

Logistics may consume class time. The completion of some skills requires cleanup of certain areas or restocking, refilling, or replacing certain items. The organization may require or the type of lesson may call for performing such duties as the following:

- Clean a spill.
- Return the classroom to order.
- Park vehicles in a designated spot.
- Clean mannequins.
- Refill self-contained breathing apparatus (SCBA) cylinders.
- Repack hose.
- Restock kits with supplies.
- Recharge AEDs.

Instructors must calculate the time it takes to do these duties. Some duties are done on class time so the next group of learners can proceed to practice; some duties are done after class time so the next class can begin promptly. In large classes with high equipment needs, such as live fire training, CPR and AED training, or driver training for groups of 25 to 30 participants, the logistical needs may require several hours of preparation and wrap-up.

For many instructors, logistics means making arrangements themselves to have the materials and support they need to deliver training. Emphasis is placed on *themselves* so that instructors understand where the responsibility lies. Who is teaching the course? The instructor is — not the training academy staff, the training supervisor, the scheduling office, or the logistics department. The instructor is the one responsible for ensuring that all materials and equipment needed are planned and arranged ahead of time. Placing the blame elsewhere by saying *"They didn't have it ready for me"* is not an appropriate excuse. *They* are not teaching the class and, therefore, do not have the final responsibility of ensuring that everything is in place at the right time.

For those instructors who work in a training facility, there may be staff members to assist with maintenance, inventory control, and scheduling of equipment. But staff members do not always have the same motivation as instructors to get things ready for a class. The final responsibility always rests with the instructor who must take the time to ask for assistance, follow procedures when making requests or reservations, and follow up on those requests. Arriving early to pick up or set up equipment and materials gives an instructor the opportunity to check for missing items, review operations, arrange room layout, find replacements or make repairs, or revert to a *contingency plan* (also know as *Plan B*). Instructors must always have at least one contingency plan per lesson.

Maintaining Continuity

[NFPA 1041: 2-3.2, 2-3.2.1, 2-3.2.2, 2-4.2, 2-4.2.1, 2-4.2.2, 2-4.4, 2-4.4.1, 2-4.5, 2-4.5.1, 2-4.5.2; 3-2.6, 3-2.6.1, 3-4.2, 3-4.2.1, 3-4.2.2]

Many fire and emergency services training programs vary in length of time, lasting from only several hours to several weeks or months, and they often have specialty lessons for which outside experts or more experienced instructors are called to teach. A challenge for the course instructor is to maintain some form of continuity and consistency within the program so the learners can reasonably follow it. *Continuity* refers to something that flows without interruption or change; *consistency* refers to having a harmonious regularity or continuity throughout. In a program with multiple instructors, multiple activities, various learner needs, potential schedule conflicts and changes, equipment breakdowns, and other similar challenges, instructors may find it difficult to maintain continuity. For these reasons, many programs include rules, procedures, and expectations for both instructors and learners that provide a foundation to rely on and return to when even normal and expected class proceedings meet with disruptions.

Many factors affect class continuity, but instructors can take steps to reduce their effects. Most of the steps can be included and taken care of in the processes of session scheduling, preparation, and logistics. Experienced instructors always anticipate problems and make contingency plans for such events as the following:

- Instructor changes
- Weather variations
- Equipment variations
- Safety factors
- Locating resources and appropriate testing locations
- Learning style differences

Instructor Changes

There are times when an instructor is not available to teach a scheduled class. Ideally, that instructor or instructor's supervisor contacts someone familiar with and experienced in teaching the lesson. Every time a different instructor teaches, it causes some extent of discontinuity in the class. Another instance that causes discontinuity is when additional instructors must assist the course instructor in teaching a skills session. All instructors have their own personal perceptions, views, beliefs, and methods of teaching, and it is likely that their ideas and methods are different from those of the course instructor. How can the course instructor maintain continuity in these situations? Some suggestions are as follows:

- ***Know your fellow instructors*** — One of the many benefits of in-service training, meetings, conferences, and seminars is that instructors meet, exchange ideas, and make assessments of each other. If an instructor must call on another to substitute or help teach a class, it will be one who the instructor knows and trusts, who has similar ideas and methods, and who has similar or better experience.
- ***Prepare your learners*** — The ideal does not always occur, and the desired instructor substitute or assistant may not be available. Learners become dependent on and familiar with the methods, attitude, and personality of *their* instructor and may not relate as well to a different one. Prepare learners for different instructors by giving them some background on the individuals, including their experiences, knowledge, and teaching methods. When the instructor arrives, introduce him or her, and outline to the learners, in front of that instructor, what the lesson plan is so both learners and instructor know what is expected.
- ***Meet with instructors to preplan the class*** — Every instructor involved in teaching a class must prepare for it. An instructor should never just show up and figure out what to do upon arrival. The course instructor who arranges for a substitute or for assistants should meet with the instructor(s) and perform the following activities (Figure 7.7):
  - — Outline what must be accomplished in the lesson.
  - — Assign specific duties or skills.
  - — Show and orient each person to the area where he or she will be teaching.
  - — Provide directions or assistance in locating and setting up equipment.
  - — Coordinate break time and clean-up time.
  - — Perform any other duties needed to have a successful lesson.

Practicing the described actions regularly has the benefit of also providing continuity in instructor preparation and delivery. Classes run more smoothly, instructors perform more effectively, and learners have the consistency they need to participate successfully in the learning process.

Weather Variations

The weather can create havoc with schedules. Programs that are scheduled during the times of year when inclement weather is possible must have some flexibility built into them. There has always been some debate on whether instructors should teach certain practical sessions in the extremes of hot, cold, wind,

Figure 7.7 Instructors must meet and discuss plans for delivering lessons. *Courtesy of Robert Wright, Maryland Fire and Rescue Institute.*

rain, or snow. One side of the issue declares that the reality is *"We fight fires and go on EMS calls in this weather, so why not train in this weather?"* The other side of the issue rebuts with the argument, *"Learning can't take place if the weather causes discomfort, concern, distraction, or safety hazards."*

There is merit to both sides of the issue, but instructors should not expect their classes to learn skills effectively in any extreme condition. When teaching basic skills, the learning environment should not distract the learners from attending to the objectives of the lesson. Instructors must also consider safety factors if they lean toward the reality side of the argument. Learning environments that include weather toward the extreme ends of the scale can become safety issues.

If the scheduled lesson falls during extreme weather conditions and it is not possible to reschedule, instructors must make arrangements for additional breaks and refreshments and for appropriate rest areas and facilities. If not, instructors are gambling with the safety of their learners and will most certainly have a problem keeping the attention of those who are tired, hot, or cold.

Equipment Variations

The section on logistics discussed the importance of having equipment available for classes. When scheduling equipment, an important factor in maintaining the continuity of the program is using the same type of equipment in the learning sessions that is used in the testing session and on the job. If a group is to be tested on tying knots using a certain type and size of rope, for example, participants should also practice tying knots with that type and size of rope. This procedure is not only fair, but it also makes testing valid and reliable (see Chapter 9, Testing and Evaluation). If a group uses a certain type of SCBA or AED on the job, the group should also be trained on the same or generically similar equipment.

Safety Factors

Safety is an action taken to avoid risk and prevent injury. Attending to safety issues while planning and delivering instruction is one of the most important concerns of fire and emergency services instruction (see Chapter 2, Safety: The Instructor's Role). Program participants learn by the examples that instructors *set*, not just by what instructors *say*. If the lesson requires that participants wear a protective hood during live fire training, then all instructors involved in teaching live fire training *must wear one* also. The same rule applies to using personal protective equipment. If instructors or standards require that participants wear personal protective equipment on the fireground, then participants need to wear it when they train. Instructors must consistently demonstrate by example every safety feature, every time, under every condition if they want participants to learn and perform every safety practice.

Safety is also an attitude. The attitude projected by instructors in how they behave toward meeting safety standards, following safety procedures, and requiring safety compliance contributes to shaping the attitudes and resulting behaviors of the learners. Recall that the title of this major section is Maintaining Continuity. Safety practices must be practiced with consistency and continuity, not for convenience. Safety that is practiced conveniently takes risks that lead to injury.

Locating Instructional Resources

If logistics is defined as providing the materials and support needed to deliver training, then *resources* can be defined as the locations from where these materials and support are provided. Instructors need to know where to find audiovisual aids and equipment, where to get the tools and props for skill practice, and where to get the learning aids (handouts and texts) necessary to deliver the lesson(s).

During their teaching experiences, instructors collect information and materials from journals, conferences, other training programs, and other instructors to use or adapt for use in their own programs. When using these types of information and materials, instructors must make sure that all sources can be cited to the appropriate credible resource or standard and that it is accepted by the sponsoring training organization.

When instructors need sources of information, they need to know where to look and who to ask for assistance. There are many types of resources. A library or media center can provide referenced and credible sources of information. State and provincial fire training organizations often have a resource center at their training academies. Professional journals and the Internet are other sources of information. See Appendix A, Instructor Resources.

Locating Appropriate Testing Locations

When a program requires testing participants on skills normally performed in particular places under certain conditions, the participants need to practice the skills in the same place under generically similar conditions during the training program. Ideally, testing conditions are similar to learning conditions, which are similar to job conditions. All should meet the requirements of the objectives, which are based on job requirements (Wlodkowski, 1993; Mager, 1984). It is important for instructors to arrange for testing areas or facilities that are familiar to the program participants. Familiarity builds comfort. Adequate comfort levels enable participants to demonstrate what they know, not worry about how they will perform because they are unfamiliar with the terrain, facilities, or environment. Organizations who have established valid and reliable testing policies and procedures understand the importance of *testing location* to training participants. It is the responsibility of the instructors to ensure that the places where the program will be taught and tested are available and that authorization is received from the responsible persons. Instructors must also follow specific guidelines or standards for certain exercises. For example, use NFPA 1403, *Standard on Live Fire Training Evolutions,* as a guide for live fire training.

Learning Style Differences

Within every group, there are those who are ready to learn and those who are not, those who learn readily and easily and those who find it difficult to learn, and those who learn quickly and can show others and those who need more time to learn the material or skill and need to be shown the steps several times. Instructors must prepare lessons with various activities and formats so that every learner can gain some knowledge and skill.

Early teaching formats for many traditional training programs provided cognitive information in a lecture style and may have used some questioning techniques to check that participants were paying attention or completing the reading assignments. Participants were supposed to listen carefully and take notes, but not everyone learned that way. Programs began to add visual materials such as transparencies, slides, and films. With visuals, participants can also see what is being said and can probably remember more of the information, but still, not everyone learns that way. Demonstration and skill-practice classes are best for those individuals who learn by doing, but again, not everyone learns that way. Ideally, instructors prepare lessons that include a variety of teaching styles so that all types of learners are exposed to the material in a manner that enables them to learn through their preferred learning styles. This is not a suggestion to prepare multiple lesson plans on one topic with each one using a different teaching style; it is suggesting that each lesson plan include a variety of teaching methods that call for different types of learner activities and participation. Many learners do not know that there is more than one way to learn, and they may not learn successfully by using the only way they know. By using a variety of teaching styles, instructors expose learners to different and possibly more successful ways of learning.

In addition to the situations described in the preceding paragraph, there are also occasions when the abilities of the participants do not match the target audience of the material to be covered. Participants may be either ahead of the material or not ready for it. What can instructors do in these situations?

If a group is behind what the instructor expects, the instructor has to adjust the teaching pace and expectation. It may be necessary to spend more time on review or practice. Participants may need a review of basic skills or knowledge before they are ready to progress to the material in the current lesson. Reviewing previous material or skills may delay the lesson agenda or program time frame initially but can allow participants to feel more comfortable with their knowledge and abilities so that they feel ready to proceed. The alternative is to have learners unprepared and not ready to progress in the learning or to test in the knowledge or skills.

If participants are advanced for the material, review to determine the level, then assign problems or exercises at that level. If the lesson includes some form of evaluation such as a test or demonstration of skills, perform the evaluation. If there is still time, preview the next lesson and get the participants involved in discussions or exercises to determine their levels of readiness. If it is not possible to move into the next lesson because of equipment limitations, have the group create exercises or scenarios and plan for the equipment they need to perform the exercises in the next lesson. Then arrange to have that equipment available. Instructors can also have participants work together in groups to create test questions complete with answer keys and text references. These tests can be used for study, debate, and discussion. Having completed all of these suggestions, an instructor may be able to dismiss participants early. Early dismissal may *not* be appropriate if the standard or the training organization requires a minimum number of contact or skill hours for completion and certification.

The key to handling all types of learners is again preparation, *preparation,* and ***preparation!*** Preparation is the most time-consuming part of teaching but also the most worthwhile. No one may notice that instructors are prepared when it is done well, but it is unfortunately obvious when instructors are not.

Physical Setting

[NFPA 1041: 2-4.2, 2-4.2.1, 2-4.2.2]

The physical aspects of the learning environment, whether inside or outside, play a major role in delivering fire and emergency service instruction. The location of physical structures and the factors of environmental controls such as rooms and buildings, grounds, light and audio sources, temperature controls, seating arrangements, and training aids or props form the elements of the learning environment. In addition to these elements, there are other factors that affect the physical setting. These factors include anything or other activity that causes distraction and any potential or real hazard that affects safety. All these elements have an effect on how participants learn and affect how instructors plan and deliver their lessons. Instructional effectiveness is, in part, a result of how instructors make the most of the elements that affect the learning environment.

By the nature of the job performance requirements, practical skills training must be performed in settings that closely resemble the locations where the service is delivered. In other words, if the job skill is normally performed outside, the training skill is performed outside. Many training facilities are in buildings that share space with motorized or other noisy mechanical equipment. These are less-than-desirable facilities but often all that is available. The operating functions, such as heat and noise, of nearby equipment can cause distractions that affect attention and learning. Ideally, instructors examine the training facility before arriving to teach. The following steps are suggested:

- Note how the training environment is arranged.
- *Preplan* how to arrange the environment to suit the lesson.
- Note any deficiencies in learning areas.
- Note items that will cause distractions.
- Determine how to fix, change, or work around these deficiencies and distractions.
- Consider strongly making arrangements to relocate the session if adjusting the environment is not possible and if adapting to the environment detracts from learning.

By understanding the limits and capabilities of learning environments, instructors can plan lessons and activities that work with or around less than ideal situations and still provide instruction that enables learning. Many factors that contribute to the comfort of individuals and therefore to the ability of them to learn are taken for granted. Seemingly insignificant elements such as types of chairs and seating arrangements, lighting, classroom climate or temperature, noise, inside or outside locations, and use of audiovisual equipment can have a significant impact on learning. The following sections provide suggestions for adapting or arranging the physical setting to enhance instruction and provide safety for participants.

Seating

The type of seating and how it is arranged has an impact on how well participants learn. Comfort is as important to learning as how the participants can see and interact with the instructor and with each other.

Instructors rarely have the option of selecting the types of chairs used in training rooms. Some facilities are able to provide padded, comfortable chairs, but typically training rooms are filled with metal folding chairs. These types of chairs are comfortable only for a limited time. A long session in a metal chair is a type of distraction. The individual becomes more concerned for comfort and relief than for attending to instructor information. The purpose of breaks, usually after every 45 to 50 minutes of instruction, is to allow participants to stand, move around, stretch, and attend to other comfort needs. Trying to delay the break time for another 10 minutes in order to finish a segment of instruction may not accomplish anything. Participants are no longer interested in learning if they are distracted by comfort needs.

Related to the types of chairs is the arrangements of those chairs. When the chairs are not fixed to the floor, as they are in auditoriums, chair arrangements can and should vary. If the seating arrangement is not effective for the planned lesson, change it. Many training facilities set up their classrooms in the format of traditional rows and require that the room be returned to that format. Instructors and program participants must respect the wishes and rules of the organization to return the room to its original arrangement when the lesson is finished.

Seating arrangements can influence how people learn and should match the instruction type. For example, when providing cognitive information in a lecture or illustrated lecture format, an arrangement with *traditional* or *chevron* (angled) rows allows for learner attention to remain focused on the front of the room toward the instructor. These traditional types of rows are not good arrangements for lessons that have group work or discussions where participants need face-to-face contact. Having seats arranged in *group clusters* or in a *U-shape pattern* is more effective when participants need to interact with each other. See Figure 7.8 for various seating arrangements.

Another issue related to type of seat is the type of desktop or table provided. How much surface space do participants have for writing? Is there a place to store as well as open and use textbooks and notebooks? Again, instructors rarely have the opportunity to choose the type of table or desktop surface for the classroom, and unfortunately, some training organizations provide desks or folding chairs with small writing surfaces that do not accommodate an open notebook. By inspecting the physical setting ahead of time, instructors can anticipate, preplan, and make arrangements for alternatives.

Lighting

Training facilities typically have lighting, but is it adequate for class activities? Ideally, the light switch is one that allows bright-to-dim adjustment, and the lights themselves provide adequate illumination on reading surfaces without glare. Some training sessions are held in makeshift rooms such as bingo halls and apparatus bays, both of which may provide either too little or too much nonadjustable light for reading and the proper use of audiovisual materials such as overhead transparencies. The best situation is a classroom that has adjustable blinds, room-darkening shades or curtains, and lights on a dimmer switch. With these features, instructors can easily regulate the amount of light in the room. Without these features, instructors must plan ahead of time to make adjustments in their presentations and learning activities.

Physical Climate

The physical climate of the classroom has to be a compromise between hot and cold. This compromise is not always easy to achieve because many heating and air-conditioning systems do their jobs with an intensity that is not adjustable. Part of the instructor preparation role may include operating the heating, ventilating, and air-conditioning (HVAC) equipment. The location of a unit may be such that its operation provides a noisy distraction during the lesson, and it must be turned off when an instructor is trying to present information. Under this condition, it is difficult to keep the physical climate at a comfortable level. Learning environments that are too hot or too cold tend to preoccupy participants with staying cool or warm rather than with learning. Instructors need to check climate controls prior to teaching in the facility, find out whether they can be adjusted and how, and advise participants on how to dress for comfort if they cannot be adjusted.

Noise

Fire and emergency service providers often work in environments with high noise levels from diesel engines, pumps, sirens, radio transmissions, and shouting while attempting to be heard above these noises. Many fire and other emergency service or rescue-type departments issue hearing protection for use *on the job*. While these personnel may often have to work in a high-noise environment, they should not have to train in one. Noise that drowns out the instructor distracts learners from the desired outcomes. A high

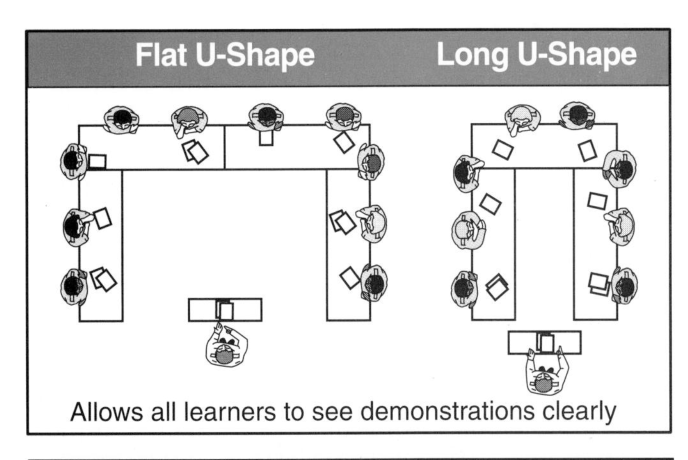

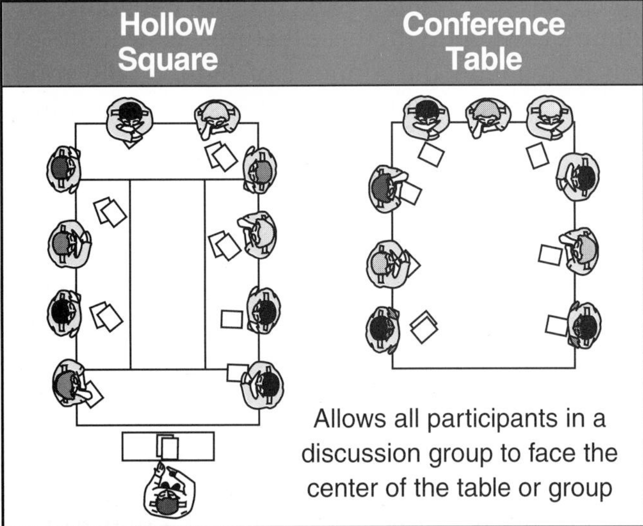

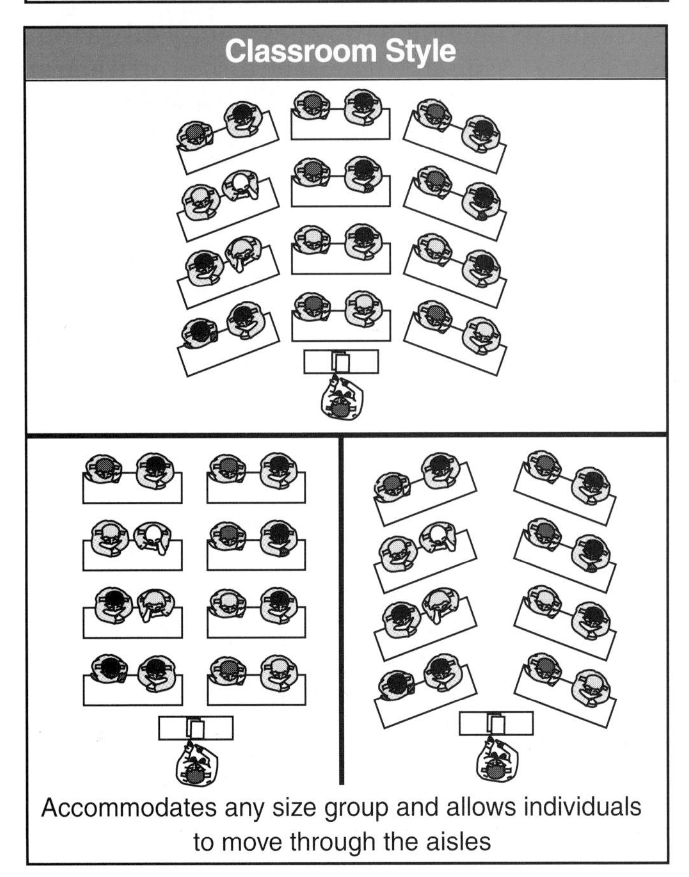

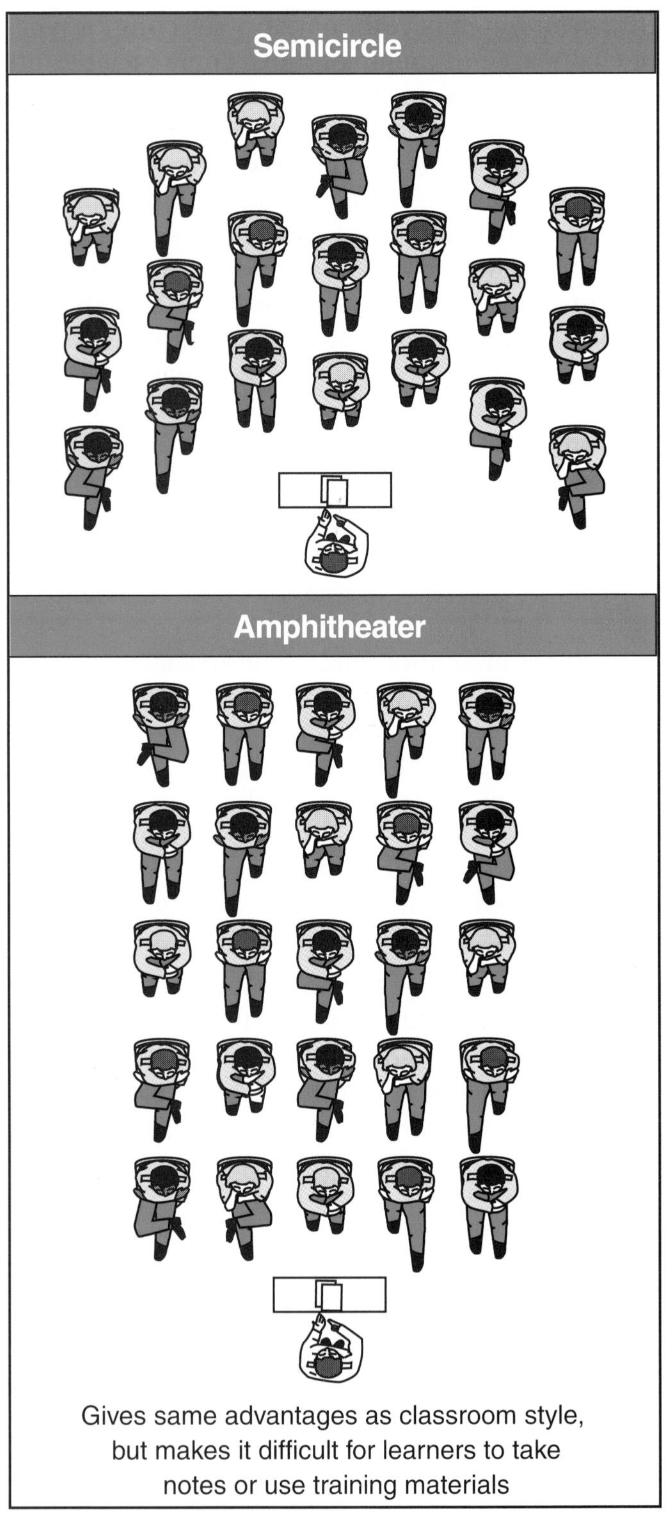

Figure 7.8 Seating arrangements.

level of noise is not only an annoyance and a distraction, it is a safety factor when working with power tools or when giving instructions on how to operate equipment.

When inspecting the training facility, instructors can take actions to find, monitor, and eliminate or limit noise levels that originate from inside and outside the classroom. These actions include turning down the volume on loudspeakers, radios, and pagers or turning them off. For participants who are *in-service* or *on call* and responding to pagers while attending the program, instructors can make arrangements with them before class sessions on how they will be alerted. Instructors can prepare other participants for potential interruptions by informing them that some individuals may have to respond to calls during lessons. This arrangement enables those in-service participants to leave classes with as little distraction to others as possible.

Outside Instruction

The outside learning environment is a challenging place to teach. All the climate-related factors of heat, cold, precipitation, and noise are in abundance in the outdoor classroom. If these elements become distractions, instructors must make adjustments for them. Adjusting to weather conditions is not always easy, but it is necessary not only for learner comfort but for safety as well.

Begin the lesson with a skill overview and explanation in an apparatus bay or classroom before taking the class outside. When demonstrating skills in extreme weather conditions, show as much of the skill as possible in the comfort of an inside area or under some protection in an outside area such as in the shade of a tree or building for hot weather or under some protective tarp, tent, or lean-to for cold or wet weather. Allow the participants to get out of the weather often and provide them with rest-area facilities and refreshments. Extreme weather conditions require longer and more frequent rest breaks.

There is a greater potential for distractions outside than inside. Enclosed or secluded training academy grounds may have less potential for distraction than on-site training in the community, but neither are immune from ordinary noises. At training academies, there may be other training activities occurring at the same time. At community sites there will be motor-vehicle and possibly train traffic passing by, aircraft flying overhead, and passersby stopping to look or ask questions (Figure 7.9). For uncontrolled noises such as diesel truck, train, and aircraft traffic, stop speaking until the noise has stopped, then continue. If the noise is lengthy, review the last few points before continuing with the instruction or demonstration. If noise is persistent or consistent, consider making arrangements to move the class to another location for the next outside session.

Figure 7.9 Potential distractions for outside instruction include noise from passing traffic (or in this case a passing ship). *Courtesy of Terry L. Heyns.*

Audiovisual Equipment

The idea of preparation to avoid or eliminate distractions also applies to using audiovisual equipment. Many instructors have had experiences as program participants in which instructors fumbled with audiovisual equipment. These instructors trying to figure out how the equipment worked contributed to the restlessness and annoyance of the group, and it also created a major distraction and concern about the instructor's ability and credibility. To give an example, a bomb-squad expert who was teaching a class on the intricacies of handling and dismantling bombs, unsuccessfully attempted to operate a videocassette recorder (VCR). The group was distracted from the topic by considering the ability and credibility of that instructor who could not handle a VCR but professed to be adept at handling a bomb.

Instructors who plan to use audiovisual equipment in their presentations must arrange ahead of time for its use. Make sure needed equipment is available and functioning before the session. Take time to ensure that the equipment is in the classroom, that it is set up and projects so all can see, and that it works (Figure 7.10). Preview a few of the overheads or slides to make sure that they are in focus and can be seen from all parts of the room. Make sure a videotape is tracking properly and set to run at the beginning of the episode. Opening credits can be distracting and time-consuming. If the videotape must be credited or referenced, give participants handouts.

Become familiar with the operation of the unit to be used. Some questions to answer include the following:

- Where is the on/off switch?
- How does the focus work (automatically or manually)?
- How is tracking adjusted?
- Where are spare bulbs?

Instructors must always preview a film, a videotape, or slides. View them before showing the class in order to check for accuracy, look for errors, find points to emphasize and discuss, and identify features critical to the lesson. Just before showing the item in class, give the learners a preview or overview of it. Tell them what it is about, what to look for, and what questions to answer afterward. The film, videotape, or slide presentation must have a point that relates to the lesson objectives. If it does not, it only serves as a distraction to the learning task.

Instructors must also eliminate the major distractions of accidents and injuries by planning and preparing ahead of time for using appropriate electrical outlets, cords, and adapters. Answer the following questions:

- Where are outlets located?
- Are extension cords needed?
- Do extension cords have grounds, or are adapters needed?
- Are cords safe, and are they free from fraying and exposed wires?
- Where will cords be placed?
- Do cords need to be taped down? Constantly stepping over extension cords is a distraction to the instructor and a safety hazard for both instructor and participants.

Safety

Most fire and emergency services programs require safety training. Instructors must train participants in safety topics as well as train them safely. Training safety includes more than just following written guide-

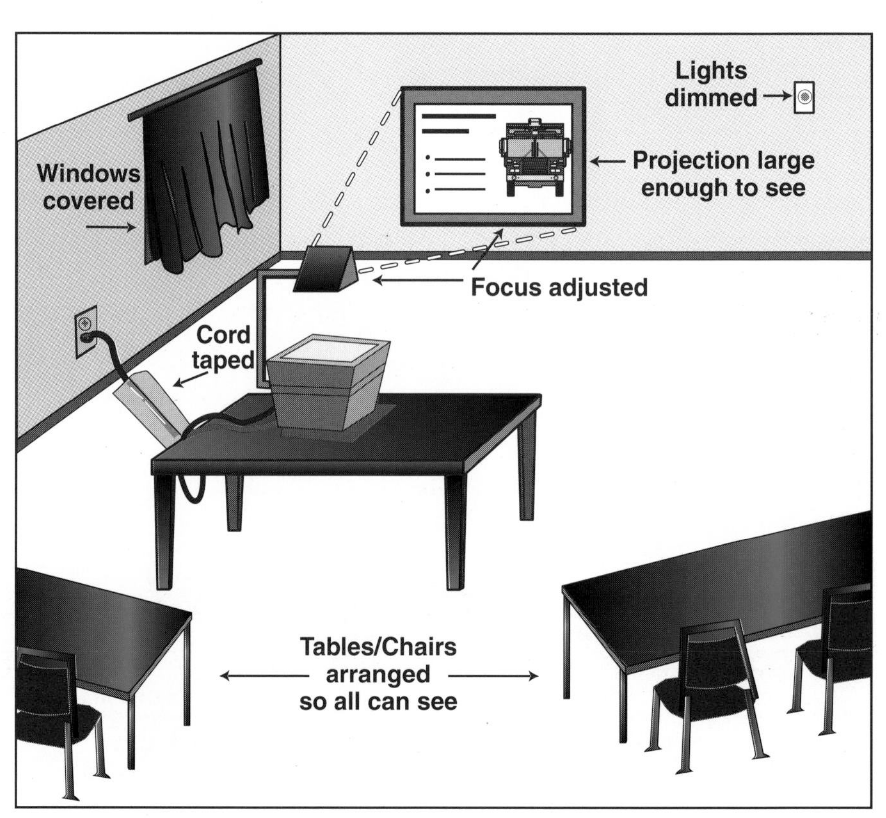

Figure 7.10 Ways to avoid distractions when arranging audiovisual equipment.

lines — it must become a mind-set, a habit, and an attitude. Training safety also means that instructors set examples, which reinforce and determine the behavior of program participants long after they leave the program, not just during it. It is worth repeating that training and safety allow for zero tolerance of disruptive behavior — no horseplay is allowed, nor is violating safety policies acceptable under any circumstances.

To ensure safe behavior during training sessions, safety rules must be provided in a written handout for reference and explained as participants read them prior to beginning the session. Instructors who show disregard for written safety guidelines and standards cannot expect participants to follow them. If an individual is injured as a result of following an instructor example, the instructor may be placing both himself or herself and the organization in a position of liability.

Remember that safety is an attitude that instructors display. Safety also should be a set of rules or policies that instructors follow and require participants to follow as well when training. Working safely must be a behavior that is observable and practiced in training *and* on the job long after training is completed.

Presentation Techniques

[NFPA 1041: 2-4.3, 2-4.3.1, 2-4.3.2, 3-4.2, 3-4.2.1, 3-4.2.2]

Through a variety of teaching methods, instructors present information through the three learning domains: cognitive (knowledge), psychomotor (applying knowledge to skills), and affective (attitude, values, and beliefs). The instructor's presentation techniques show through public speaking, skills demonstration, and attitude/values display. In each of these formats, a variety of methods convey the knowledge, skills, and attitudes that instructors are trying to impress on learners.

Public Speaking

Instructors communicate knowledge through a variety of presentation techniques; some are subtle and some are obvious. Though knowledge can be presented in very subtle and effective methods, a common form of presenting knowledge to groups or individuals is by the obvious public speaking format.

Instructors who want to be effective public speakers, must have excellent verbal communications. They must speak clearly and with expression. Public speaking requires good verbal skills, which are among the strongest tools of effective teaching. Verbal skills include the following abilities:

- Speaking with correct grammar
- Outlining major discussion topics consistently
- Following up discussion topics clearly with orderly details
- Composing appropriate responses to questions carefully
- Directing discussion effectively
- Summarizing important points concisely

Effective public speaking requires that instructors project their voices. An instructor's voice can be an effective tool in the classroom. By changing the speed of delivery, adding a pause for effect, or raising or lowering the volume, instructors can provide emphases that leave lasting impressions on learners. Too many instructors are afraid of silence after they ask questions and quickly fill silent spaces with answers. Realize that it may take a few moments for the participants to gather their thoughts, formulate a response, and finally respond to a question. Coupled with how instructors use their voices, they can also add effect, gain attention, and maintain interest with different facial gestures. Gesturing can also add emphasis and help keep attention directed at the instructor. The best public speaking effect is to use these techniques often but subtly. Then to really emphasize a point, an overt change in the normal delivery style makes an important issue memorable. Be careful not to overdo gestures and expressions because they can become distractions.

Repeated gestures and expressions are as distracting as repeated words. Overused words and phrases include such common "crutches" as *"OK," "uh"* or *"um," "yeah," "you know,"* and *"you all."* Instructors often use these crutches to fill silent spaces, and they often repeat them to distraction. Learners are soon paying more attention to how many times the word is used than to what is being taught. Some instructors catch themselves using certain words frequently; others are not aware of their habits. To catch and correct presentation distractions, instructors can videotape themselves or have other instructors watch and offer critiques of lessons. Learn to place pauses during the lesson and become comfortable with them. Pausing and speaking slowly allow learners to process information.

Many beginning instructors tend to speak quickly because they are nervous; some speak quickly because they are running out of time. Instructors can overcome both situations with the simple remedies of preparation, planning, and practice. These remedies enable instructors to be aware of sentence structure and grammar so that they can present sensible, intelligent, and credible lessons. In this current relaxed society, people may not be critical of slang and profanity in everyday, casual conversations, television shows, or films, but this is not the case with instructors. Sometimes it is effective to use slang like *"ain't"* or fad terms like *"cool,"* but they need to be used sparingly and with discretion. Learners do not take the instructor who uses a lot of slang seriously. Consider that profanity is *always* unprofessional and may offend participants and other instructors. Profanity can be construed as an element of a *hostile environment.*

The following list recaps important presentation techniques that help instructors speak and communicate effectively. These methods help instructors identify distracting areas that need improvement, eliminate many annoying habits, and identify tendencies that distract from delivery of the message.

- Practice delivery of a lesson.
- Videotape a presentation and review it for distracting actions and speech patterns (see Table 7.2). Videotaping also enables instructors to experiment with different ways to present materials, which can increase instructional effectiveness.
- Speak clearly and distinctly; enunciate and clearly pronounce each word. Do not mumble or slur words together.
- Use expressive voice inflection and add emphasis to words, both of which transmit interest and excitement about what is expressed. Do not speak in a monotone.
- Govern the speaking pace or speed. Begin slowly with new information to new learners and gradually pick up speed as information becomes familiar and learners understand it. Slow down for important points and when learners need to take notes. Restate important points, and emphasize their importance with voice tone.
- Pause periodically so learners can catch up with their thoughts, think about what they heard, and ask questions. Do not feel pressured by silence. Learners need time to think about and format responses.
- Use correct grammar. When using technical terms, abbreviations, and acronyms, define and explain them in context with the lesson. Avoid slang and expletives.
- Relax and speak in a conversational tone so listeners feel at ease and ready to receive information.
- Make eye contact with learners while speaking. Eye contact reinforces the feeling that the speaker is interested in the learners and concerned that they understand the message. Speakers who look at the wall or ceiling or out the window rather than at the learners are not communicating with them.
- Use appropriate gestures to illustrate mental pictures or to emphasize key points.
- Avoid distracting mannerisms such as the following:
  - Pacing
  - Playing with or tapping pens, pencils, and other items
  - Jingling keys or change
  - Chewing gum, fingernails, matchsticks, or toothpicks
  - Repeating worn phrases such as *"OK," "you know,"* and *"um"* or *"uh"*
  - Using the word *"I"* excessively
  - Overusing pet words or fad phrases

Table 7.2
Reviewing a Videotaped Presentation

Review Methods	Speed/Sound	Focus
Complete content review	Normal speed/sound on	Speech patterns and complementary gestures, eye contact, and use of visual training aids
Quick cue review	Fast forward cue/sound off	Gestures, repetitive motion, and movement habits
Audio review	Normal speed/sound only	Speech patterns, repetitive word usage, and voice projection and enunciation

Skills Demonstration

Just as knowledge can be imparted subtly or overtly, so can presenting skills, but instructors usually demonstrate skill procedures very deliberately while explaining the steps. Also, just as important in presenting skills demonstrations (as in presenting knowledge) is the need for instructors to prepare, plan, and practice. An effective method for presenting a skill is to give the first demonstration in *real time* — that is, show the entire procedure from start to finish as it would normally be done on the job. Then, slowly show each step. Allow participants to practice these steps while being guided and coached by instructors. Allow participants to practice more as they critique and coach each other. Once or twice during a practice session, demonstrate the entire skill again if necessary, and at the end of the session, show the entire skill one more time. Instructors must show the skill steps correctly and in sequence. An instructor who forgets a step and has to backtrack or who incorrectly performs a step and has to correct it not only distracts the learning process but loses credibility.

Attitude/Values Display

Coupled with the important elements of presenting knowledge and skills is the ability to convey a positive and enthusiastic attitude. To present the affective domain, which includes attitudes, values, and beliefs, the subtle approach is usually more effective. People are usually "turned off" by those who want to force a personal value or belief on them.

The first area, attitude, is conveyed overtly by the overall actions and appearance of the instructor. Actions and appearance that demonstrate attitude include facial expressions, body posture or body language, type of clothing and how it is worn, and grooming. Appearance, particularly clothing and grooming, is not about winning a beauty contest — it is about presenting a professional appearance (Figure 7.11). What is a professional appearance? It is an instructor who arrives to teach in a clean uniform or clothing appropriate for the training session. For practical skills evolutions, it is wearing protective clothing that fits and meets standards and is not worn, torn, or burned. How instructors present themselves speaks volumes to learners. An instructor's appearance often speaks louder than the message being delivered.

The other two areas within the affective domain are the values and beliefs that instructors portray. Instructors must show belief in and the value of knowledge and skills by acting out and practicing outside the classroom what they teach inside the classroom. If instructors teach a session in the classroom on giving clear, concise instructions on the emergency scene, then they must practice giving clear, concise instructions on actual scenes. Likewise, if instructors teach the proper use of safety equipment, they must wear and use it properly on the job. It is easier for instructors to demonstrate values and beliefs in the classroom if they actually perform the skills consistently on the job.

Instructors who clearly present these aspects of the affective domain — attitude, values, and beliefs — convey important information and skills and set positive examples for learners. Learners can then internalize the information and develop the necessary skills while training and then correctly perform what they have learned later on the job.

Concepts of Instruction

[NFPA 1041: 3-4.2.1, 3-4.2.2]

Instructors use many different instructional methods to reach learners who have a variety of learning styles and needs. It is a challenge to match teaching styles to learning styles. Matching styles does not require that the instructor teach to each individual style, but an instructor should provide a variety of methods so that individuals have opportunities to learn by being exposed to styles that are familiar and comfortable. By being exposed to many teaching styles, learners often find that they become comfortable with other methods and can expand their abilities to learn. Fire and emergency services programs are

Figure 7.11 An instructor presents a professional appearance through posture, dress, and grooming. *Courtesy of Jarett Metheny, Midwest City (OK) Fire Department.*

often designed to present information in the following various formats:

- *Cognitive* areas of introducing, sharing, discussing, discovering, and questioning information
- *Psychomotor* areas of applying, touching, experimenting, showing, and practicing skills from learned information
- *Affective* areas of developing, changing, accepting, and contemplating attitudes, values, and beliefs

With appropriate planning and preparation coupled with knowledge and experience, instructors routinely demonstrate a variety of teaching methods by introducing new information, demonstrating skills, and providing time for practice in progressive steps and in easy-to-complex scenarios. While working with the learners in the first two domains, effective instructors subtly accomplish the affective domain by being consistent as they practice and perform what they teach.

The following sections provide answers to the following questions on these instructional concepts and guide instructors in methods of preparing lessons that enable and enhance learning:

- How do I go about presenting this material?
- What is the sequence of information?
- How do I maintain interest?
- What can I do to encourage learners and help them be successful?
- Are there special techniques to get all learners to fully participate?
- What are the best ways to reinforce learning?
- Are there effective ways to use questions?
- What types of questions are effective?

Presenting New Material

Instructors spend a lot of teaching time presenting new material. Their goal should be to present the material in a manner that gives many learning opportunities in a variety of formats. Determining how to provide a variety of learning opportunities effectively and to meet the program requirements along with the needs of the learners are challenging situations. The challenges do not necessarily get easier with experience but persist with every lesson instructors prepare for presenting new material to a different group of learners. One of the first considerations in presenting new material is determining the order in which it is taught. At what point does an instructor begin when there is so much information to teach? Organizing the lesson so that it is presented in a logical sequence that makes sense to the learner is the first consideration.

The guidelines for teaching in *logical sequence* or from *the simple to the complex* are good but varied. The sequence can start with what came first followed by what came next, or the sequence can start with theory followed by practice. Many times what the instructor believes is important as the first step is not what interests the learners. Teaching the history of power tools will bore a group of learners whose primary interest is getting their hands on the new equipment and learning how to use it. Mager (1988) suggests the following steps for presenting new material:

- ***Begin the lesson with the topic of highest interest***— It is important to first get learners' interest. Give them what they really want to know in the first half hour, but tell them that there will be more about the background (history, rules, etc.) later. Continue to sprinkle high-interest topics throughout the lesson.
- ***Move from the big picture to the details***—This is the *whole-part-whole* concept (discussed in the next section, Applying Sequencing Methods). Show learners the final product, or tell them the final outcome first. Then get into the details of how to get to that point. Conclude with another look at the final product or review of the final outcome.
- ***Give learners many opportunities to decide what they want to work on, question, or discuss*** — If you give learners the opportunity to choose topics, they will find (and let you know) the areas in which they need help, clarification, or more practice. This freedom keeps learners motivated. Instructors can provide guidelines of areas that learners can move into or must satisfy before moving to other areas.

Even though many instructors must teach to strict organizational requirements or to specific standards, they can provide some leeway in their lessons, which enhance the learning process. In the book, *Adult Learning Methods,* edited by Michael Galbraith (1990), Associate Professor of Adult Education at Temple University, it states, "*. . . don't be so rigid with your (lesson) organization that you fail to change your game plan when needed. Sensitivity to the needs of the learners always takes precedence . . . .*" Even so, when presenting new material, instructors must have a plan or outline from which to work and by which they provide instruction in logical progression. A later segment of a lesson may be presented first or an earlier segment after a later one, but the order of each segment is presented in a logical sequence.

Applying Sequencing Methods

A review of Chapter 4, The Psychology of Learning, is helpful in linking the memory process with the sequencing methods. When planning the instructional sequence, instructors typically use the following three common and generally accepted sequences for delivering new material:

- ***Known-to-unknown*** — Start with material that the learners are familiar with — or what they know — and then venture into the unfamiliar — or what they do not know. This method is effective because it gives learners an opportunity to find relationships in their memory schema. When introducing new tools or equipment, most learners *know* that tools are aids to doing a job and the new tool aids in doing a particular job. How to use the tool to accomplish a job is *unknown* or new to them. Consider Mager's theory given in earlier paragraphs: Rather than *hearing* the theory first, learners most likely want to get their hands on the tool to *feel* and *see* how it works. Instructors can first show them how it works while demonstrating and emphasizing appropriate safety precautions, and then let them try it under supervision. The theory, history, and appropriate standards can be described and emphasized once the instructor has hooked learner attention and interest with a dynamic demonstration and a hands-on introduction.
- ***Simple-to-complex*** — Start first by teaching the basic knowledge or skills, and then add more difficult or complex knowledge as the lesson progresses. Learning is a process, not a product (Merriam and Caffarella, 1991). Instructors must begin with the basic knowledge and first steps of a skill, which are necessary foundations to mastering the more complex knowledge and skills. Learners cannot progress to algebra without learning and practicing the concepts of basic math. Similarly, for example, instructors must be sure that learners have mastered basic rope skills such as tying basic knots before moving on to more complex uses of rope such as setting up rigging and hauling systems. Without learning the simple (basic rope skills) first, learning the complex (setting up a hauling system) is not possible.
- ***Whole-part-whole*** — (1) Start with an overview *(whole)* of the topic or a demonstration of the skill in real time; (2) then discuss, describe, or demonstrate each individual topic or step *(part);* and (3) finish with a review and summary of main points or a final demonstration of the skill *(whole)* (Figure 7.12). Another way to express this is *introduction-body-review.* A familiar phrase that some instructors use is *"Tell them what you're going tell them* (whole), *tell them* (part), *and tell them what you told them* (whole)." Most textbooks use this format. Notice that the first chapter of this and many instructional texts gives an overview of the following chapters. Instructional techniques are based on the whole-part-whole process because it gives learners direction with a preview of what to expect, a breakdown of information into manageable segments, and closure. Whole-part-whole is an effective technique with both cognitive and psychomotor types of lessons.

Instructors commonly use all these three organizers to present new material because they provide a solid foundation for the learning process. These organizers also help instructors outline the points that are essential to understanding the topic and mastering a skill. In their text, *Educational Psychology: A Realistic Approach,* Good and Brophy (1977) suggest that instructors use several checkpoints when preparing a lesson that enable them to check for understanding during the lesson. A short list of checkpoints from their text is as follows:

- Analyze the presentation to make sure that it is logical in its sequence.
- Provide physical examples of unfamiliar objects or demonstrations of unfamiliar processes.
- Diagram a complex set of ideas held together by some kind of structure on the chalkboard, on a handout, or through overhead projection.
- Use demonstrations and modeling where possible, particularly cognitive modeling in which you "think out loud" while performing the physical activities of demonstrating.
- Build motivation by pointing out the interest value or application possibilities of the new material. Stress aspects that interest learners.

Figure 7.12 The "whole-part-whole" theory of instruction.

In all lessons, instructors must ensure that they introduce learners to the key points, that these points are stressed in the related parts of the lesson, and that they are reviewed and summarized at the end of the lesson. Even though a lesson segment may be presented out of order from the outline, the lesson or segment must flow logically, that is, from simple-to-complex and known-to-unknown information or skills.

Generating Interest

Fire and emergency services training programs include adults who bring their previous significant experiences to the learning situation. These adults want to relate their experiences to the training, and wise instructors plan lessons that permit learners to integrate these experiences with new information. By getting learners involved in the instruction, instructors stimulate and maintain their interest.

Stimulating interest is more than just having learners pay attention — it is having them *desire* more information and understanding about the lesson and *want* to participate in the learning process. Interested learners are open and responsive and want to concentrate on what they are learning and retain it (Wlodkowski, 1993).

Maintaining interest can be accomplished by following several strategies. Fire and emergency services training can be repetitive because it often involves review of skills for accurate performance and for recertification requirements. Instructors need to show learners a personal connection with the lesson. Wlodkowski (1993) suggests the following strategies for stimulating interest:

- ***Relate learning to adult interests*** — Material relevant to learner experiences is important. By offering material that pertains to their goals and tasks, instructors expose learners to experiences that arouse their curiosities and desires for understanding.
- ***State clearly or demonstrate the resulting advantages of the learning activity*** — Learning that offers the possibility of gaining an advantage in life or on the job is interesting.
- ***Use humor appropriately*** — Humor has many attractive and stimulating qualities: It offers enjoyment, unique perspective, and unpredictability; it makes people laugh; it makes learning interesting. Use it spontaneously; laugh *with* people, not at them, and laugh at yourself.
- ***Stimulate emotions*** — People become interested in events that cause strong emotions such as anger, delight, affection, and sorrow. The experiences that learners bring to the training sessions evoke all these emotions and can arouse interest when these experiences are incorporated into the lesson.
- ***Explain and illustrate with examples, stories, analogies, and metaphors*** — Examples allow learners to focus on new learning so that they can picture in their minds how to perform. Carefully chosen examples help learners understand. Well-told and imaginative stories that are related to the topic can captivate learners. Analogies and metaphors are ways of suggesting colorful likenesses between two completely different things. For example, an instructor might say, *"Firefighters doing overhaul often act like a bull in a china shop. What can we do during overhaul to help preserve the victim's property?"*
- ***Use questions to stimulate interest*** — Questions are opportunities to stimulate learners either with interest, attention, and excitement because the learner knows the answer or with fear, uncertainty, curiosity, and again interest because the learner does not know the answer but wants to find out what it is. Questions also stimulate thinking and participation. With practice and experience, instructors learn to pose well-timed, thought-provoking, appropriate, and quality questions that stimulate thought, encourage participation, and develop understanding.
- ***Use unpredictability and uncertainty with a sense of security and enjoyment*** — Used appropriately, the unexpected can arouse interest, and unpredictability can be stimulating. The news media often use uncertainty and surprise to keep audience interest. Anticipating the unexpected is exciting when it is done so that it makes the audience feel safe and know that no one will be hurt. Instructors must plan these unexpected or surprise events to ensure safety and enjoyment during the learning process.

The term *stimulation* was used several times in the preceding paragraphs. *"Stimulation is any change in our perception or experience with our environment that makes us active"* (Galbraith, 1990). People are attracted to such stimuli as vivid colors, strange or different sounds, and loud sounds. When an event or individual presents a surprise, it triggers emotions. All these stimulating experiences can be interesting or frustrating and invigorating or irritating, but they get attention and keep people actively interested and involved in the situation. Interest and activity enhance learning.

Encouraging Success

Encouraging success is emotional coaching that assists learners in completing tasks. Most people want to succeed and usually arrive at a training session with the desire to complete the program successfully. Reasons for taking a class include many kinds of motivators that fuel a desire for success such as keeping a job, getting a promotion or raise, gaining recognition, feeling important, or joining a group (organization or club).

People want and need to succeed. Fear of failure is a major physiological block that prevents people from taking risks. This same fear may hold learners back or cause them to feel that they *can't do a task.* Learners find it difficult to dislike a class in which they are successful. Instructors can positively influence learner attitudes and encourage their successes by offering the following instructional enhancements (Galbraith, 1990; Wlodkowski, 1993):

- ***Quality instruction that helps learners who try to learn*** — Make the first experience with a new subject or topic safe, successful, and interesting. First impressions are important and have a lasting impact on future learning.
- ***Evidence that learner efforts make a difference*** — Stress to learners the importance of the amount and quality of effort needed for success in learning tasks before they begin. Learners can control the effort they put forth. Emphasizing the importance of effort, without threatening, accomplishes the following:
  - — Establishes learner responsibility
  - — Reduces feelings of helplessness
  - — Increases perseverance
  - — Generates feelings of pride and accomplishment
- ***Continued feedback about learner progress*** — Make the learning goals and the evaluation criteria clear. Knowing what to expect and how they are progressing help learners succeed. They need to know exactly what they are to learn and how well they are performing in preparation for testing. Instructors who provide learning and testing criteria, continuous and constructive feedback, and appropriate coaching and encouragement have confident and successful learners.

Fostering Participation

Instructors can stimulate interest through a variety of methods that foster and encourage participation. Keeping learners involved in the lesson can be a major task, but it is the most important part of learning. Certain strategies can assist instructors in fostering participation. Of these strategies, possibly the first and most important one may be that the instructor expects, encourages, and sets the stage for participation at the beginning of the program. To begin this strategy and still have new participants feel safe and comfortable, instructors can begin by asking for opinions and examples of topic-related experiences, decisions, results, and conclusions. The instructor encourages discussion by providing positive reinforcement — by emphasizing that no opinion is wrong and that the activity provides a forum for sharing and learning. When using this type of strategy, instructors must guide the discussion so that all learners, even the quiet and shy ones, are able to participate.

As instructors foster participation through discussion techniques, there are many opportunities to ask questions. The instructor as well as the learners can generate questions as their interests and curiosities are stimulated by the discussion activities. Questions open new pathways that lead to discovering solutions through debates, researches, analyses, and drawing conclusions. Strategies for actively applying knowledge can include such activities as assigning groups to plan strategies, solve problems, accomplish tasks, or perform skills. Groups can then explain or demonstrate their results (Figure 7.13).

Questions provide feedback and give the instructor an idea of how much or how little the learners know. Questions also give learners an idea of how much they

Figure 7.13 Fostering participation: Instructors can stimulate learner interest by keeping class members involved in lesson activities. *Courtesy of Jarett Metheny, Midwest City (OK) Fire Department.*

know. Using questions for feedback is not the same as calling on someone who is not paying attention or being disruptive. (**Note:** Various questioning techniques will be discussed in the Applying Questioning Techniques section.) Calling on some individuals to respond gives reserved or reticent individuals an opportunity to answer questions. These same people, while introverted in front of the whole class, may be more comfortable working in small groups. The key issue in fostering participation is to make participants feel comfortable and accepted rather than threatened. Learning is an active process, not a passive one: Doing is remembering.

Reinforcing Learning

Consider reinforcement in two ways: (1) repeat in order to reemphasize and review and (2) encourage in order to maintain or increase correct responses. When instructors repeat information or skill demonstrations, they do so to ensure that learners grasp and retain the importance of it and have opportunities to place information or skill steps into memory. Repetition prepares learners for testing. When presenting material, emphasizing and repeating key points help ensure that learners recognize its importance. Do not be afraid to say *"This is important; we won't go on to the next step if you don't understand."* If the material or skill is important enough to be emphasized repeatedly, learners take this as a cue to practice and learn for successful testing. Through various types of written tests (quizzes or comprehensive exams) and skill-ability tests, instructors can observe the level of knowledge or skill that learners acquired and can reinforce low score or ability areas through more review and reemphasis. Instructors may emphasize importance by telling learners they will see certain material or similar problems again on a test. Reinforce the material by providing ample opportunities to practice and apply similar skills and knowledge so learners have the confidence to test. This suggestion does not imply that instructors tell the learners what the questions are before a test; it means guiding the learners to the important areas to study or practice for successful testing.

While reinforcing the learned material, instructors also encourage learners through *positive reinforcement* of their attempts to master the material. If a step is performed correctly, compliment the individual; if a step is performed incorrectly, coach and encourage the individual as he or she tries again. Compliment every effort that attempts to master knowledge and skill. Remember that some individuals require more time and opportunity to learn, but given both, along with positive reinforcement, anyone can learn anything.

Applying Questioning Techniques

An earlier part of this chapter mentioned using questions as a method to foster participation. There are many types of questioning techniques, and it is helpful to be familiar with and understand how to use them. This section describes and discusses oral questions. Written questions are covered in Chapter 9, Testing and Evaluation.

Questions are useful teaching and learning instruments. They can do the following:

- Stimulate and maintain communication.
- Act as control devices for troublesome individuals.
- Provide a measure for understanding and learning.
- Generate thought.
- Arouse curiosity.
- Exchange ideas.
- Mold opinion.
- Provoke interest that triggers related questions.

Instructors must learn to use questions effectively. Effective questions have the following qualities:

- ***Are open-ended*** — Questions should require more than a *yes* or *no* response.
- ***Do not suggest the answer*** — The question should not hint at or lead to a certain response. Ask questions that require some thought.
- ***Seek information*** — Questions should not make learners feel ignorant. Instructors should not use questions that show off how much they know or how little learners know. People attend class to learn. Ask questions that perform the following activities:
  - — Relate questions to the material discussed.
  - — Solicit opinions about the topic.
  - — Enable learners to discover or realize they know the answer.

Instructors can stimulate thinking, instill confidence, and encourage success by using the various questioning techniques in a nonthreatening manner. Questions should never be used to intimidate or embarrass learners. Individuals who give wrong answers and are criticized will not learn and will cease to participate for fear of making other mistakes. If the answer is wrong, there are several ways to tactfully handle the situation so the learner is not embarrassed. First, never humiliate, ridicule, or laugh at an individual who gives a wrong answer. Instead, instructors can look for the "right" parts of the answer and ask the group to come up with other "right" parts to add to the answer. If the question is a difficult one, prompt the individual or group with alternatives that lead them to the right answer. An instructor who has established a comfort level with the group may also respond to a wrong answer with, *"Tell me how you arrived at that answer."* This technique helps the individual review the thought processes used and examine the perceptions developed to arrive at the answer. It often aids this person in discovering the correct answer and is a *positive* reinforcer (see Encouraging Success section). Finally, hold the group accountable for the answer. The instructor cannot let a wrong answer stand without clarifying or correcting the response. If needed, review the information or steps, telling learners that the review is to help them understand the material.

Effective questions can serve many positive purposes. They can emphasize key points, reinforce material, provide feedback, and maintain interest. Instructors can also encourage success and improve participation with the use of good questioning techniques and various types of questions (Figure 7.14).

Identifying Types of Questions

Instructors can use the suggestions in the previous section in various types of questioning formats. These formats each have special effects and can be used for various purposes and to generate different types of responses. Types of questions include the following:

- ***Rhetorical*** — It is not the intent of this type of question to have one correct answer nor to necessarily have an oral response. Rather, rhetorical questions are used to stimulate thinking or to motivate participants. This type of question is addressed to the entire group and often serves as an effective attention-getter. For example, an instructor might open a safety lesson by saying, *"What are the most important pieces of equipment you will use on an emergency scene to protect yourself from injury or exposure? By the end of this lesson, you will know how to answer this question."*
- ***Direct*** — This type of question is directed at one person in the class who is expected to respond. Direct questions can put the individual "on the spot," and they are not often used with adult learners because they can be threatening. To give an example though, an instructor calls on Terry: *"Terry, what is the engine pressure for the attack pumper in*

Figure 7.14 Qualities of questions and types of questions.

Discussion Formats

- Guided
- Conference
- Case Study
- Role-Play
- Brainstorming

Figure 7.16 Using different types of discussion formats arouses interest and stimulates participation. *Courtesy of Robert Wright, Maryland Fire and Rescue Institute.*

other. Having eye contact with each other is essential to conducting an effective discussion. Instructors must preplan all discussion formats. Any type of discussion can be time-consuming and can use that time with no measurable results if not guided carefully.

Guided

In a *guided discussion,* an instructor presents a topic to a group. The members of the group discuss ideas in an orderly exchange controlled or guided by the instructor. The intent of this type of discussion is to gain knowledge from other group members, to modify their own ideas, or develop new ones. As facilitator, the instructor's role is to keep the discussion on the topic, add pertinent details, ask thought-provoking questions, or analyze conversations to ensure understanding — all of which are designed to *guide* the discussion and meet the lesson objective(s).

Conference

The *conference* method of discussion is used to direct group thinking toward the solution of a common problem. A successful conference has a clearly stated end result. This method is effective in bringing about changes in participant thinking and attitude as participants compare experiences, techniques, and beliefs with others. When a participant hears the interpretation others have made of a similar experience, that person is often willing to adopt the attitude taken by the group. Participants must be willing to share ideas and to trust that the consensus of the *group* is better than the ideas of any *one* member. A conference cannot be spontaneous. It must be planned, have an agenda, have a set time and place, and have a goal. Instructors influence the direction and outcome of the discussion by providing background information, and their goal is to have the group develop understanding and recognition of the issue. It is also helpful for participants to know the limits, scope, and purpose of the conference. In addition, other instructor roles may include controlling or eliminating bickering and irrelevant discussion, reconciling differences of opinion, and uniting participants. Instructors are no longer teachers in conference situations. They do not tell the group how or what to think nor steer the results of the group's thinking in a personally preferred direction. The instructor must not enter into the discussion except to state or restate problems, ask questions, state cases, or summarize.

Case Study

A *case study* is an examination of real or hypothetical events. The next section discusses role-play, which can be based on a hypothetical event followed by a case study to analyze situations in preparation for real events. Typically a case study reviews and discusses detailed accounts of past events with the purpose of suggesting possible solutions. This method helps develop a learner's ability to analyze situations and examine facts to reach a conclusion. It provides opportunities to form and discuss ideas and do problem-solving exercises by using the problems faced by the parties in the event. The instructor must create an environment that encourages participants to review the case and express ideas without fear of ridicule. Instructors may not agree with the participants' discussion, but they must allow participants to complete their ideas and draw their conclusions. Emphasize the reasoning that participants use, then the conclusion.

Role-Play

In *role-play,* individuals act out roles in scenarios. This method provides opportunity to practice actions that prepare individuals for situations such as a supervisor dealing with a subordinate, emergency medical services (EMS) or law enforcement personnel handling a violent spouse on a domestic response, or instructors counseling a disruptive or failing learner. Role-play can be used in training personnel for public safety telecommunication positions (dispatching and receiving calls) or for public information officers. The aspects of the affective domain (values, beliefs, and emotions) are easily covered in role-play methods. By placing learners in roles that are new or unique to their experiences, they have opportunities to understand what it is like to "walk a mile in someone else's shoes."

There are areas to consider when planning a role-play lesson. Instructors must ensure that scenarios are pertinent to course materials and clearly defined. Select participants who can play the roles easily, and plan to keep all participants involved in some way — acting, critiquing, note-taking, or watching. Because it is only possible to supervise one group at a time, the actual role-playing is limited to a few who play the roles while the rest of the group are assigned activities directed toward a summary discussion and conclusion. For a large group, having several "plays" provides many, if not all, individuals the opportunity to act out a role. Some people are natural actors; many are not performers and are uncomfortable playing a role. Do not force anyone to act.

Set the following guidelines for the role-play:

- Control the group.
- Emphasize that learners must stick to the script and avoid getting "carried away" with performing.
- Set time limits for the scenes.
- Set time limits for a summary discussion that guides the role-play back to the purpose of the lesson.

Brainstorming

Brainstorming is an effective discussion method if participants have adequate knowledge of the subject matter. Through this method, a class (or groups within a class) is given a problem or situation and then given time to determine a solution to it. These groups are often called *buzz groups* because they display a constant noise or buzz that reflects the talk and interaction among the participants. Effective brainstorming sessions require that the following activities are performed (Figure 7.17):

Figure 7.17 Brainstorming steps.

- ***Record all ideas*** — The purpose of brainstorming is to allow individuals to think creatively within a group and express their ideas.
- ***Allow everyone to speak*** — Everyone should participate and stay on topic. Do not let one or two strong personalities dominate the process or wander from the issue.
- ***Encourage far-fetched ideas*** — No idea is wrong; ideas open up discussions.
- ***Encourage piggy-backing*** — An individual's suggestion may spark an idea for other persons, which eventually enables the group to produce a solution to a problem.
- ***Evaluate and prioritize issues after ideas are exhausted*** — Select a top five or ten issues to discuss or resolve.

Illustration

What is called lecture in most classrooms is really an *illustrated lecture* — a showing method that uses both the senses of sight and hearing. In this format, the instructor uses drawings, pictures, slides, transparencies, videotapes, films, models, and other visual aids that help clarify details or processes. Instructors often improperly use the illustration method as a substitute for demonstration. An illustration may show a picture of a piece of equipment, but it does not show how to perform the operation. Illustrations can supplement a demonstration but cannot take its place.

Instructors often use visuals to illustrate the key points of a lesson. Whether on slides, transparencies, easel pads, or chalkboards, the visible outline shows the framework of the lecture. Learners find that note-taking is easier when they can see the key points and these points remain visible while they write down information as the instructor fills in the outline. Also, if a subsequent point is not clear, learners can easily refer back to previous points for clarification.

The following guidelines are helpful when using visuals such as pictures, charts, diagrams, models, or mock-ups as illustrative teaching aids:

- Illustrate a certain lesson objective in each visual.
- Make the visual large enough to be easily seen. Rearrange the room if necessary.
- Keep visuals out of sight until the lesson topic calls for them to be displayed. Sometimes, the instructor wishes to display a visual early to arouse curiosity, but often a visual that is displayed early is a distraction to attention.
- Show series charts or drawings one at a time when they are designed to show steps or an operation. Displaying them all at once is a confusing distraction.

Demonstration

A *demonstration* is the act of showing how to do something. It is a basic means for teaching manipulative skills, physical principles, and mechanical functions. It can be used effectively to compare products or equipment and show the results of their use. The instructor demonstrates a task while explaining how and why it is performed. This method communicates to both sight and hearing senses. When participants practice the skill, they add the sense of touch to their learning experiences.

There are always limitations to the demonstration method. Some are as follows:

- Instructors must plan for extensive preparation and cleanup times, especially when using such items as power tools, hose, breathing apparatus, and CPR mannequins.
- Careful lesson planning is important because setup and practice often take up most of the class time.
- Large groups of participants require extra equipment for practice as well as additional instructors for supervision, coaching, and safety.
- Skills that require outside practice are at the mercy of the elements. Instructors must remember to arrange for facilities and equipment.

To offset limitations, consider the following positive advantages to the demonstration method:

- Participants can receive feedback immediately.
- Instructors can readily observe a change in behavior.
- Learners have a higher level of interest when participating.
- Instructors can determine what objectives have been met.
- Carefully supervised skills that participants learn correctly in a safe environment give them the confidence to operate on the job.

The only way to learn, practice, and perform a skill is the safe way. Because of the hazardous nature of fire and emergency services, instructors must stress safety as every step of each skill is demonstrated. Many learners want to be able to perform a skill quickly when they are first learning. Skill and speed only come with practice, and trying to perform new tasks without carefully learning the steps or developing coordination risks safety.

Demonstrating a task may appear simple, but instructors must follow a few guidelines in order to use this method effectively. The following lists are separated into two critical areas: preparing for a demonstration and demonstrating the skill.

1. Preparing for a demonstration

✔ Know clearly what is to be demonstrated and its objective.

✔ Be proficient in every step of the demonstration by practicing ahead of time with all instructors who will be involved.

✔ Acquire all equipment and accessories, ensure they work, and arrange them for use.

✔ Arrange the room so that all participants can see the demonstration.

2. Demonstrating the skill

- ✔ Begin the demonstration by linking new information with the learners' current knowledge.
- ✔ Explain what the demonstration will show the group how to do.
- ✔ Demonstrate the skill once at normal speed.
- ✔ Repeat the demonstration step-by-step while explaining each step slowly.
- ✔ Repeat the demonstration again while a class member or the group explains each step.
- ✔ Ask for a volunteer to demonstrate the skill while explaining the steps. The instructor reassures the individual that he or she will be coached and guided as necessary through the process. The instructor tactfully offers suggestions or corrections during the demonstration.
- ✔ Provide the opportunity for participants to practice, and allow them to supervise and correct each other as they become skilled.
- ✔ Reassemble the group and demonstrate the skill one more time at normal speed and/or one more time slowly as the group explains the steps as a summary. Relate the skill(s) to the objective(s) and performance on the job.

Team Teaching

Team teaching is an arrangement in which a group of instructors, often from different jurisdictions or organizations, cooperate so that all their classes have contact with more than one of the instructors during a lesson. It is an effective educational tool for instructors and a unifying force among organizations. Usually, the instructor with the most knowledge or experience in a particular topic teaches the cognitive and demonstration portion of the lesson to all the classes and then joins with the other instructors in supervising practice. The purpose of team teaching is to combine the knowledge and expertise of several instructors with the content, requirements, and materials of the lesson to meet the program objectives.

Team teaching has many advantages. It is an instructional delivery method that works well when the topic is broad. Team teaching affects several jurisdictions differently and requires varied approaches or perspectives. By using different instructors with expertise in the different topics, all approaches can be covered. Different instructors can keep the attention of the group and maintain participants' interest with varying methods and teaching styles and by each individual's voice, pace, and personality (Figure 7.18).

In order to be successful at team teaching, instructors find that the process requires more advanced planning than solo teaching. To prepare for team teaching, consider the following suggestions:

- ***Choose instructors whose teaching styles contrast yet balance one another*** — For example, match a high-energy instructor with one who has a more relaxed teaching style. This provides a change of pace and enhances anticipation for participants.
- ***Agree with the chosen instructors on who will teach what topics*** — This suggestion allows each

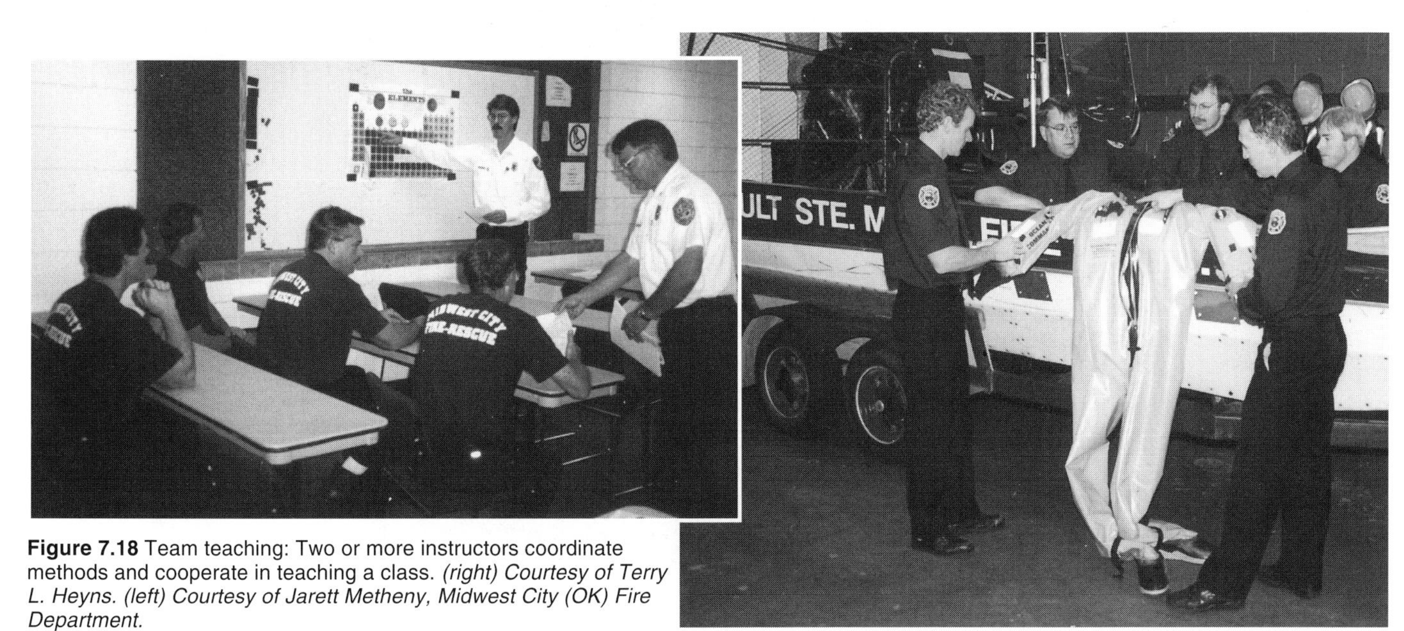

Figure 7.18 Team teaching: Two or more instructors coordinate methods and cooperate in teaching a class. *(right) Courtesy of Terry L. Heyns. (left) Courtesy of Jarett Metheny, Midwest City (OK) Fire Department.*

instructor to prepare to present his or her own topic ahead of time.

- ***Share and review instructor outlines together to know what each instructor is saying and doing*** — This suggestion avoids disagreement and questionable credibility during the lesson in front of participants.
- ***Agree on and use the same format for audiovisual materials*** — For example, agree that all transparencies and slides will be horizontal or that handouts will be computer-generated versus handprinted and drawn. This suggestion provides the lesson with continuity and an appearance that looks planned rather than thrown together.
- ***Designate one lead instructor per topic*** — This suggestion avoids stress and tension between instructors by knowing who is in charge at what point.
- ***Determine time commitments and require each instructor to adhere to them*** — This suggestion enforces the importance of starting and finishing on time so that all instructors have the opportunity to teach their topics and meet their objectives. In addition, review procedures for practicing simple courtesies such as not interrupting each other or devising a signal when someone needs to insert an important point or safety procedure.
- ***Agree that all instructors will be present for the entire program, not just when it is their turn to teach*** — This suggestion accomplishes several things. Instructors can help with the following tasks:
  - Setting up equipment
  - Distributing handouts
  - Noting gaps in information
  - Attending to class or equipment problems
  - Adding different perspectives in response to class questions
  - Assisting with feedback during breaks or after class

Mentoring

A *mentor* is a trusted and friendly adviser or guide, especially for someone who is new to a particular role or job. To be a mentor implies that a special bond exists between an individual and the mentor. The mentor becomes a source of advice and counsel for an individual through the course of a training program or orientation to a job. In addition, individuals may also seek guidance from mentors when they make a transition into new roles or jobs (Figure 7.19).

Mentors set tasks for individuals to accomplish that help them learn and become familiar with the aspects of their jobs. As individuals perform the tasks, they can call on their mentors with questions and for guidance and advice to ensure that they are proceeding correctly. The purposes of these tasks are to give individuals ways to gain knowledge through experience and to relate the meaning of the tasks to the jobs through reflection and analyzing results or consequences. To meet these purposes, mentors must have clear ideas of what tasks to assign that will assist individuals in moving ahead in their training. Tasks should not be assigned as "busy work" but should stimulate thinking and planning in order to accomplish a goal. Mentors provide specific and positive feedback during every meeting with the individuals and to every question from them.

Mentoring is a special kind of instructional method. Mentors provide a personal connection in a learning or new-job situation, which can often be impersonal and threatening. Mentors may be the course instructors, or they may be other instructors or an experienced person on the job. If an instructor or a training program offers mentors, they must be available to all participants and not just selected individuals. Offering mentoring to all avoids the possibility of discrimination when all participants have the same opportunities to learn and benefit equally.

Learners are not the only individuals who benefit from mentoring. It is a method that also benefits new instructors. There are many questions and uncertainties that new instructors face, even though they

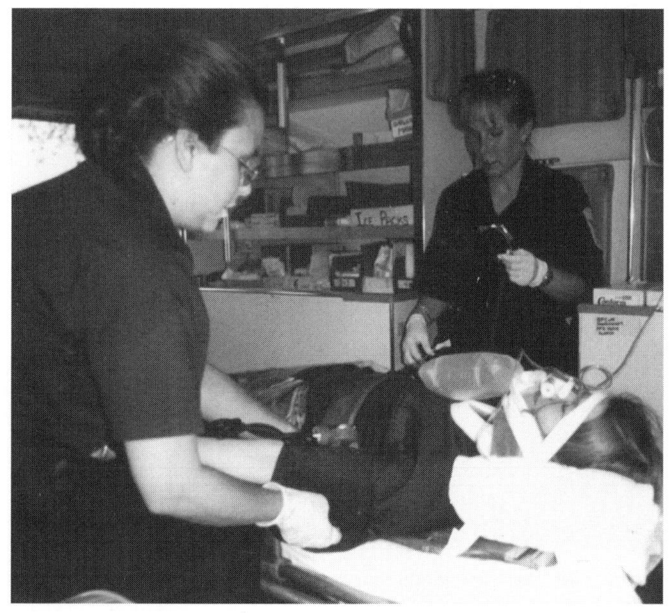

Figure 7.19 A mentor is a trusted advisor for an individual in training. *Courtesy of Dan Gross, Maryland Fire and Rescue Institute.*

have just completed a thorough and comprehensive instructor-training program. Every organization and every job has its unique requirements, rules, procedures, and paperwork. Many organizations assign a mentor (though they may not call the person that) to new instructors who go through an orientation period under that person's guidance and supervision. Many organizations have policies that require new instructors to do their first teaching assignments with a seasoned instructor. This process "walks" them through all the procedures of teaching a program and completing the paperwork from start to finish. If the employing organization does not routinely assign a mentor, new instructors should find experienced instructors who can give them the wisdom that comes from experience, provide them with the rules and procedures of the organization, and tell them the "quirks" of the program. New instructors must find out these things from somewhere or someone; they should not be afraid to ask for advice from or share good and bad experiences with other instructors.

Alternative Methods of Teaching

Though there are many variations to delivering instruction, the typical and traditional methods are performed with groups in classroom settings. There are several more instructional methods that take learners out of classrooms yet provide them with unique formats for accomplishing program objectives. The following sections describe these methods, and there are potential variations for all of them.

Consider that an alternative method of instruction must have some system of credibility and accreditation in order to allow participants to meet standards and certification or licensing requirements. Having *accreditation* means that a training program has received the recognition and approval of academic standards of an educational institution that is an external and impartial body of high public esteem. Training organizations do not allow individuals to participate in an alternative type of instructional methods without first ensuring that the method is designed to accomplish the lesson objectives and meet required qualifications and standards.

Jurisdictions often take the time and effort necessary to meet accreditation requirements for several important reasons:

- Ensures professionalism among personnel and departments
- Responds to insurance coverage standards or qualifications
- Allows for transfer of certifications and credits to advanced programs or to other jurisdictions (reciprocity)

Individualized Instruction

Recall that learners have a variety of learning styles. For very real educational reasons, some of those styles may not enable a learner to progress with the group. An alternative teaching method may be *individualized instruction:* the process of matching instructional methods with learning objectives and individual learning styles that enable a learner to achieve lesson objectives (Figure 7.20). Individualized instruction offers a major advantage to learners: It provides a successful learning format for individuals who could not normally succeed in the typical learning environment but could be successful given time and opportunity to learn.

Individualized instruction is a tool for the instructor to use in combination with other traditional methods, but it is not a substitute for the instructor. Individualized instruction does not imply *solitary* learning. These individuals need to work with others to meet certain requirements and share experiences. They also work under the supervision of an instructor or mentor who ensures that they meet lesson or chosen individual objectives.

Variations of individualized instruction can match instructional methods with learning objectives. For example, the instructor prescribes the objectives, but the learner selects personal methods or techniques of accomplishing them. In another example, the instructor chooses fixed methods and techniques for meeting the objectives, and the methods and techniques chosen by the learner are optional.

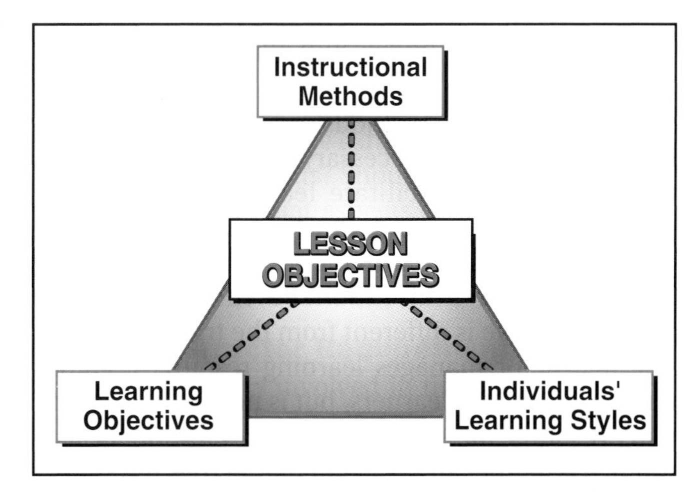

Figure 7.20 Individualized instruction: Instructors work with individuals to match instructional methods, learning objectives, and learning styles to achieve the objectives.

Figure 7.22 An example of high-hazard training is rappelling. Instructors must ensure all safety precautions are in place.

supporting instructor must be qualified to teach and supervise participants in high-risk training exercises. For each evolution, the lead instructor gives all instructors clearly defined roles and responsibilities, and they must rigidly follow standards and safety guidelines. For example, NFPA 1403 sets the learner-teacher ratio at five to one in all live-fire scenarios. This standard also specifies an instructor in charge and a safety officer. The general requirements of this particular standard (small learner-teacher ratios, a safety officer, and an instructor in charge) can be applied to many other similar types of high-hazard training evolutions such as using power tools or practicing confined-space evolutions.

All supporting instructors must be briefed on the instructional plan well enough ahead of each evolution to make their own individual training plans, acquire all safety equipment and have it in place, and inspect the environment for any safety hazards. Appropriate preparation requires that instructors preplan the evolution and discuss what their roles and responsibilities encompass. It is equally important for instructors to know what they are *not* supposed to do. Improvising can be dangerous in high-hazard training.

References and Supplemental Readings

Arends, Richard. *Learning to Teach,* 3rd ed. New York: McGraw-Hill, Inc., 1994, p. 268.

Badt, Milt. *How to Handle the Classroom Sharpshooter,* comp. by David Zielinski, *Adult Learning in the Classroom,* 3rd ed. Minneapolis, MN: Lakewood Books, 1996, p. 30.

Galbraith, Michael W. (ed.). *Adult Learning Methods.* Malabar, FL: Krieger Publishing Company, 1990, pp. 107, 111-112, 176.

Good, Thomas L. and Jere L. Brophy. *Educational Psychology: A Realistic Approach.* New York: Holt, Reinhart & Winston, 1977, pp. 71-75.

Mager, Robert F. *Measuring Instructional Results.* Belmont, CA: Lake Publishing Company, 1984, Chapter 5.

Mager, Robert F. *Making Instruction Work.* Belmont, CA: Lake Publishing Company, 1988, Chapter 16.

Merriam, Sharan B. and Rosemary S. Caffarella. *Learning in Adulthood.* San Francisco: Jossey-Bass Publishers, 1991, p. 124.

Wlodkowski, Raymond J. *Enhancing Adult Motivation to Learn.* San Francisco: Jossey-Bass Publishers, 1993, Chapters 4 and 8.

Job Performance Requirements

This chapter provides information that addresses the following job performance requirements of NFPA 1041, *Standard for Fire Service Instructor Professional Qualifications* (1996 edition). Colored portions of the standard are specifically covered in this chapter.

Chapter 2 Instructor I

2-3.2 Review instructional materials, given the materials for a specific topic, target audience and learning environment, so that elements of the lesson plan, learning environment, and resources that need adaptation are identified.

2-3.2.1 *Prerequisite Knowledge:* Recognition of student limitations, methods of instruction, types of resource materials; organizing the learning environment; policies and procedures.

2-3.2.2 *Prerequisite Skills:* Analysis of resources, facilities, and materials.

2-3.3 Adapt a prepared lesson plan, given course materials and an assignment, so that the needs of the student and the objectives of the lesson plan are achieved.

2-3.3.1 *Prerequisite Knowledge:* Elements of a lesson plan, selection of instructional aids and methods, organization of learning environment.

2-3.3.2 *Prerequisite Skills:* Instructor preparation and organizational skills.

2-4.2 Organize the classroom, laboratory or outdoor learning environment, given a facility and an assignment, so that lighting, distractions, climate control or weather, noise control, seating, audiovisual equipment, teaching aids, and safety, are considered.

2-4.2.1 *Prerequisite Knowledge:* Classroom management and safety, advantages and limitations of audiovisual equipment and teaching aids, classroom arrangement, and methods and techniques of instruction.

2-4.2.2 *Prerequisite Skills:* Use of instructional media and materials.

2-4.3 Present prepared lessons, given a prepared lesson plan that specifies the presentation method(s), so that the method(s) indicated in the plan are used and the stated objectives or learning outcomes are achieved.

2-4.3.1 *Prerequisite Knowledge:* The laws and principles of learning, teaching methods and techniques, lesson plan components and elements of the communication process, and lesson plan terminology and definitions.

2-4.3.2 *Prerequisite Skills:* Oral communication techniques, teaching methods and techniques, utilization of lesson plans in the instructional setting.

2-4.4 Adjust presentation, given a lesson plan and changing circumstances in the class environment, so that class continuity and the objectives or learning outcomes are achieved.

2-4.4.1 *Prerequisite Knowledge:* Methods of dealing with changing circumstances.

2-4.5 Adjust to differences in learning styles, abilities and behaviors, given the instructional environment, so that lesson objectives are accomplished, disruptive behavior is addressed, and a safe learning environment is maintained.

2-4.5.1 *Prerequisite Knowledge:* Motivation techniques, learning styles, types of learning disabilities and methods for dealing with them, methods of dealing with disruptive and unsafe behavior.

2-4.5.2 *Prerequisite Skills:* Basic coaching and motivational techniques, adaptation of lesson plans or materials to specific instructional situations.

Chapter 3 Instructor II

3-2.6 Evaluate instructors, given an evaluation form, department policy, and job performance requirements, so that the evaluation identifies areas of strengths and weaknesses, recommends changes in instructional style and communication methods, and provides opportunity for instructor feedback to the evaluator.

3-2.6.1 *Prerequisite Knowledge:* Personnel evaluation methods, supervision techniques, department policy, effective instructional methods and techniques.

3-3.2 Create a lesson plan, given a topic, audience characteristics, and a standard lesson plan format, so that the job performance requirements for the topic are achieved, and the plan includes learning objectives, a lesson outline, course materials, instructional aids, and an evaluation plan.

3-3.2.2 *Prerequisite Skills:* Basic research, using job performance requirements to develop behavioral objectives, student needs assessment, development of instructional media, outlining techniques, evaluation techniques, and resource needs analysis.

3-4.2 Conduct a class using a lesson plan that the instructor has prepared and that involves the utilization of multiple teaching methods and techniques, given a topic and a target audience, so that the lesson objectives are achieved.

3-4.2.1 *Prerequisite Knowledge:* Use and limitations of teaching methods and techniques.

3-4.2.2 *Prerequisite Skills:* Transition between different teaching methods, conference, and discussion leadership.

3-4.3 Supervise other instructors and students during high hazard training, given a training scenario with increased hazard exposure, so that applicable safety standards and practices are followed, and instructional goals are met.

3-4.3.1 *Prerequisite Knowledge:* Safety rules, regulations and practices, the incident command system used by the agency, and leadership techniques.

3-4.3.2 *Prerequisite Skills:* ICS implementation.

Chapter 8

Practical Training Evolutions

Practical training evolutions and instructions are key elements in any fire and emergency services organization's training program. *Hands-on* training is required by applicable NFPA standards, the Occupational Safety and Health Administration (OSHA), professional associations, and local jurisdictions. The need for job-performance objectives (such as those outlined in NFPA 1001, *Standard for Fire Fighter Professional Qualifications*), behavioral requirements, and measurable skill levels needed by fire and emergency service responders is clear. The fire and emergency services instructor plays a key role in formulating, supervising, and conducting practical training evolutions. The instructor must provide such training in a controlled environment where learning a new skill or improving a skill is possible for those who are participating. Practical instruction, hands-on training, and various evolutions in which fire and emergency response personnel participate provide the opportunity to reinforce what has been learned in the classroom and to put that knowledge into practice. For experienced firefighters, for example, live fire and other evolutions give opportunities to develop additional skills and increase skill performance levels. In addition, realistic practical training evolutions promote enthusiasm, morale, and *esprit de corps* among fire and emergency service responders.

Many different kinds of practical evolutions are important to individual fire and emergency service responders and their respective departments or organizations. The responsibilities of fire service, law enforcement, emergency medical service, and emergency planning and preparedness organizations have expanded in recent years to include many areas of response activities, including the following:

- Technical rescue
  - Belowground and below-grade rescues
  - High-angle rescue
  - Rope rescue
  - Surface-water rescue
  - Vehicle and machinery rescues
  - Confined-space rescue
  - Structural collapse rescue
  - Trench rescue
- Tactical emergency medical training
- Counterterrorism operations

All these organizations and services have to be integrated into a total response effort. In addition, other personnel who may not be directly connected with fire and emergency services are sometimes present and working at a scene. These individuals also have to be integrated into a practice evolution.

Practical training evolutions, for example, can either involve one firefighter or an entire battalion of firefighters. However, many activities are "joint" in that other emergency response or support organizations play key roles in the exercises. In small fire and emergency service organizations, the entire membership may be involved. But, no matter how many are included, an instructor has to prepare well ahead of time. This means following a process that is very similar to that used in preparing for classroom instruction. Many of the considerations discussed in Chapter 5, Planning Instruction, hold true for practical evolutions. Specific goals and learning objectives must be established for each exercise. Once these goals and learning objectives are clear, the instructor must communicate them to each learner. All participants must know what is expected of them, otherwise they will not learn and benefit from the experience. For more complicated evolutions, each participant may have a different set of behaviors to learn. The instructor's task is to ensure that all individual learning objectives and job performance requirements are successfully accomplished in order to attain the overall exercise objective.

This chapter helps an instructor prepare for a practical evolution. Preparation must include an adequate lesson plan with appropriate learning objectives. Advanced planning ensures the correct scheduling of equipment and personnel. Safety is a consideration throughout the evolution. But safety concerns are best addressed during the initial planning stage because at this point, safety considerations can be built into the practical evolution from beginning to end. The Incident Management System (IMS) applies to a training evolution just as it applies in actual field operations. If training evolutions include live fire, for example, instructors must ensure that an adequate number of safety officers are present. The use of acquired structures (donated buildings), Class B fires, and wildland fire training evolutions deserve special consideration. Finally instructors must not forget safety when training with power tools and in specialized rescue evolutions. References and supplemental readings and job performance requirements are given at the end of the chapter. A suggested way to start preparing for a practical evolution is to examine the training setting.

Figure 8.1 Training facilities cover a broad range of types and design. Permanent structures vary in cost. Facilities range from the more modest structure shown to very elaborate simulators with computerized instrumentation and controls. *Courtesy of Terry L. Heyns.*

Training Settings

[NFPA 1041: 3-4.3.1]

A wide range of facilities, resources, and circumstances are involved in and available for practical evolutions (Figure 8.1). Large fire and emergency service organizations with adequate budgets often have permanent facilities equipped with computer-aided training props, burn simulators, buildings, and laboratories. The advantage of these facilities is that instructors can exercise considerable control and thus reduce the possibility of surprises and unpredictable events. Many of these newer facilities and structures are designed with features that enhance instructional possibilities with built-in safety devices. Instruction and practical evolutions take place on a regular basis. As long as these permanent facilities are used for their intended design purposes, there are maximum possibilities for learning in safe environments. While not all organizations can afford such facilities, instructors still have responsibilities to conduct safe, practical evolutions.

Fire and emergency service organizations with limited budgets and resources face a different challenge when it comes to practical evolutions. For these organizations, inherited facilities, acquired structures that are scheduled for demolition, and limited equipment are the norms. Such facilities may be the only alternatives available if practical evolutions are going to take place. Instructors face special challenges in working with settings that have not been designed for practical training evolutions.

While climate conditions affect every training evolution, the impact on temporary facilities can be extreme. Wind speed and direction, humidity, temperature, and time of day all affect the training environment. Weather extremes are a given in many locations, whether it is harsh below-freezing cold or debilitating high temperatures with high humidity. Hazards such as lightning, hail, and high winds may also present additional dangers when severe storms approach. Noise made by equipment, light levels, and even the types of vegetation at or near the training ground are important factors that can affect the outcome of a training exercise. If all these factors combine to result in environments that are detrimental to learning or are inherently unsafe, instructors may have to limit the training exercises or not allow such exercises as a live burn for example. Canceling an exercise due to weather may be more difficult than it first appears, especially if the evolution has been scheduled for some time, involves people who have traveled long distances, includes a great number of fire and emergency service responders, or is a requirement imposed by a funding source. Instructors must have alternative plans for postponing large-scale training operations, including a system for notifying all participating personnel and organizations.

Live fire exercises that are conducted in temporary settings such as an acquired structure pose special problems for instructors. Acquired structures are usually scheduled for demolition. Sometimes organizations are invited to burn these structures because this may be the least expensive way for the owner to eliminate a building that has outlived its usefulness. The training officer must first determine whether it is even possible to use the structure for live fire training. Environmental laws, for example, may prohibit the burning of a structure due to its location. The structure may also have been designated as a historical landmark. If live fire training is not possible, perhaps the structure might be suitable for forcible entry training, ladder evolutions, search and rescue operations, or a class on building construction. In any event, the building must undergo a complete and thorough inspection. See Appendix G, Structural Live Fire Training Forms, for sample structural inspection checklists.

Considerable preparation is sometimes necessary to ensure that these structures provide safe and effective settings to accomplish learning objectives. An inspection will reveal the extent of preparation necessary. Preparations might include the following items:

- Removing asphalt shingles or vinyl siding
- Clearing vegetation
- Shoring parts of the structure
- Arranging for asbestos abatement
- Installing railings and safety dividers
- Reinforcing stairs
- Removing excess furniture, debris, and trash

The local building and land-use organization may require that the building owner use certified personnel to remove asbestos or other hazardous materials. Instructors should learn local regulations. This ensures compliance with the law and may relieve the instructor from the direct responsibility of removing asbestos or hazardous materials.

Certainly, some structures have to be rejected as unsuitable and unsafe for use in any type of training evolution. Other structures might be suitable, but only after considerable work is completed such as clearing vegetation, cleaning interior and exterior areas in and around the structure, and making modifications.

When using an acquired structure, instructors conducting live fire or other training evolutions must consider the impact on the surrounding neighborhood. Refer to Appendix G, Structural Live Fire Training Forms, for sample forms that may help instructors prepare letters and public announcements. Sample checklists for live fire training situations are also given. The instructor's responsibilities include the following:

- Distribute a notice (letter or brochure) to each resident, informing them of the date and time of the event, a description of the training activity, and its impact on the surrounding area such as street closures.
- Plan carefully the placement of hoselines and apparatus, and consider how they restrict access.
- Notify the water department if hydrants are involved.
- Flush water mains so that rust and sediment do not cause problems for pumping operations or for surrounding households if an acquired structure is located in an area where there has been little flow recently through the water distribution system.
- Prepare water supply and flow analyses. Training officers must know the required fire flow, including safety margins.
- Videotape or photograph surrounding structures, vehicles, and grounds. If neighboring property is damaged by the evolution, these videotapes and photographs help document the conditions present before the training activity began. Videotape documentation may be important if legal claims arise later.

NFPA 1231, *Standard on Water Supplies for Suburban and Rural Fire Fighting,* gives important guidance for water supply and fire fighting purposes in rural and suburban areas. These requirements also apply to fire attack during training evolutions. Many acquired structures are in locations that are not convenient to hydrants or other water sources. While these locations provide an excellent opportunity for training evolutions on water shuttle, portable dump tank, and relay operations, such training should be completed before a live fire evolution starts. There must be reliable water sources for the entire duration of any live fire evolution. In this sense, water supply operations during a live fire exercise become the "real thing."

Once the setting has been determined and inspections completed, an instructor then considers the type of training best suited for that set of circumstances. There will always be a tendency to overreach and to set goals and learning objectives that cannot be achieved given the limits imposed by the facility. The number of instructors available, the total number of learners, safety considerations, equipment available, and type of training itself establish limits. Therefore, advanced planning is critical to the success of any practical training evolution.

Planning Practical Evolutions

[NFPA 1041: 3-4.3]

Just as in classroom instruction, a lesson plan is important for a practical evolution (see a sample in Appendix G, Structural Live Fire Training Forms). The activity or series of activities is aimed at accomplishing the learning objectives for that practice evolution. The performance involved and expected should be clear to each participant. In addition, individuals must understand how their own learning objectives or skill performances tie into the overall goal for the exercise. Much of what is involved in planning practical evolutions has been discussed in Chapter 5, Planning Instruction. However, keep in mind that practical evolutions are designed to teach and develop psychomotor skills. At the basic level, the series of behaviors designed for beginning fire and emergency service responders is outlined in the materials contained in other IFSTA manuals. Successfully reaching the highly proficient or advanced level of learning requires significant time and energy on the part of both learner and instructor.

Practical evolutions for experienced fire and emergency service responders fall into a category of training that requires highly proficient performance that is usually acquired through advanced-level training and extensive on-the-job experience. This type of evolution is a challenge for an instructor to plan and provide. The role of the instructor moves away from that of a *teacher* and more to a role that is best described as a *monitor* or *facilitator*. In this context, the instructor arranges a training evolution so that fire and emergency service responders may develop their own abilities, test new ideas, reinforce or develop new skills, and avoid behaviors that are best eliminated (for example, bad habits and questionable shortcuts). Experienced fire and emergency service responders respond best to training evolutions that help them overcome problems they have encountered in the "real world." Their approaches are different from those of new recruits who initially concentrate on mastering individual skills and then learn how to blend these skills into coordinated responses with others. Experienced fire and emergency service responders are generally task-oriented, enjoy working as a team, and are cooperative learners. They enjoy working as a group and accomplishing an assigned task with others. This orientation has important implications for planning practical evolutions.

Consider each of the following factors when planning a practical evolution and when establishing the desired learning outcomes:

- Give each participant the opportunity to have an input and to influence the outcome. This means each participant must understand his or her role and what must be accomplished.
- Do not assign too many participants to specific tasks. Keep all participants busy, and eliminate or greatly reduce "stand-around time."
- Maintain a suitable instructor-to-participant ratio. The lead instructor needs help when it comes to practical evolutions. The exact instructor-learner ratio varies with the nature of the evolution. Assign a safety officer to monitor ongoing activities.
- Design the practical evolution so that a positive outcome is possible. Assigning a task that is very difficult or impossible to accomplish provides a limited learning experience.
- Provide a thorough evaluation and critique of performance after the exercise is complete. Point out positive behaviors as well as those that could be improved. There are many ways to accomplish a task, and there may be several different ways to handle a situation. Many times experienced firefighters, police officers, rescue personnel, and emergency medical service responders will suggest alternate techniques based on their experiences. Instructors may integrate these alternative suggestions into the discussion if they are appropriate to the topics or local protocols and procedures.
- Provide a summary of what has been learned and what can be carried into the operational environment and actual emergency setting. See Appendix H, Incident Action Plan, for an example.

Another important aspect of planning has to do with the coordination and scheduling of personnel, equipment, facilities, or special needs with other organizations. Many actual incidents now involve multiple-organization responses. There is a need to include these organizations in practical evolutions or disaster exercises.

Safety is another issue in planning. Just as there should be no "freelancing" at an actual emergency, there also should be none on the training ground. Therefore, participants must have clear ideas of their roles and understand how their behaviors fit into the entire exercise.

Safety Issues

[NFPA 1041: 3-4.3.1]

In achieving the exercise objectives, safety is considered in both the planning and execution stages. There are certainly dangers to all participants in live fire and other training evolutions. Indeed, practical evolutions are best accomplished using a safety officer and the Incident Management System discussed in NFPA 1521, *Standard for Fire Department Safety Officer.* Another excellent reference source is NFPA 1500, *Standard on Fire Department Occupational Safety and Health Program.*

Safety officers play important roles in practical exercises, fire stream evolutions, vehicle extrication training, high-angle rescue training, belowground or below-grade operations training, and other areas of technical rescue. The lead instructor is concerned with the learner ratio, which varies with the number of participants, but cannot forget that there is also a need for an adequate "safety sector" that can observe all participants through the entire course of the evolution. With a small number of learners, the instructor can serve as the safety officer, just as a company officer serves as the safety officer at small incidents. However, with increasing numbers of participants and a more complex set of performance/learning objectives that need attaining, the instructor must appoint a separate safety officer.

As stated earlier, safety starts with a complete survey of the setting, especially in acquired structures. However, a complete review of fixed facilities is also necessary. Complete an inspection of all equipment permanently installed or brought to the site of the exercise. A problem with equipment might not only interfere with the evolution, but also has the potential to inflict injury (Figure 8.2). Inspections of personal equipment (protective clothing), personal alert safety system (PASS) devices, self-contained breathing apparatus (SCBA), and other items required by a participant's organization are always incorporated into the lesson plan. One of the behavioral objectives can be the proper use of safety equipment.

One goal is to provide an ingrained set of behaviors that emphasizes safety (Figure 8.3). Understanding the use and care of required personal protective

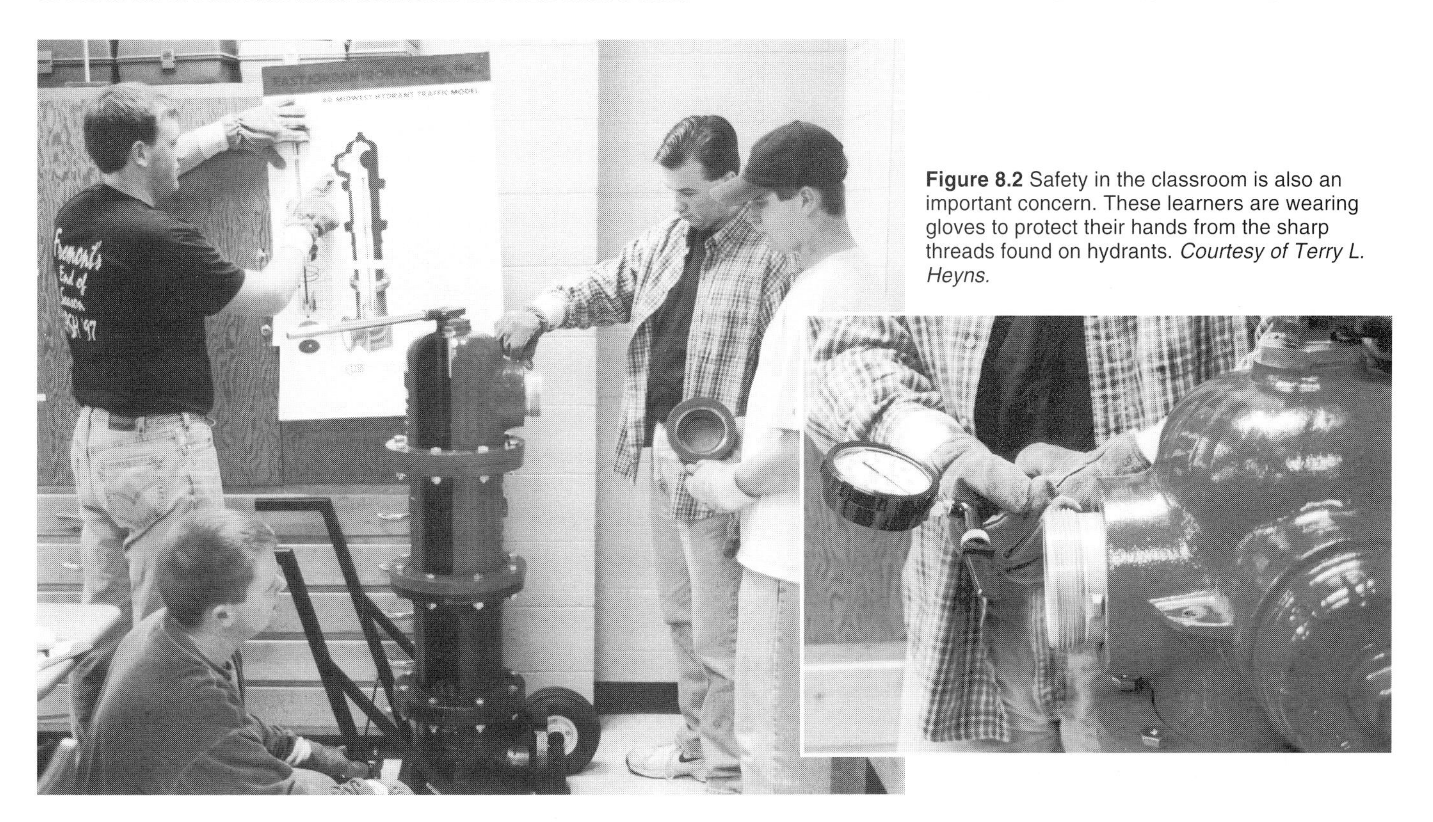

Figure 8.2 Safety in the classroom is also an important concern. These learners are wearing gloves to protect their hands from the sharp threads found on hydrants. *Courtesy of Terry L. Heyns.*

equipment (PPE) is a good place to start in establishing a safety behavior pattern. Certainly an important item for personal equipment inspection is a tag, fastener shield, or other identifying device that is used for accountability (Figure 8.4). These issues and situations lead us to learning and understanding a system that must be employed for safety: incident management.

Incident Management Systems

[NFPA 1041: 3-4.3.2]

Any practical exercise that involves multiple teams of fire or emergency service responders automatically demands the use of an incident management system. With such a system, instructors, participants, and observers are accounted for and kept under control. Exercises that are more complex, involve more participants, or may involve dangers inherent in the exercise itself demand incident management. There are several systems that address the key issues of safety and accountability that are appropriate to use on the training ground.

The Incident Management System is one example that might be adopted and used for practical training exercises. NFPA 1561, *Standard on Fire Department Incident Management System,* also provides guidance and direction for safety officers, supervisors, and lead instructors. A safety officer, staging officer, and functional line groups are assigned by task. Each of these groups might be headed by an assistant instructor. Staff functions might include a communications sector and a logistics sector. The logistics sector could handle refilling air cylinders, providing a rehabilitation (rehab) area, and maintaining a classroom or assembly area where participants could come together to receive instructions, directions, briefings, or after-exercise critiques. Other sectors or divisions could be created as required by the lesson plan for that exercise.

The use of an incident management system during a practical training exercise also has an additional benefit of acquainting participants with the operation of the system. Participants can take this training experience and apply what they have learned at the scene of an actual emergency. Instructors may wish to adapt a variant of an incident management system to fit the requirements of the particular exercise. In the case of any evolution that holds the potential for injuries, the use of an incident management system is an absolute necessity. Useful references are any books in the *Model Procedures Guide* series developed by the National Fire Service Incident Management Consortium and published by Fire Protection Publications.

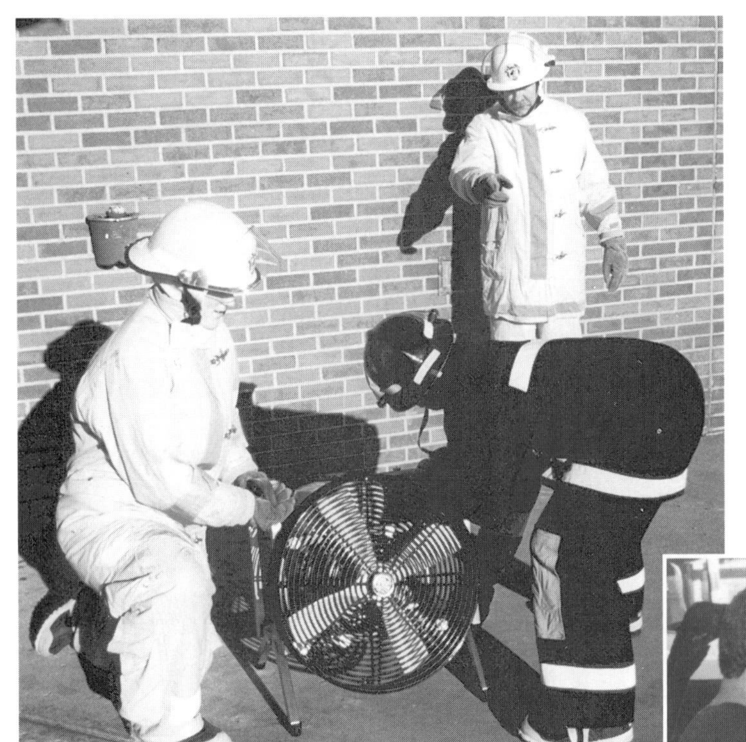

Figure 8.3 Proper lifting techniques prevent injuries. Instructors must immediately *stop* any unsafe action and correct any situation that might cause injury. *Courtesy of Terry L. Heyns.*

Figure 8.4 An accountability system aids in ensuring the safety and survival of personnel on a working incident. Accountability helps prevent freelancing, enables the incident commander to know the location of personnel, and locates personnel in the event of an emergency situation that requires a personnel accountability roll call. *Courtesy of Jarett Metheny, Midwest City (OK) Fire Department.*

Live Fire Training Evolutions

[NFPA 1041: 3-4.3]

Live fire training is an important part of both recruit and experienced firefighter training. New skills are learned; previously learned skills are developed, improved, or altered. In many communities with strict code enforcement and modern construction, the number of structure fires is being reduced significantly. With the expanding role of firefighters in the emergency medical system, actual fire experience may be only a small part of their responsibilities. Thus, experience in actual fire conditions is accomplished primarily on the training ground. But even while training, the potential for an actual emergency is real. Indeed, NFPA 1403, *Standard on Live Fire Training Evolutions,* equates live fire training with operations conducted at the scene of actual fires. NFPA 1403 gives information for establishing procedures for live fire exercises. This standard clearly establishes the requirement that all participants be trained by individuals who are qualified to provide instruction in their particular areas of expertise or in subjects that are covered by the evolution.

While the objectives for the overall exercise are shared, each participant in a live fire evolution also has his or her own specific learning objectives. Pump operators, for example, are concerned with responsibilities designed to provide effective fire streams. Firefighters who are making initial fire attacks are learning effective fire suppression techniques. Fire officers have opportunities to orchestrate coordinated fire attacks. Thus, the outcome of an exercise gives each fire officer, pump operator, or firefighter the opportunity to critique performances and see how accomplishing his or her own lesson objectives contribute to the success or outcome of the overall operation.

Live fire evolutions may include several fire training areas: structure, Class B fuel, and wildland. Exercises for these areas are discussed in the sections that follow.

NFPA 1403, *Standard on Live Fire Training Evolutions*

Chapter 2, Acquired Structures, Section 2-1.2: Prior to being permitted to participate in live fire training evolutions, the student shall have received training to meet the performance objectives for Fire Fighter I of the following sections of NFPA 1001, *Standard for Fire Fighter Professional Qualifications:*

Section 3-3 Safety

Section 3-5 Fire Behavior

Section 3-6 Portable Extinguishers

Section 3-7 Personal Protective Equipment

Section 3-11 Ladders

Section 3-12 Fire Hose, Appliances and Streams

Section 3-16 Overhaul

Section 3-19 Water Supply

Structure Fire Exercises

The increasing use of fire simulators makes live fire exercises safer because conditions are carefully controlled. When allowed by code and other circumstance, the acquisition of old structures scheduled for demolition or removal is another option. However, the use of old structures and houses for live fire training also requires an increased concern for safety. NFPA 1403 provides valuable information for an instructor who is planning to use an acquired structure for live fire training. In areas where live fire training is possible with an older structure, environmental regulations immediately come into play. Limiting the products of combustion and controlling/containing "runoff" are subject to federal and local guidelines. Among the most important of many questions are those involving structural integrity, fire behavior, and water supply, which were discussed in the Training Settings section.

With increasing emphasis on making fire training environmentally cleaner and safer for participants, gas-fueled training systems are becoming more popular. The degree of realism of a gas-fueled simulator has less to do with the level of technology than it does with the ingenuity of design. Fire training organizations should ensure that the system they purchase meets their short-term and long-term training needs by being expandable, flexible and user-friendly. The technology used to monitor safety, particularly in interior simulators, must be reliable and tested regularly. No gas-fueled system is a complete substitute for actual fire fighting.

A dedicated live fire training facility that has built-in safety features, that is governed by standard operating procedures, and that is used on a regular basis provides the best and safest situation in which to conduct live fire training. The issues of water supply and structural integrity are already solved by the facility's design. Fire behavior can be controlled with vents, including automatic safety devices sensitive to excess temperature. Donated structures such as old houses and buildings provide no such safety design features. Therefore, an instructor needs to conduct careful research to determine the suitability of a donated structure for live fire training.

An on-site inspection or survey of the structure and surrounding area is very important (Figure 8.5). This survey may reveal that the structure is unsound, poses environmental difficulties, or has hazardous materials such as asbestos that would pose a danger to participants. In some cases, the condition of an acquired structure might preclude its use in training evolutions. Many of these structures are filled with trash and old furniture. If abandoned for a long time, vegetation may impede access/egress routes and cause visibility problems.

Are there abandoned wells near the structure? The instructor needs to complete a thorough water/hydraulic analysis. This analysis would include calculations of both gallons per minute (gpm) or liters per minute (L/min) needed and available. The analysis also designates a standby water source for use in emergency situations. If tender shuttles and portable dump tank operations are required, all arrangements should be rehearsed well ahead of the live fire exercise. Adequate water supply during the actual live fire is critically important.

Figure 8.5 An inspection of an acquired structure that may be used in training is especially important for the safety of all participants. The structure is inspected from top to bottom with special attention aimed at the roof, basement, chimneys, and structural supports. Any and all characteristics that might cause unsafe conditions are noted. *Courtesy of Terry L. Heyns.*

Once the decision is made that a structure is suitable and useful for live fire training, a great deal of work must be done well ahead of the scheduled exercise. For example, remove brush and grass, finish the water survey and supply plan and rehearse it, and make a structural safety check. Pay particular attention to any alterations, especially those involving ventilation. Many times a house is used for other types of training such as forcible entry, search/rescue, and building construction. Have these previous training sessions affected the safety of the live fire evolution? Once these and other issues are addressed in the survey, the live fire exercise can move to the planning stage.

Class B Exercises

In addition to live fire training in simulated structures or acquired structures, the use of petroleum/hydrocarbon fuels in pits and aircraft fire simulators provides important training opportunities. Because of environmental concerns, the use of liquefied petroleum gas (LPG) is increasing because there are fewer runoff problems. With proper installation of control valves, the fuel can be regulated and turned off if necessary. While there may be less life-threatening danger to participants and less damage to the environment, there is still the danger of burns or possible injury to participants. Therefore, all equipment must be inspected before and after use, participants must be given clear learning objectives, and an incident management system (including a safety officer) must be established.

LPG systems are also used in aircraft fire and rescue training simulators. Like modern live fire buildings, these simulators have built-in safety features and computer-controlled operations. Instructors may simulate a wide variety of circumstances with these simulators. Some systems can bubble up through a water pool to simulate flammable liquid fires.

The use of petroleum/hydrocarbon fuel pit fires raises a number of issues that address both safety and environmental concerns. Once the liquid is burning, there is no way to stop combustion — this situation makes for special safety concerns. An adequate water/foam supply is a must, and experienced firefighters must be available with backup lines. Various pit fire exercises also involve sweeping the fire off the surface of the burning liquid and practicing several different foam application techniques. These evolutions require careful coordination and emphasize close teamwork. Indeed, an appreciation for teamwork is one possible learning objective. However, such exercises also impose special responsibilities for the instructor(s) and safety officer. They must exert positive control over the participants and ensure that there are wide safety margins because it is not possible to regulate the intensity of the fire.

Petroleum/hydrocarbon fuel fires also may not be allowed in certain areas due to environmental regulations. If these fires are allowed, there may be a number of requirements that must be satisfied well ahead of the scheduled exercise. Issues such as air, ground, and water pollution are all involved with this type of fire. Federal and state/provincial requirements are very stringent, and it may not be possible to conduct fire training using petroleum/hydrocarbon products without adequate controls. Instructors must be sure that such fires are legal and that adequate safety considerations are built into the exercises.

Wildland Fire Training

Wildland fire training puts firefighters outdoors where they can construct firelines, conduct controlled burns, and practice backfire operations. These exercises may take place in an open field behind fire stations or in a wilderness area. Certainly, it is important to coordinate such training with representatives of a forest service agency. In addition to weather and climatic concerns, there are other issues that are important to the safety of the exercise. Some personal considerations are as simple as using sunscreen on the face, removing jewelry, wearing proper protective clothing (structural fire fighting gear may not be appropriate), safely using both handheld and power tools, using insect repellent, and maintaining adequate water intake. Other issues are more complex such as contour map interpretation, the interaction of terrain and fire behavior, and the use of navigation aids such as a global positioning system (GPS). As with other exercises, the use of an incident management system is required, especially if the exercise takes place over a large area where a team or crew may become lost or disoriented.

In the case of a controlled burn or backfire, safety considerations become more important. As with live fire structural evolutions, there are potentials for accidents and transitions to actual emergency situations. Sudden wind shifts could cause a fire to spread or change direction. Adequate fire fighting resources must be on hand to deal with such situations. The fact that inexperienced personnel are on the fireline may also contribute to situations that could be more dangerous than otherwise might appear. As stated previously, the importance of an established and effective incident management system and the provision for a safety officer are critical to the successful outcome of these exercises.

Training to control fires in the wildland/urban interface is complex. In addition, responders must receive training in the use of appropriate personal protective clothing for wildland fires because structural protective clothing is not suitable and, in fact, may be dangerous in wildland situations. Training requirements for these areas are described in the sections that follow.

Wildland/Urban Interface

Fire training in the wildland/urban interface (area where structures mix with wildland fuels) is a distinct discipline. Fire and emergency service instructors must familiarize themselves with their state/provincial and the federal office land management agencies that coordinate wildland fire programs. These agencies are a valuable resource for access to the latest information on wildland/urban interface training and education. For example, the National Wildland Fire Coordinating Group (NWCG) based in Boise, Idaho, provides a series of training courses specifically designed for wildland/urban interface environments. The courses are divided into a series of skill and management levels and cover topics from basic wildland fire fighting to advanced wildland fire behavior. Fire and emergency responders should only participate in actual wildland fire responses after completing training in the S-190 "Introduction to Fire Behavior" and the S-130 "Introduction to Wildland Fire Fighting" skill-level sessions of the wildland/urban interface program.

Wildland Personal Protective Clothing/Equipment

Personal protective clothing used for structural fire fighting is generally too bulky, too hot, and too heavy to be practical in wildland fire fighting. NFPA 1977, *Standard on Protective Clothing and Equipment for Wildland Fire Fighting,* provides specifications for wildland fire fighting personal protective clothing (often called *brush gear*) and equipment. Descriptions of wildland personal protective clothing are as follows (Figure 8.6):

Figure 8.6 Wildland firefighters are equipped with personal protective equipment designed especially for wildland fire conditions. *Courtesy of Dana Reed, San Jose (CA) Fire Department.*

- ***Gloves*** — Wildland fire fighting gloves are made of leather or other suitable materials and provide wrist protection. They need to be comfortable and sized correctly to prevent abrasions and blisters.
- ***Brush jackets/pants/one-piece jumpsuits*** — The cuffs of sleeves and pants of protective clothing are designed to close snugly around the wrists and ankles. The fabric is treated cotton or some other inherently flame-resistant material. Underwear of 100 percent cotton, including a long-sleeved T-shirt, is worn under brush gear.
- ***Head/neck protection***—To protect the head, a lightweight hard hat or helmet with chin straps is preferred to a structural helmet. The helmet is equipped with a protective shroud to protect the face and neck. Goggles should have clear lenses.
- ***Footwear***—Acceptable footwear varies in different geographical regions for wildland fire fighting, but some standard guidelines apply in all areas. Lace-up safety boots with lug or grip-tread soles are used most often. Boots must be at least 8 to 10 inches (200 mm to 250 mm) high to protect the lower leg from burns, snakebites, and cuts and abrasions. Socks should be made of a natural fiber.

CAUTION: *Never* wear synthetic materials at a fire. These materials melt when heated and stick to a wearer's skin, which greatly increases the likelihood of major burn injuries.

As part of their protective equipment, wildland firefighters must carry and know how to deploy a fire shelter — a special protective aluminized tent that reflects heat and provides a volume of breathing air in the event of entrapment. Different types of respiratory protection are also available for wildland firefighters. Though not considered safety equipment, containers of drinking water to prevent dehydration are necessary items for wildland firefighters.

Nonstructural and Technical Training Evolutions

[NFPA 1041: 3-4.3, 3-4.3.1]

A number of other situations demand practical training exercises in order to keep skill levels and performance competencies at high levels. While not all could be listed here, examples would include the following:

- Vehicle extrication and agricultural rescue operations
- Swift water rescues
- Trench and building collapse operations
- Hazardous materials operations
- Above/below-grade operations
- Ice rescues
- Power tool operations
- Confined-space entry operations
- Aircraft rescue and fire fighting (ARFF) operations

There is one common element in each of these special situations and that is the need for qualified instructors who are specialists in their particular areas of expertise. In such cases, the duty of the lead instructor is to arrange for suitable training sites with the advice and assistance of these recognized experts. In previous sections of this manual, prior planning has been emphasized. However, in the case of specialized instruction, making prior arrangements is paramount.

The instructor ensures that technical experts are available during the proposed training period. Another important consideration is selecting a time most convenient for the participants who will enroll in the course. Schedule training around seasonal activities, such as harvest and game hunting, and avoid conflict with major holidays. Also important is the setting needed for specialized training. For example, rescue evolutions inside tanks or other confined spaces should *NOT* expose participants to dangers expected in actual emergencies. Tanks used for confined-space operations should have an end or side cut away to facilitate immediate access to a participant who experiences difficulty and to provide for emergency egress. Training in ARFF also requires a particular type of training ground. Some regions may have aircraft mock-ups available.

For transportation emergency exercises, some training organizations have specialized rigs that provide hands-on training. These training rigs can be scheduled in advance and could involve several neighboring organizations. By coordinating their training schedules with other organizations, training officers can help maximize the effectiveness of their individual programs and provide instruction that a single organization might not be able to obtain on its own.

Each evolution, however, involves its own set of safety requirements. These requirements must be clearly established and satisfied before actual training begins. The instructor or group of instructors who provides this specialized training is also responsible for coordinating all safety measures with the lead instructor in a cooperative effort (Figure 8.7). Additional federal or state/provincial regulations may apply for this specialized training. For example, a permit required for confined-space operations can also be used in training situations. This permit provides a checklist that helps to ensure safety.

Power Tools/Equipment Training

[NFPA 1041: 3-4.3, 3-4.3.1]

Fire and emergency service responders must also know how to maintain and operate power tools/equipment such as fans, saws, hydraulic rescue devices, generators, monitoring instruments, medical devices, and other equipment associated with their duties. The product supplier may provide necessary instruction, but the training officer should review the supplier's material prior to the presentation to ensure that it meets organizational policies and procedures. For some equipment, such as infrared sensing devices, the manufacturer may require training from an approved instructor. In most cases, the use and maintenance of a tool are regulated by the manufacturer's instructions and guidelines. Firefighters, for example, understand this situation, because it is similar to the training they receive in the use of SCBA, PASS devices, and other PPE. All fire and emergency service responders should understand the safety rules involved in the use of specialized equipment. If these rules are not followed or the device is used incorrectly to accomplish a task for which it was not designed, manufacturer warranty may be voided (Figure 8.8).

Figure 8.7 Instructors need to inform all participants about the layout of the area in which a training evolution is held. This includes briefing with a structural diagram followed by a tour of the facility. Participants must know where exits are located. *Courtesy of Terry L. Heyns.*

Figure 8.8 When using equipment, follow safety rules and use the device according to the manufacturer's guidelines. *Courtesy of Bill Partridge, Utah Fire and Rescue Academy.*

General Considerations for Practical Evolutions

The fire and emergency services instructor has to resolve a number of issues that involve all practical evolutions. Some general considerations are as follows:

- Determine that each participant is in good health. A medical examination should be required before a person can participate in strenuous training evolutions.
- Ascertain that each participant is in good physical condition. A participant who is "out of shape" may not be able to finish the training due to exhaustion. Someone who becomes tired runs a greater chance of incurring injury or of inflicting injury on another participant.
- Monitor participants carefully during strenuous practical training evolutions. An exercise should not be so demanding that learning is difficult or impossible. Even if the physical demand is carefully measured, the effect of extreme heat or cold on participants must be considered.
- Do not allow the length of an exercise to interfere with participants' abilities to learn and succeed at their tasks. Monitor intake of food and fluids. Provide rehabilitation and staging areas.
- Use an incident management system to control the movement of personnel. The system limits the number of trainees in certain areas that pose special dangers such as the interior of structures used for live fire training.
- Make sure that those who engage in the training have the necessary skill levels and knowledge needed for that particular exercise. A participant who is not familiar with the specialized knots needed for high-angle rescue, for example, must acquire additional training before participating in such a class. In other words, prerequisites may have to be satisfied before learners are allowed to participate in an exercise. A complete review of each individual's training record reveals those who are qualified to participate in advanced exercises. This review encompasses job performance requirements and prerequisite skills and knowledge.

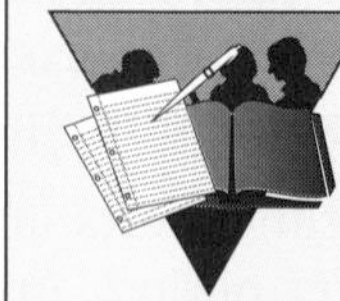

Case Study: Milford, Michigan, Incident: Deaths in Acquired Structure Live Burn Training

Introduction

What follows is a summary of the Milford fire-training incident in 1987. NFPA 1403, a standard that specifies procedures for live burns in acquired structures, was adopted in 1986. A worthwhile exercise for instructors who are working with live fire exercises is to obtain the original Milford report and read it while referring to appropriate NFPA standards.

Goals of the Exercise

The exercise was designed to provide training for four area volunteer fire departments. The goals were to familiarize the participants with arson fires and provide experiences in structural fire attack. Accordingly, the plans were to set a fire using liquid accelerants, have attack teams extinguish the resulting fire, and allow participants to examine possible evidence associated with flammable and combustible liquids.

The Structure

A farm house, perhaps over 100 years old, was donated by the owner who wished it to be destroyed. The first floor contained a kitchen, entry/storage room, and two other rooms. The second floor had a corridor and two rooms. Construction was wood frame with lath and plaster walls and ceilings with attic spaces. The structure was in a state of disrepair with holes in floors and ceilings. An accumulation of trash and debris littered the house. The only access to the second floor was a stairway with approximately 35 feet (10.7 m) of travel distance from the bottom of the stairs to the exit door. One of the six windows on the second floor allowed access to the kitchen roof. Trees and shrubs were standing close to the structure, and some of the windows were covered by plant growth.

Fuel

While specific placement is not known, the flammable liquids used were gasoline, stove fuel, and kerosene. To simulate an arson fire, sheets of waxed paper were sprinkled with gasoline to form trailers (materials used to spread fire from one point to another), rubber gloves were filled with flammable liquids, and couches were soaked with additional flammable liquids. Flammable liquid pools were also placed in several areas of the first and second floors.

Ignition

At first, participants noted that there was little fire. Though several spot fires were noticed, they were not spread by the trailers. This condition persisted for several minutes and generated moderate smoke. An assistant chief and a firefighter reentered the house to ignite the fires on the second floor. A four-member attack crew that had been standing by was directed to go inside the house to observe how the fires were developing. These four crew members went to the second floor, which now was occupied by six people.

During this initial period, an observer outside the house was asked to break two upper windows on the south side. The windows were broken by throwing rocks. The six firefighters on the second floor then noticed a sudden increase in smoke, which caused them to attempt leaving the structure by the stairs. However, they found the exit was blocked by flames. With zero visibility, they lost track of each other. Three of the six reached the second-floor window and escaped onto the kitchen roof. Flames were now coming out of the second-floor windows. Rescue attempts were not successful. Three firefighters died.

Questions for Discussion

1. As an instructor, what would you have done differently to prepare for this exercise?
2. Which positions must be assigned/staffed to meet NFPA 1403 requirements?
3. What effect would a presurvey of the building have had on preventing the fatalities? What other preparations could have been made to prevent injuries/fatalities?
4. What is the proper instructor/learner ratio in such an exercise?
5. What fuels would you use to ignite an acquired structure?

What Happened: Lessons Learned

Liberal use of flammable liquids and conditions consistent with a ventilation-controlled fire may have caused flashover. Fiberboard tile and wood walls also contributed to the overall fuel load. The cause of the deaths seems to be related to incapacitation as a result of inhaling carbon monoxide. Although firefighters were wearing SCBAs, the straps may have melted, causing masks to pull away from their faces. It was also noted that one SCBA facepiece had a missing lens, one facepiece had a broken lens, and two SCBAs had severe damage to the low-pressure lines.

(Continued)

Case Study (Continued)

By following the recommendations of NFPA 1403, the training officers could have avoided this tragedy. The report finds that the following safety precautions should have been considered:

- Use caution with flammable liquids and limit their use — they are dangerous when combined with highly combustible interior finishes.
- Remove trash from the building.
- Repair holes in walls and ceilings.
- Prepare roof openings for emergency ventilation.
- Preplan an evacuation signal.
- Provide an instructor with each group of participants.
- Provide a briefing to participants on the structure's interior layout and escape routes in the event evacuation becomes necessary.
- Use an incident management system to provide for control and accountability of all participants and to have better communications between interior and exterior sectors.
- Assign a safety officer to monitor events and keep the incident commander informed of any threats to the health and well-being of instructors and participants.

Conclusion

Those who were in command were well-motivated and trying to provide training to their members and experiences that would help all those participating in the exercise. Their overall objective was to improve the skills of their firefighters, thus increasing the protection afforded to their communities. However, the results were tragic. Instructors can learn from this incident. Use applicable NFPA standards in all training exercises, and be especially careful when conducting live fire training.

References and Supplemental Readings

Davis, Larry. "A Call for a Fire Service Reality Check on Making the Fireground as Safe as the Training Ground." *The Voice,* International Society of Fire Service Instructors (ISFSI), February 22, 1993.

International Society of Fire Service Instructors (ISFSI), "Instruct-O-Gram," Volume X, Issue 12, December 1989. (Provides a reference dealing with the planning, preparation, and reporting of the requirements of NFPA 1403.)

Kramer, William M. "Training and Education." *The Fire Chief's Handbook.* Joseph R. Bachtler, and Thomas F. Brennan (eds.). Saddle Brook, NJ: Fire Engineering Books and Videos, 1995, pp. 335-351.

Norman, John. "Live Fire Training: Dos and Don'ts." *Firehouse,* March, 1988, p. 79. (Also includes a discussion of a loop oil pan for petroleum/hydrocarbon fuel fires.)

Teele, Bruce W. (ed.). *NFPA 1500 Handbook.* Quincy, MA: National Fire Protection Association, 1997.

United States Fire Administration (USFA). "Three Firefighter Fatalities in Training Exercise, Milford, Michigan (October 25, 1987)." Technical Report Series, Report 015, 1987.

Wieder, Michael A. "Smokehouse and Fire Training Building Safety." *Speaking of Fire,* International Fire Service Training Association (IFSTA), Oklahoma State University, Fall 1991.

Job Performance Requirements

This chapter provides information that addresses the following job performance requirements of NFPA 1041, *Standard for Fire Service Instructor Professional Qualifications* (1996 edition). Colored portions of the standard are specifically covered in this chapter.

Chapter 3 Instructor II

3-4.3 Supervise other instructors and students during high hazard training, given a training scenario with increased hazard exposure, so that applicable safety standards and practices are followed, and instructional goals are met.

3-4.3.1 *Prerequisite Knowledge:* Safety rules, regulations and practices, the incident command system used by the agency, and leadership techniques.

3-4.3.2 *Prerequisite Skills:* ICS implementation.

Chapter 9

Testing and Evaluation

Training programs are not complete without some way to measure that learning took place. Common measurement terms and processes include *testing* and *evaluation*. Selecting a term is often based on the type of measurement.

Evaluation refers to methods of identifying the effects and judging the effectiveness of learning experiences, courses, or complete curricula. Evaluation is terminal or a measurement of the end product of learning. The term evaluation is sometimes used interchangeably with the term assessment. *Assessment* refers to the processes used to find out the knowledge, skills, and abilities of a learner by observation and special activities including quizzes, examinations, oral tests, and similar testing devices. Assessment can also include measuring learner levels before *and* after the learning experience—a valuable part of an evaluation process. Assessment is a broad term and reflects the kind of testing and measuring that instructors must do.

A *test* is a means by which the absence, presence, amount, or nature of some learner quality or ability is observed or inferred and appraised or measured. The measurement terms have the same goal: measuring learner knowledge, attitudes, and skills. To some extent, tests assess changes in attitudes, values, and beliefs.

Traditionally testing implies that the testing instrument is a paper-and-pencil measurement tool. Paper-and-pencil testing is often used to assign a grade. Coupled with skills check-off by observation, it is also a means to determine whether a learner is ready to move onto the next set of topics. The current trend in testing is to determine learner status with a wide variety of measuring devices. This text emphasizes the various learning styles that may exist in a group and the importance of using a variety of teaching methods. A variety of testing methods also measure the many important kinds of learning that cannot be measured appropriately by paper-and-pencil tests alone (more accurately referred to as assessment).

Many organizations use tests that have been professionally prepared and are part of purchased programs. These prepared tests are validated and checked for reliability, but they may not always accurately test specific jurisdictional criteria and policies or reflect instructional methods and learning styles. For these reasons, instructors may often find that they need to develop test questions or entire tests that appropriately measure the learning that has taken place in their courses.

This chapter first discusses the purpose of testing and describes the classification of tests. Creating a test is not an easy process, so this chapter discusses the importance of knowing how to plan tests, including what to consider when creating tests, and how to select and develop the appropriate evaluation tool from the many common instruments. This chapter also briefly discusses testing in the affective domain and describes the common formats used. Other important aspects of this chapter are how to interpret test data once it is collected and how instructors use the data once the test is given and graded. Finally this chapter discusses the purpose of evaluation and the importance of evaluating the course and instructional design. References and supplemental readings and job performance requirements are given at the end of the chapter.

Purpose of Testing

[NFPA 1041: 2-5.2, 2-5.3, 2-5.4, 3-5.4, 4-2.7, 4-5.2]

Testing is a method of measuring learning results (evidence), and it is an integral part of instruction and constructively doing something other than practicing a skill they have already mastered.

Progress

Progress tests are often viewed as quizzes, pop tests, or question/answer periods in class that are given throughout the course or unit of instruction. These tests typically *measure improvement* and give the instructor and learners feedback on learning progress. When measuring improvement, the test answers the question, *"Is the learner achieving the objectives?"* Tests include the most important behavioral objectives or all of them if possible. Each test item matches the level of difficulty of the corresponding behavioral objective and thus is criterion-referenced.

Instructors who encounter individuals with learning problems during the course should check with the organization. It may be the organization's policy to consult with appropriate authorities and use a specific type of test to *diagnose learning difficulty.* When trying to diagnose causes of the learning problem, the test tries to answer the question, *"If the learner is not achieving the objectives, what are the specific causes?"* Test items are easy and provide a sample of tasks in which the learner has experienced difficulty. The test is designed so the evaluator performing the diagnosis can pinpoint specific causes or difficulties.

> ***Instructor Tip:** Instructors often recognize individuals in class who seem to have difficulty learning such as those who have difficulty reading. To assist these individuals, first check with the training organization and follow its policy on aiding or directing learners to appropriate help. Then, if the organization does not have internal assistance, check local agencies such as the board of education, the health department, and high schools for resources and/or references for assisting learners.*

Comprehensive

Comprehensive tests measure learner achievement in an entire area on a number of topics covered over a long period of time such as a semester or other major segment of the course. They test general mastery of a broad academic field and are typically given in the middle (midterm) or at the end of instruction (final). When measuring comprehensive knowledge and skill, the test answers the question, *"Has the learner achieved the course objectives?"* Examples of comprehensive tests given for a training program are the written and/or practical exams given at the midpoint or end of emergency medical technician (EMT), basic fire fighting, and rescue programs. The learners who are tested must demonstrate comprehensive knowledge and skills learned in all areas from the beginning to the testing point (midterm or end) of the program.

Administration Method

There are various physical methods of giving tests that instructors and their organizations can use that give some options to enable all learners to experience success. For one reason or another (such as good or bad testing experiences or successes or failures), learners develop a preference for a certain type of test — written, oral, or performance — much the same way they develop a preference for a certain learning style. Program developers should consider including a variety of test methods throughout the program and in the program final evaluation so that they may get an accurate learning measurement of all individuals. The following sections describe each type of test.

Oral

During *oral tests,* the learner generally gives verbal answers to spoken questions. Oral tests are not commonly used in the fire and emergency services, but under certain circumstances, the instructor may find them necessary. An example of this would be testing a learner who cannot read at the level to which a test is written. Oral tests may also be used to supplement performance tests to determine whether a learner knows the reasoning behind the jobs performed. Oral tests are usually given one-on-one with an instructor. Oral tests may also be a component part of a comprehensive examination program for officer candidates. In this case, the oral test may be given by a board of experts to each individual officer candidate.

When giving oral tests, the instructor or the board of experts must listen carefully to an individual's responses because people phrase answers differently, though they mean or are trying to say the same thing. When giving oral tests, instructors or examining boards should consider the test's purpose and carefully create appropriate questions. Oral tests should never be used as the *sole* means of evaluating learners for terminal performance or officer candidates for promotion.

Written

Written tests evaluate the accomplishment of cognitive and affective objectives. Written tests may be subjective (based on individual perceptions) or objective (based on facts), and they are useful in measuring retention and understanding of technical information. Fire chemistry, laws and ordinances, hydraulic principles, and medical protocols are examples of technical subjects for which learners are given written tests (Figure 9.2).

Objective test items. Test developers must carefully construct these types of test items. Often developers limit objective test items to the lower cognitive learning levels of recall and recognition because these levels lend themselves easily to the objective test format. But objective test items can also be used to measure higher levels of cognitive learning such as interpretation and analysis. For example, an objective that asks the learner to choose the correct function for a tool from among four listed functions requires only that the learner recognize the correct function. However, the test item that presents a problem and asks the learner to choose the best tool for solving the problem requires the learner to interpret and analyze.

Subjective test items. This type of test item has no set response. "Correct" responses vary with each learner's solution to the problem. Because they allow the learner the freedom to organize, analyze, revise, redesign, or evaluate a problem, subjective tests are effective for measuring the higher cognitive levels of analysis, evaluation, and interpretation. In the fire and emergency services, subjective tests, usually in essay format, are used most often in officer training. At this level, most of the training involves intellectual, problem-solving skills, and participants must be able to demonstrate higher cognitive skills to meet higher level job performance requirements.

Performance

In *performance tests* learners are actually required *to do* (perform) something rather than talk or write about it. Candidates are required to perform a skill or action as it would be performed on the job, and they are tested on their present abilities or attainments rather than their potentials. Test developers must base performance tests on standard criteria and performance objectives. As they create the test, developers must be mindful of the following factors that are important to successful testing:

- Determine the materials or equipment the learners need to perform the skills or activities. Also ensure that materials can be provided in appropriate quantities for the testing situations.

Figure 9.2 Written tests are useful in measuring retention and understanding of technical information. *(left) Courtesy of Jarett Metheny, Midwest City (OK) Fire Department. (right) Courtesy of Larry Ansted, Maryland Fire and Rescue Institute.*

Assessment Center

Assessment centers may be used by fire and emergency service instructors to give learners a chance to practice what they have learned. As learners work through each assessment center exercise, instructors can observe and evaluate. The assessment center process is specific, and instructors must learn all of the elements, how they fit together, and how to administer the exercises to ensure good results. This brief overview is to encourage instructors to obtain further information before using the assessment center process with learners, not to merely mimic portions that they have observed.

The assessment center is a process in which a candidate participates in several structured exercises. The exercises require candidates to perform realistic tasks in a controlled situation. The exercises are evaluated using a standardized method that compares the candidate's behavior against a predetermined set of behaviors.

Assessment centers were originally designed as a management training process. Assessment centers are now used to obtain information for career development and to help managers make personnel decisions. They are valuable as promotional examinations because the standardized method gives candidates an equal opportunity to demonstrate behaviors and eliminates much of the bias found in traditional examinations or interviews.

The assessment center's standardized method is comprised of specific elements. The major elements include the following:

- Inclusion of multiple exercises with at least one job-related simulation
- Observation by multiple, trained assessors
- Identification of specific areas of behaviors or dimensions (for example, decision making, initiative, and communications) that are evaluated
- Evaluation by assessors of observed behaviors
- Discussion of observations by assessors and formation of consensus on their judgments

There are other elements that comprise the standardized method. Assessment center organizers must properly utilize all of the elements in order to administer a proper assessment center that is fair to the candidate and reduces the opportunity for bias.

Each exercise includes work activities that a candidate would normally perform in the actual position. The selection and design of exercises are based on the behaviors and dimensions that the organization believes are important to the position.

Examples of assessment center exercises are as follows:

- **In-Basket Exercise** — The candidate is directed to review a stack of memos, letters, schedules, or other documents. The candidate must prioritize items and decide whether to act on an item himself or herself, delegate it to a subordinate, or present it to a superior. The candidate must write the actual response to a memo or letter.
- **Cooperative Group Exercise** — The candidates meet as members of a committee or task force to accomplish a task or organize a solution to a shared problem.
- **Role-Play Exercise** — The candidate is given a specific role in a job-related situation. The candidate may have a meeting or telephone discussion with a person portraying a citizen, subordinate, or superior with a specific grievance, question, request, or problem.
- **Tactical or Strategic Exercise** — Command decisions are made by candidates as they analyze and respond to "emergency" conditions depicted by photographs, written documents, "radio reports," or other means. Changes are presented as the simulated emergency progresses, and the candidates must respond to those changes. This exercise is frequently used in the fire and emergency services.

Identifying a candidate's specific strengths and weaknesses in job-related exercises provides valuable training and development information. A summary of the assessors' observations shows the candidate's performance in the specific dimensions. Recommendations for further training or experience that will help candidates improve their skills may be made.

Test Planning Sheet

Course: *Breathing Apparatus*

Test Type: *Written*

Test Purpose: *Quiz*

Total Test Items: *30*

Job or Topic	Instruction Level Taught	Number of Questions: Knowledge	Understanding	Application
Overview of Respiratory Standard	2	1	1	0
Respiratory Hazards	2	1	3	0
Principles of Operations	2	1	4	1
Major Components, Purposes, Use	2	1	4	1
Inspection and Servicing	3	1	4	0
Donning and Doffing	3	1	0	0
Emergency Procedures	3	0	2	4

Remember!

1. Identify learning levels.
2. Prioritize materials.
3. Test all objectives.

Test Planning Sheet

Course: ______________________ Test Purpose: ______________

Test Type: ______________________ Total Test Items: ______________

Job or Topic	Instruction Level Taught	Number of Questions: Knowledge	Understanding	Application

Figure 9.3 Test planning sheet — a valuable aid for test construction. Reproduce the blank form and use it to plan your tests.

specifications with individual test items. Tips given in the following sections help in writing effective tests.

Match Test Items to Objectives

Instructors should draft test items that provide a measure of the intended learning outcome stated in the lesson objectives. If the objective is to convert metric measurements (International System) to U.S. measurements (Customary System), an item requiring the learner to identify metric numbers or to add a set of metric numbers would be testing on items other than what is required by and included in the objective.

Eliminate Testing Barriers

Compose test items that eliminate or at least minimize barriers to taking the test. One way to do this is to use words that the learner would use during training or on the job. Some barriers to test taking include the following:

- Higher reading level than the learner audience possesses
- Lengthy, complex, or unclear sentences
- Vague directions
- Unclear graphic materials

Avoid Giving Clues to Test Answers

Write items that do not give clues on how to answer the question correctly. Some areas to ***avoid*** include the following:

- Word associations that give away the answer
- Plural or singular verbs or use of the words *a* or *an* that hint at the answer or eliminate an answer
- Words that make some answers more likely (such as *sometimes*) or less likely (such as *always* or *never*)
- Answers consistently placed in the same location (such as the Choice B answer in multiple-choice questions) or consistently longer correct answers (such as making all true statements longer than false statements)
- Words and materials routinely copied verbatim from the textbook
- Stereotypical answers
- Test items that give the answer away in other items

Select Proper Level of Item Difficulty

Since a function of criterion-referenced testing is to describe the precise learning outcome of a given learner, item difficulty must match the course objectives. Norm-referenced testing, on the other hand, is designed to rank learners against others in the class. Test items should range in difficulty to allow appropriate discrimination among learners.

Decide Appropriate Number of Test Items

Although the test planning sheet indicates how many test items are to be used, there are practical constraints and educational considerations that dictate the number of test items on a test. Educational recommendations for criterion-referenced tests are ten items per learning *level*. This number ensures that each level of learning is adequately measured. Do not confuse the number of items per level with the suggestion of creating at least one test item per learning *objective*. (Many objectives are written at the same level of learning.) For practical considerations, the number of items must match the time allowed for testing. Generally, an adult can average one multiple-choice, three short answer/completion, *or* three true-false items in 1 minute. Remember that some learners in the class will take more time than others.

The number of items is the key to comprehensiveness in test writing. A test should be constructed so it can measure the learners' abilities in all phases of the course. It should be complete without unnecessary details.

Ensure Ease of Testing and Scoring

A test that burdens the instructor with the mechanics of administration is an obstacle to effective testing. Therefore, tests must have the following characteristics:

- ***Easy to give*** — Construct a test so the instructor can concentrate on monitoring learners and answering questions during the testing period.
- ***Easy to take*** — Include directions that are clear and complete so the instructor, as well as the test takers, know exactly what they are to do. Provide sample test items to show the class how they are to answer the items.
- ***Easy to score*** — Construct a test that is specific in what it is asking; that is, it is not vague or ambiguous in the test items or directions. Provide an answer sheet or simple test form that aids the instructor in quickly scoring the tests. Other easy-to-score methods include electronic scoring devices. See Short Answer/Completion portion and examples in the Written Tests section later in this chapter.

Build in Validity and Reliability

The two most important conditions of a well-designed test are validity and reliability. *Validity* is the extent that a test measures what it is supposed to measure. It is built into the test by selecting an ample number of test items for each learning level and content area. The best way to ensure validity of the course content is to take the following steps:

- Identify the content of the course and the behavioral objectives to be measured.
- Develop a table of specifications, which specifies the sample of test items to be used.
- Design a test that matches the specifications.

Reliability is the consistency and accuracy of test measurement. A reliable test is free of ambiguous items or directions, vague scoring criteria, environmental distractions, and opportunities for cheating or guessing. A reliable test is one in which separate scorers would give the same score to the same learner's test. Devoting attention to each of the test characteristics, analyzing a test each time it is given, and discarding or rewriting items that do not meet requirements aid in improving test reliability. Any test that cannot give consistent and accurate scores cannot be measuring what it said it would measure. This characteristic of reliability is an essential condition of validity, but it is not the only condition.

Common Considerations for All Tests

[NFPA 1041: 2-5.2, 2-5.3.1, 2-5.4, 2-5.5, 3-5.2; 3-5.3]

When designing tests, some points are commonly overlooked but are considerably important for smooth testing. The most significant consideration for test design is that all test items must be referenced to a learning objective. In turn, these learning objectives are teaching points that are found in credible sources such as national standards, organization policies, and textbooks. Test designers must also give consideration to test format, arrangement, and instructions. Methods of ensuring test security, ease of test administration, and reporting test results are also important.

Format

Certain format considerations make test administration, test taking, and test scoring simple. The following should be included in any test format (see Tools for Testing Learning section):

- Spaces for the date and the learner's name on the test sheet unless separate answer sheets are provided
- Test title or label
- Numbered tests or different test versions, which aid in reporting scores and maintaining security
- Clear, easy-to-follow instructions at the beginning of the test and each test section that begins a different test format (multiple-choice, matching, true-false, or fill-in-the-blank)
- Sample test item demonstrating how the test is to be taken
- Consecutive numbering of test items
- Single spacing within test items
- Double spacing between test items
- Point value of each test item, for example, multiple-choice items are 1 point each, short answer items are 2 points each, and true-false items are 1 point each

Arrangement

After the test items are prepared, they must be arranged in a logical way. They can either be grouped by learning outcome (such as basic, intermediate, or advanced) or by type of test item (such as multiple-choice, matching, or short answer). In either case, it is recommended that the items be placed in sequence of increasing difficulty. This allows learners to answer some easy items as they begin, gain confidence, move on to the moderately difficult items, and then be challenged by more difficult items toward the end of the test or section.

Instructions

All tests should include instructions on how to complete the test. Instructions should be brief and to the point. They should explain the purpose of the test, how the learner is to record answers (whether on the test itself or answer sheet), and whether to guess when in doubt of the answer. It is helpful to learners to know how long they have to complete a test. Include instructions on how much time is allowed. In tests with multiple parts where questions are grouped by their type (such as sections on multiple choice, matching, true-false, etc.), indicate how much time the learner can spend on each section. Give specific instructions at the beginning of each section. It is good policy for instructors to read all instructions to the class, point out any variances with different sections of the test, and then ask if there are any questions.

Security

Cheating on tests presents special problems for test developers and instructors — problems that they may reduce by careful attention to a few details. Protect test security by regularly revising tests and by exercising care in typing, duplicating, and storing test materials. Most cheating is a result of advance knowledge of specific test content caused by inattention to these details. To improve test reliability and validity, instructors can ensure that test scores are the result of the learner's own unaided efforts by creating special seating arrangements in the testing room and monitoring the learners during the test. A method to ensure that all test booklets are returned is to number them and ask that they be returned with the answer sheets.

Administration

Administration of a test begins before the test is given. Tell learners about the test type and content, how they can prepare for the test, and what they must bring to class for the test (paper, pencil, pen, notes, books, etc.). Notify them in an informative and motivating manner well in advance of the test date. On the day of the test, ensure that the environment aids learners in testing effectively and that the quality of their performance is not influenced by adverse conditions such as poor lighting, uncomfortable seating, noises, extreme heat or cold, and other causes of discomfort and distraction.

Though environmental factors are the usual concern for instructors and learners at test time, motivation also affects test performance. Instructors must make every effort to alleviate stress and anxiety while providing motivation and positive points for taking the test.

Reporting

Report test results and use them to decide on further learning. For norm-referenced tests, report overall general results to the class while maintaining individual confidentiality. To report general results, instructors can provide the range of scores (such as *68 to 97*) and the *A-B-C* or *pass/fail* categories (such as *70 and above* is passing or *90–100 = A, 80–89 = B,* and so on). Provide individuals with *their scores,* the *range of scores,* and the *number of passing scores* (or *A's, B's, C's,* and so on) so everyone knows their general position in relation to other class members.

For criterion-referenced tests, compare individual results with course objectives and the total number of possible points. Let learners know the total possible points, the minimum score, and what they must do to earn the minimum points and to meet course objectives if their scores were unsatisfactory. Always review tests so that learners know what test items they missed and why and that they have the opportunity to review and reinforce the material in order to learn it appropriately.

Tools for Testing Learning

[NFPA 1041: 2-5.2, 2-5.3, 2-5.4, 3-5.2, 4-5.2]

The three learning domains — cognitive, affective, and psychomotor — each have several levels of learning that enable instructors to teach in progressive steps and learners to proceed from simple to advanced knowledge, attitudes, and abilities. There are also different levels of training courses — beginning, intermediate, and advanced — each of which require different levels of knowledge and skills. These factors make the task of developing appropriate tests a challenge, yet it is important that test developers and instructors select the right testing tool for the levels and types of programs, objectives, and learners. There are three basic types of tests: (1) performance tests that test skill ability, (2) oral tests that test communication ability, and (3) written tests that test knowledge. Each type has advantages, disadvantages, and test construction considerations.

Performance Tests

Performance tests measure an individual's proficiency in performing a job or evolution such as achieving a psychomotor objective. This type of test holds the test-taker to either a *speed standard* (timed performance) or a *quality standard* (minimum acceptable product or process standard) or both. Performance tests are the most direct means of finding out how well an individual can do a job. Some examples of performance tests include those that require demonstrating care of tools and equipment, driving and operating apparatus, and performing emergency care steps and techniques (Figure 9.4).

There is a difference between performance tests and drills. The purpose of a *drill* is to give learners an opportunity to practice skills (on which they may later be tested) in a supportive environment. The purpose of performance tests is to give learners an opportunity to demonstrate their proficiency under controlled conditions after appropriate practice sessions. Only when testing conditions are controlled can instructors make valid and reliable judgments about learner performance.

Test Construction Considerations

Test developers and instructors must take several steps when creating performance tests. Consider the following:

Figure 9.4 Performance tests are used to measure individual proficiency in performing a job.

- ***Specify the performance objectives to be measured, then construct test items based on those objectives*** — Each test item should require the performance of a *number* of basic skills, which allows a broad sampling without consuming the time necessary to test performance of *each* basic skill. For example, a test item that requires ventilating a pitched roof also requires the learner to demonstrate use of ground and roof ladders, cutting tools, safety ropes, and hoselines.

- ***Select rating factors on which the test will be judged and design a rating form*** — Rating factors for performance tests include such items as (1) the learner's approach to a stated job or procedure; (2) the care shown in handling tools, equipment, and materials; (3) a demonstration of accuracy; and (4) the time required to complete a job or procedure safely. Rate learners against a standard, not against the performance of other learners (Figure 9.5).

- ***Prepare directions that clearly explain the test situation to learners*** — A written set of instructions supplemented with an oral explanation and an opportunity for learners to ask questions are necessary so learners understand what is expected of them.

- ***Try a new performance test on other instructors before using it on learners*** — A trial test contributes to test validity and may uncover problems that can be corrected.

- ***Use more than one evaluator*** — Request other instructors or officers to be evaluators. Avoid using the instructors who taught the class to evaluate the groups that they taught. Provide instructions to

evaluators about what they are to look for during the test and how to use the rating scales and forms. Calculate an average score from all evaluators for each learner or learner team. For example, a learner or team who receives three scores of *97, 94,* and *94* from three evaluators receives an average score of *95.* Scoring for performance tests that are criterion-referenced are typically *pass/fail* or *satisfactory/unsatisfactory,* though this scoring may be coupled with a *minimum points-earned* requirement.

- ***Follow established procedures during performance test administration*** — Instructors must have all necessary apparatus and equipment ready before beginning the tests. A person assigned as an evaluator must use the same equipment throughout the test for testing learners, follow the same sequence of jobs for all learners, and rate each individual on the same basis. All distractions must be eliminated from the testing area so that evaluators can concentrate on observing and evaluating learner performances and learners can concentrate on demonstrating proficient performances.

- ***Make a score distribution chart after tests have been administered and evaluate learners with low scores*** — Those learners who have difficulty in performing manipulative skills must receive immediate attention.

- ***Rotate team members to every position for team evaluation ratings*** — It is important that each learner is observed and evaluated in each position in an evolution.

Chapter 4 — Firefighter Personal Protective Equipment
Lesson 4B — SCBA

Performance Test

Choose the following problems or set up similar situations in which the candidate will have to use collectively those psychomotor skills performed independently.

When creating your own performance test problems, make the problem as realistic as possible by presenting it within a framework that simulates real-life situations that will be encountered by the candidate after completing training.

To pass the performance test, the candidate should be able to determine required processes in a given situation and to perform those processes to a competency level of at least 2.

Competency Rating Scale

3 — Skilled — Meets all evaluation criteria and standards; performs task independently on first attempt; requires no additional practice or training.

2 — Moderately skilled — Meets all evaluation criteria and standards; performs task independently; additional practice is recommended.

1 — Unskilled — Is unable to perform the task; additional training required.

N/A — Not assigned — Task is not required or has not been performed.

✔ **Evaluator's Note:** Formulate and inform the candidate of the standards for this performance test (time allowed and number of attempts, if applicable). Observe the candidate perform the task, note whether the performance meets each of the evaluation criteria, and then use the rating scale above to assign an overall competency rating. If the candidate is unable to perform any step of this task, have the candidate review the materials and try again.

Student's Name ______________________ **Date** ____________

Performance Task 1 ☐ Assigned ☐ N/A

Don and doff protective clothing. ***(NFPA 1001, 3-1.1.2)***

Performance Criteria	Yes	No
Donning		
Trouser legs over boots	☐	☐
Hood on properly (no exposed hair)	☐	☐
Coat completely fastened	☐	☐
Coat collar up and fastened	☐	☐
Helmet secured under chin	☐	☐
All articles of protective clothing donned within 1 minute	☐	☐
Doffing		
Gloves in coat pocket	☐	☐
Helmet stored with front facing back of storage unit	☐	☐
Hood in coat pocket	☐	☐
Coat on storage hook with collar up and front completely unfastened	☐	☐
Boots and trousers doffed as one	☐	☐
	☐	☐
Other		
1. ______________________	☐	☐
2. ______________________	☐	☐

Competency Rating ____________

Figure 9.5 Performance test rating sheet.

Advantages

- ***Validity*** — A performance test is the only valid method of measuring learner achievement and ability to perform manipulative skills.
- ***Reliability***—A properly constructed test using specific criteria is a reliable measure of performance when coupled with an appropriate rating scale.
- ***Observation*** — Evaluators can observe individual differences in judgment and approach to problems. Some individuals may not be able to express themselves orally or in writing. However, they may be able to perform a job as well as or better than other learners.
- ***Learner motivation*** — Performance tests are an excellent means of motivating learners. Knowing that they are expected to demonstrate skill ability in a performance test usually motivates learners to spend in-class and out-of-class time practicing productively to prepare and develop comfort levels for their skills.
- ***Sense of accomplishment***—Learners who successfully complete well-prepared and carefully administered performance tests are proud of their accomplishments.

Disadvantages

- ***Unreliability*** — Scores may be unreliable because of evaluator subjectivity. Reliability is proportionate to how well the evaluator identifies the skills observed and compares them with the criteria specified. Lack of definitive rating criteria adds to evaluator subjectivity and reduces reliability.
- ***Economy of test taking*** — Performance tests are time- and resource-consuming. This type of test also requires more instructors/evaluators to observe, monitor, and evaluate learners.
- ***Difficulty*** — It is difficult to test each individual in team evolutions, to eliminate subjectivity, and to provide consistent conditions for each individual or group because the test progresses into long hours.

Oral Tests

Oral tests were described in the Administration Method section. Many of the advantages and disadvantages of performance tests also apply to oral tests. Evaluators must make every effort to put individuals at ease and in comfortable situations so that performances are not influenced by adverse emotions and environments. Three areas that provide more detail on oral tests include the following:

- ***Development*** — Base oral questions on standard criteria and performance objectives. Developers must ensure that the questions clearly state what the learners are to describe.
- ***Validation***—Validate oral tests by trying the questions on other faculty, experts, and learners in other training programs prior to using them on actual testing candidates. Revise inappropriate questions and update questions as needed to meet changing or updated criteria.
- ***Evaluation***—State the question, carefully listen to and consider the response, assess knowledge based on objectives and standards, and score the response based on what is expressed compared to the knowledge required.

Written Tests

The purpose of *written tests* is to measure learner retention and understanding of technical information and evaluate learner accomplishment of the cognitive objectives. There are several types of written tests that are considered objective such as multiple-choice, true-false, matching, and short answer/completion. They are described in the following sections. In addition, a brief description of essay tests, which are considered subjective, is included.

Multiple-Choice

A *multiple-choice* test item consists of either a question or an incomplete statement, commonly referred to as the *stem*. Following the stem is a list of several possible responses referred to as *choices* or *alternatives*. The learner is asked to read the stem and to select the correct response from the list of choices or alternatives. The correct choice in each item is known as the *answer*, and the remaining choices are called *distracters*. The obvious purpose of distracters is to confuse those learners who are not sure of the correct answer but not to trick or mislead test takers (Figure 9.6).

A well-constructed multiple-choice test is one of the most versatile of the objective test types. This type of test can measure a variety of learner abilities and can be adapted to most types of subject matter. Multiple-choice tests are not easy to construct. In order to perform as described, they must be unambiguous and test to the learning level desired. Once constructed, they are quick and easy to grade.

Test construction considerations. When creating multiple-choice tests, it is important to include the following components:

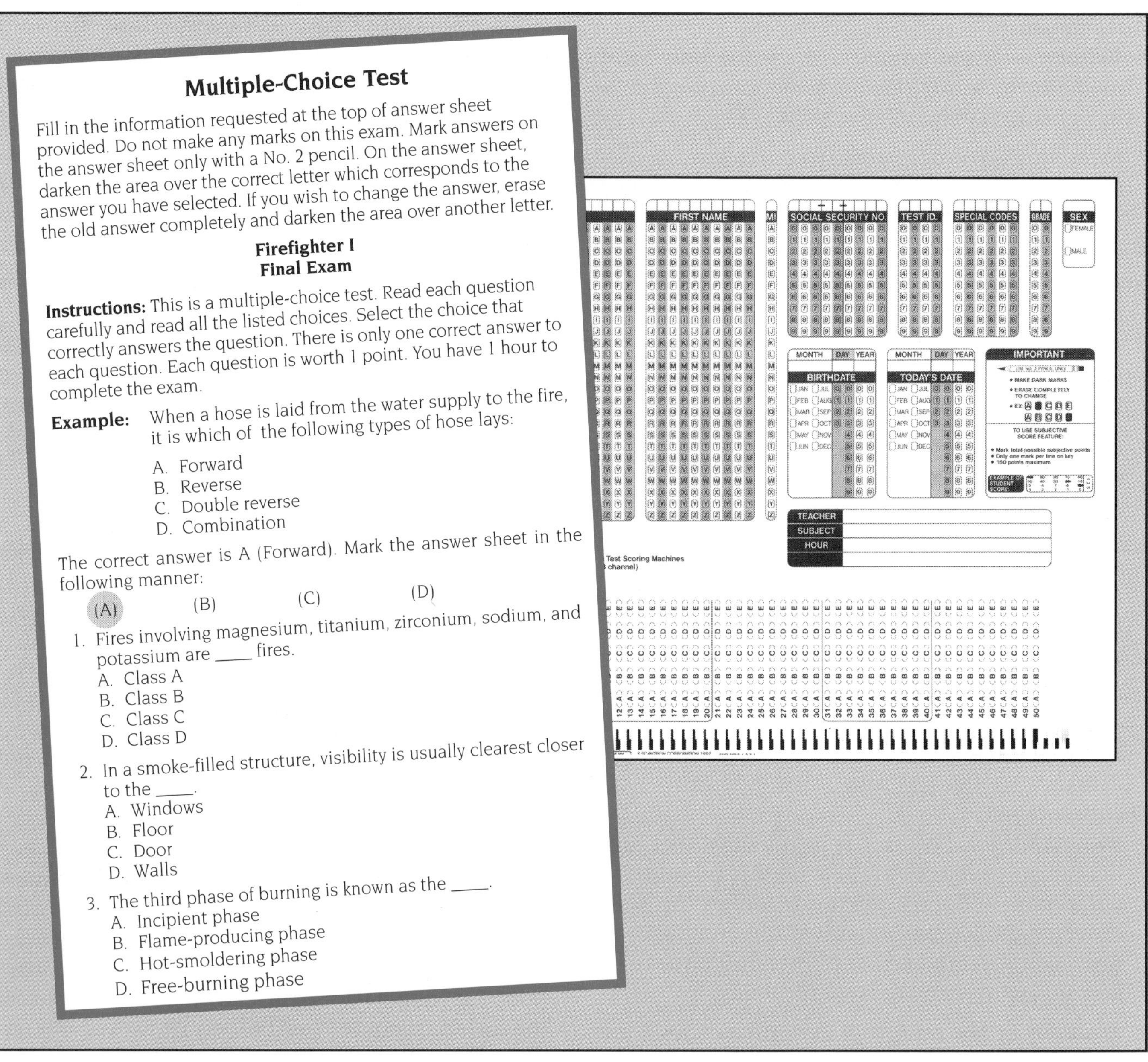

Multiple-Choice Test

Fill in the information requested at the top of answer sheet provided. Do not make any marks on this exam. Mark answers on the answer sheet only with a No. 2 pencil. On the answer sheet, darken the area over the correct letter which corresponds to the answer you have selected. If you wish to change the answer, erase the old answer completely and darken the area over another letter.

Firefighter I
Final Exam

Instructions: This is a multiple-choice test. Read each question carefully and read all the listed choices. Select the choice that correctly answers the question. There is only one correct answer to each question. Each question is worth 1 point. You have 1 hour to complete the exam.

Example: When a hose is laid from the water supply to the fire, it is which of the following types of hose lays:

A. Forward
B. Reverse
C. Double reverse
D. Combination

The correct answer is A (Forward). Mark the answer sheet in the following manner:

(A) (B) (C) (D)

1. Fires involving magnesium, titanium, zirconium, sodium, and potassium are ____ fires.
   A. Class A
   B. Class B
   C. Class C
   D. Class D

2. In a smoke-filled structure, visibility is usually clearest closer to the ____.
   A. Windows
   B. Floor
   C. Door
   D. Walls

3. The third phase of burning is known as the ____.
   A. Incipient phase
   B. Flame-producing phase
   C. Hot-smoldering phase
   D. Free-burning phase

Figure 9.6 Multiple-choice test and answer sheet. *Courtesy of Maryland Fire and Rescue Institute.*

- Write the stem in the form of a direct question or an incomplete sentence that asks and measures only one learning outcome.
- Write a clear, brief stem that contains most of the wording (rather than placing repeated words in the choices).
- Write positive statements, but if and when negative statements are used, underline (or italicize) or emphasize negative words in bold.
- List at least four choices: the answer and three plausible, attractive distracters.
- Create choices with phrases in parallel form and grammatically consistent with the stem.
- Place correct answers in varied positions among the A, B, C, and D choices.
- Place each choice on separate, indented lines and in a single column.
- Begin responses with capital letters.

Other possible formats for laying out multiple-choice questions depend on the training organization's style preference and whether the organization uses a computer-generated test creator program, which automatically sets spacing between the choice letters (A, B, C, and D) and the choice sentence. Placing periods after responses may depend on style preference or whether the response finishes an incomplete sentence or fills in a blank in the stem.

While assuring that all the components listed are included in multiple-choice tests, test developers should be aware of the following points:

- Do not include choices that are obviously wrong.
- Construct the stem and choices to avoid grammatical clues to the correct answer such as using the words *a* and *an* or singular and plural word endings at the end of the stem that direct learners to choose a response that matches the tense and case of the stem.
- Make all choices close to the same length; the correct response should not be longer than the distracters.
- Rarely, if ever, use the phrases *all of the above* and *none of the above* as the fourth choice.

Advantages. When constructing multiple-choice items, consider the advantage of their wide ability to measure achievement and complex learning outcomes and various types of knowledge. These measurements are advantages only if the test is constructed with clear, unambiguous questions, which take time to carefully plan and thoughtfully word. Keep in mind that a question that is clear to the test developer or instructor is not always clear nor may not mean the same to the learner.

Disadvantages. As useful, common, and favored as multiple-choice tests are, they do have some disadvantages. These tests are not as well-suited as essay tests for measuring some cognitive skills such as the ability to organize and present ideas. It can be very difficult to construct multiple-choice tests that include different difficulty-level questions that measure a variety of learning levels. In addition, it is often difficult to create enough appropriate and plausible distracters for each stem. It is also possible for the testing learners to guess the correct answer yet not know the material. Having at least four choices limits the possibility of correctly guessing to 25 percent instead of the 50 percent possible in true-false tests.

True-False

Probably the best known of the objective tests is the *true-false*, or alternative-response, test. The true-false item usually consists of a single statement that the learner is required to recognize as either true or false. Though considered relatively easy to construct, true-false tests are also the most abused, and their quality is often doubted. The difficulty in constructing this type of test is creating statements that are completely true or completely false (Figure 9.7)

In addition to the traditional true-false test, there are also modified true-false tests. The modified test item asks the learner to explain why an item is false if that is the answer chosen. The modified version tests learning at a higher level (Figure 9.8).

Test construction considerations. When creating true-false tests, consider the following points:

- Place the letters *T* and *F* at the left margin if answers are to be marked on the test paper.
- Write instructions that direct learners to draw a circle around the answer they select rather than have them write the letter *T* or *F*, the symbol + or -, or the word *yes* or *no*.
- Create a sufficient number of items to provide reliable results.
- Distribute an equal number of true and false items randomly throughout the test.
- Do not use specific determiners that provide unwarranted clues. Words such as *usually, generally, often*, or *sometimes* are most likely to appear in true statements; the words *never, all, always*, or *none* are more likely to be found in false statements.
- Avoid creating items that are tricky or designed to catch the learner in a mistake.
- Do not use double-negative statements; they are very confusing to the learner and do not accurately measure knowledge, only cleverness at figuring out the statement.
- Avoid using personal pronouns such as *you* in test item statements.
- Do not use command statements. These types of statements usually incorporate other "don'ts" as well. For example the statement *"You must always wear gloves when caring for a patient"* uses the personal pronoun *you* and the qualifier *always* in a command statement.
- Do not use questions that ask trivia or obscure facts. Tests do not need to measure insignificant or common knowledge.
- Develop items that require learners to think about what they have learned, rather than to merely remember it.
- Attempt to make all statements the same length. The length is often a clue. Test developers tend to write true items consistently longer than false items because they are trying to justify them as true.
- Create statements that are brief and simply stated, rather than lengthy and complex. It is not necessary

way, it is tempting to award high ratings, even when the learner really didn't earn them. Evaluators should remember that learners who are different can do a good job and that there are several effective ways to approach and accomplish any task.

- *Contrast effect* — An evaluator must often rate several learners who are very dissimilar. An average learner can look extremely good or extremely poor in contrast to a very "low-scoring" or very "high-scoring" learner. Evaluators must keep in mind that they are rating learners' actual behavior compared to the performance standard.
- *First-impression effect* — This effect is closely related to the halo effect. Sometimes a learner performs very well in the beginning and then falters, but the evaluator has already formed an opinion based on the first impression. Evaluators must consider learner performance during the entire evaluation.
- *Recency effect* — Many times the evaluator remembers events or learner performance that occurred recently. If the recent performances or events were good, the evaluator may rate the learner better than the learner actually performs. If the recent performance was poor, the evaluator may rate the learner worse. Refer to the Laws of Learning section in Chapter 4, The Psychology of Learning.
- *Stereotype effect* — This particular type of effect can be a very difficult error to avoid since it can involve closely held and sometimes unconscious personal values. Sometimes evaluators assign ratings, not because of observed behavior but because they "fit" with the behavior associated with the stereotype. The evaluator may have preconceived ideas or prejudices about the stereotyped individual, which the evaluator may use unconsciously in rating the learner.

Scoring Methods and Interpreting Test Data

[NFPA 1041: 2-5.3, 3-5.4, 3-5.4.1, 3-5.4.2, 4-2.7]

Testing is a measurement process that determines whether learning has occurred. Giving a test is only part of the process. Scoring and analyzing the results are also necessary parts of testing. The primary purpose of analyzing test results is the same as the purpose of evaluation — to improve the teaching/learning process (see Purpose of Evaluation section). Performing some simple test analyses provides the instructors, the test developers, and the organization with information on test validity and reliability. The results of a test analysis point to test questions or answers that may need altering, rewording, or restructuring to make test points more clearly understood and more significant to the information being tested. Instructors can perform some simple analyses to assess the effectiveness of tests. Some simple test analyses, scoring methods, and corrective techniques are discussed in the following sections.

Scoring Methods

Scores are meaningless unless measured against a predetermined scale. A score of *66* points is excellent if the total possible is *70,* but it is usually considered a failing score if the total possible is *100.* Whether a score is recorded as a numeric or letter grade, it must be meaningful. If learners earn scores in several areas or for different activities, instructors must have a method of combining all scores and determining a final score. Several methods are appropriate and also contribute to analyzing tests and results. Composite scoring, point systems, and class participation rating scales are discussed in the following sections.

Composite Scoring

The basis of assigning scores or grades should be a composite (mixture) of various course activities and other factors. Establish specific values for activities such as assignments, projects, quizzes, examinations, and participation. Other less obvious learning outcome factors that can be scored come from the affective learning domain such as attitude and cooperation.

A system that is more objective and works for scoring both criterion-referenced and norm-referenced tests is the *point system.* Establish point values for each course activity, then total the points for the course score.

There are several ways to use composite scoring. The following example uses a particular point spread, but any point spread may be used. The total points earned by a learner can be converted to a letter grade in norm-referenced testing. For example:

85–100 percent of possible points = A
75–84 percent of possible points = B
65–74 percent of possible points = C
50–64 percent of possible points = D

The point values can be converted to a mastery or nonmastery classification in criterion-referenced testing (also called *competency testing* and *pass-fail*). Refer to the Criterion-Referenced part of the Interpretation Method section. For example:

> 85–100 percent of possible points = Mastery (competency/pass)
>
> 84 percent or less of possible points = Nonmastery (noncompetency/fail)

Instructors must be as objective as possible in scoring and grading and must not be influenced by factors not pertinent to the learner's achievement. The instructor must be prepared to defend any score or grade given. Developing a standard system of scoring and grading and then adhering to it serves to increase objectivity, consistency, fairness, and reliability.

Class Participation Rating Scale

The *class participation scale* documents the active participation of learners in a class and rates learner class participation in an objective way. In the following example, the instructor circles the number that best describes the learner in class:

> The numbers represent the following values:
>
> *5 = excellent*
> *4 = good*
> *3 = acceptable*
> *2 = needs improvement*
> *1 = not acceptable*
>
> 1. To what degree does the learner participate in class discussions?
>
>   *1 2 3 4 5*
>
> 2. To what degree are the learner's comments related to the discussion topic?
>
>   *1 2 3 4 5*
>
> 3. To what degree does the learner pose thoughtful questions on the discussion topic?
>
>   *1 2 3 4 5*

A rating scale such as this enables the instructor to judge performance on an objective scale. Judgment by the instructor is subjective and determined by consciously observing learners during class sessions. Instructors should realize when assessing participation levels that some learners are more aggressive than others. This aggressiveness may make the docile learners appear unusually introverted. The instructor must compare and evaluate each class member as an individual while simultaneously comparing them to other class members.

Test Item Analysis

A *test item analysis* is a helpful tool for showing how difficult a test is, how much it discriminates between high and low scorers, and whether the alternatives used for distracters, for example, work to confuse learners. This process makes the instructors and test developers aware of problem test items so the items can be improved.

The simplest way to analyze a test item is to list each item with the possible answers, then count and record the number of items selected for each possible answer. The correct answer is emphasized in a colored bold italic in the following example of three multiple-choice questions:

Test Question (Number)	Choice A	B	C	D
1	5	8	***6***	1
2	2	3	2	***13***
3	***2***	17	1	0

By using the example for Question No. 1, analyzers can see that only 6 learners chose the right answer, Choice C. Since 8 learners selected the B distracter, assume that it served as an effective distracter. The A distracter worked with 5 learners, but the D distracter had limited effectiveness. Analysts may choose to rewrite the D response to make it more plausible. However, if only 6 out of 20 learners answered this item correctly, then the question may not have tested the intended learning.

In Question No. 2, the majority of the class chose Choice D for the correct answer. This question did not distinguish among learner knowledge very well, since there were few who answered incorrectly. The test analyzers have to make a decision on the effectiveness of this item. From this simple analysis, it cannot be determined whether the item was too easy or whether it accurately tested for learning.

The third test question shows the power of Choice B to distract learners from the correct answer. However, Choices C and D are poor distracters. Analysts may choose to discard them and create new distracters.

Test Result Analysis

Analyzing test results is the process of using different methods to interpret results. The process of analyzing test results also aids in determining validity and reliability — significant aspects of testing. Analysis methods include the processes of simply grading and ranking scores, determining measures of central tendency and the variability of scores, and using other more complex statistical methods. The following sections describe the relationships of validity and reliability to analysis and give some simple statistical methods of test analysis.

Validity and Reliability Components

If test results are not valid or reliable, they have no meaning. Recall that validity is the degree to which a test measures what it sets out to measure. Reliability is the consistency of test scores from one measurement to another.

Reliability is a condition of validity. In other words, if test scores differ on a test given one day from a test covering the same material given the next day, then the test scores are not reliable. If the scores differ, they cannot be said to measure what they set out to measure and are not valid. Therefore, reliability is necessary to validity.

Validity has different meanings when interpreting the results of criterion-referenced and norm-referenced tests. In both types of tests, validity means the extent to which the test measures learner achievement of the behavioral objectives. In criterion-referenced tests, validity refers to the measurement of mastery or nonmastery by an individual as compared against behavioral objectives. In norm-referenced tests, validity refers to the measurement or grade of the individual learner compared against other learners in a class.

Reliability also has different meanings when interpreting test results. Reliability of criterion-referenced tests means the consistency of results in classifying mastery or nonmastery of an individual. In contrast, reliability of norm-referenced tests means the consistency of results among a group of learners. The type of test — criterion-referenced or norm-referenced — should be specified and designed by intent, not by accident.

Statistical Methods

Some methods of analyzing test results are generally referred to as statistics. *Statistics* are nothing more than ways of organizing, analyzing, and interpreting test scores. The elementary statistical methods described here use simple arithmetic skills that instructors already possess. The only new element is the introduction of new terms, which is expected in any new learning.

Raw scores consist of the points a learner receives on a test. If a learner answers 38 items correctly out of 40 test questions, then the raw score is *38*. A raw score is easily converted to a *percentage score*. Compute a percentage score by adding the number of correct answers and dividing by the total possible answers. In the example given, the percentage score is *95 percent*, which is computed by taking the raw score of 38 correct answers and dividing it by 40 total possible answers.

The raw score is of little value in norm-referenced tests unless it is converted into some type of calculated score that shows how it compares to other scores from the same test in the same class. This method of comparing scores to other scores is called *ranking of scores*. The simplest method is to list all scores from highest to lowest (or lowest to highest), which gives a quick visual identification of how many learners scored what points. Once ranking is done, it is easy to see the *range* of scores and to see the *median* (*middle* score) and the *mode* (*common* score). To determine the *mean* (*average* score), total all scores and divide by the number of individuals testing. It is not unusual for the mean, median, and mode to be the same or close to the same number (often within 1 to 3 points). With these basic arithmetic calculations, instructors can quickly calculate basic statistics for any group of testing scores.

The following example illustrates a simple analysis:

1. List *raw* scores (21 scores): *76, 62, 84, 95, 98, 93, 84, 88, 89, 84, 84, 83, 84, 81, 78, 83, 84, 97, 79, 69, 78.*
2. *Rank* scores from lowest to highest: *62, 69, 76, 78, 79, 80, 82, 83, 83, 84, 84, 84, 84, 84, 84, 88, 89, 93, 95, 97, 98.*
3. *Range* of scores is easily seen: *62 to 98.*
4. *Total* of all scores: *1,756.*
5. Divide total of all scores by total number of scores to get the *mean* or *average* score: *1,756 ÷ 21 = 83.6* (rounded to *84*).
6. The *middle* score of the 21 scores is *84* (the eleventh score, which has ten scores before and after it). If there is an even

number, the middle score is the average of the two middle scores. If there was one less score in this example, the middle score would still be *84*. (If the middle scores were 82 and 83, the average score would be 82.5, even if there is not an actual score of 82.5. This score is an "anchor point" to determine the mean.)

7. The *mode* or most common score is *84* since there were six scores of *84*. There may be several modes in a range of scores.

In the example given, the instructor or the organization may decide that *84* is the average or *C* range. Scores above *84* are ranked as *B* and *A*; scores below *84* are ranked as *D* and *E* or *F*. Another option is to "curve" the scores and decide that the *84* is the *B* score; every score above *84* is an *A*, and scores below it are *C, D,* and *E* or *F*. For many certification classes, scores cannot be curved and *70 percent* is considered mastery or passing.

Corrective Techniques

Test scores may sometimes appear skewed (distributed mostly toward one end of a scale rather than evenly distributed). When an entire class scores poorly or the scores are skewed toward the lower end of the scale, the organization or the instructor can perform a careful analysis to determine reasons for poor testing performance. Corrective actions can be immediate or long range and also simple or complex. Some steps that instructors can use to correct skewed test scores are as follows:

- ***Throw out poor test items and refigure the score*** — Take this step after carefully evaluating the test item and determining that its content or structure was misleading to learners.
- ***Review the test analysis, adjust the test items, and give the test again*** — Adjust test items by rewording to make them more clear or change the wording in distracters, for example.
- ***Reteach the lesson and retest*** — Take this step upon the discovery that certain critical information was not taught or made clear to learners during the program.

Other possibilities are to review the instruction, circumstances of the instruction, and testing situations. Any of these factors may have confused, misled, or distracted the learners in testing effectively. Adjust the instruction, reteach, and retest as necessary if there were critical circumstances. If two or more identical classes are held at the same time, another analysis technique is to compare test results of all classes, then decide what steps need to be taken.

Purpose of Evaluation

[NFPA 1041: 2-2.1, 2-2.3, 2-5.4, 3-2.5, 3-2.6, 3-5.2, 3-5.3, 4-2.5, 4-2.5.1, 4-2.7, 4-5.3, 4-5.4]

A complete evaluation process must include a review of behavioral objectives and test results in order to make a decision about whether the results are a true measure of the objectives. This review process enables instructors and program developers to meet the primary purpose of evaluation, which is to improve the teaching/learning process. Many evaluation techniques are available to assess learning levels and to determine how instruction can be changed to enhance learning.

The evaluation process systematically collects information for the purpose of making decisions. Evaluation has three major components: criteria, evidence, and judgment. In evaluating learning levels, instructors must have well-defined behavioral objectives (*criteria*) and results of tests or observations (*evidence*). Based on these two components, instructors can make decisions (*judgments*) on whether the results (evidence) indicate accomplishment of the behavioral objectives (criteria) that were set at the beginning of the program (Figure 9.12).

Each component must be present in order to have a complete evaluation process. For example, if there is no behavioral objective, there is no way to judge the test result because there is no criterion to measure

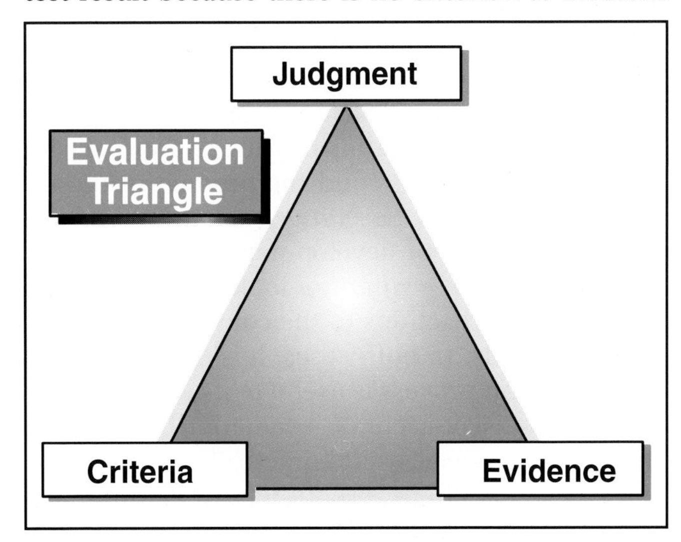

Figure 9.12 The three components of evaluation: criteria, evidence, and judgment.

evidence of accomplishment. And if no one can make a decision or judgment about how well behavioral objectives and test results match, then no effective evaluation takes place.

Evaluation is used for many other purposes such as the following:

- Discover weaknesses in learning as well as in instruction.
- Diagnose causes of learning problems or weaknesses.
- Establish guidance or recommendations for further study.
- Assign grades.
- Make administrative decisions about learners, instructors, and courses.

Evaluating Course and Instructional Design

[NFPA 1041: 3-5.2, 3-5.3, 3-5.3.2, 4-5.2, 4-5.3, 4-5.4]

Instructional design is the analysis of training needs, the systematic design of teaching/learning activities, and the assessment of the teaching/learning process. As training needs change, the instructional process changes. When the design of teaching/learning activities is no longer effective, the process must change. Change is the reason for continuous assessment of the teaching/learning process. Instructors must always be alert to ways of improving or updating instruction, or even eliminating instruction that is no longer needed. In a changing world of instructional needs, content, methods, and techniques, instructors must remain flexible.

Often, instructors use only test results for evaluation. But there are many other types of evidence that collectively give a more complete picture as to whether the teaching/learning process was successful. Two kinds of evaluation provide an approach for looking at the process and the product of the instructional process: *Formative evaluation* looks at the process of course development and instruction, while *summative evaluation* looks at the product and evaluates reactions to the course and instructional methods. In both cases, it is important that the evaluation process is useful to its users, so data collection must be accurate. Evaluation instruments should always be easy and practical to administer, and results should be easy to report. Planning the approach is also important.

Planning Considerations

To get the most from a course evaluation, the instructor must plan the approach by thinking through and answering the following questions:

- How did participants feel about the training? What did they learn? How did training affect their attitudes and behavior? What were the organizational results?
- How can items addressed be answered? Will information gathering be administered by paper-and-pencil tests, questionnaires, or surveys? Will tests require participants to demonstrate their new knowledge and skills in role-plays, simulations, or actual performances?
- What are the objectives of the training program? Are the evaluation criteria based on these objectives?
- Do the criteria indicate improvement between expected and actual performance when measured against the results of the needs analysis?
- What data sources are already available to help measure results (productivity reports, daily log sheets, and training and personnel records)?
- What alternative methods for gathering data are available (interviews and on-site observations)?
- What are the best and most cost-effective methods for measuring the results of the training? Are there less costly, more efficient ways of administering the evaluation?

In evaluating the instructional process, it is critical that the training organization make clear to anyone involved *what* is to be evaluated and *why*. Answering *what* and *why* clarifies and guides how evaluation is to be conducted and eliminates misunderstanding. Include the following six essential areas in a thorough evaluation of the instructional process. Answer the questions related to each area.

- ***Reaction*** — Were participants satisfied with the course? Were instructors and management officials satisfied with the learning that occurred?
- ***Knowledge*** — What new knowledge did the participants acquire and demonstrate?
- ***Skills*** — What new skills did the participants acquire and demonstrate?
- ***Attitudes*** — How has the training changed participants' opinions, values, and beliefs?
- ***Transfer of learning*** — How has the training affected the ways participants perform on their jobs?

- *Results* — How has the training contributed to accomplishing organizational goals and objectives?

Formative Evaluation

Formative evaluation is the ongoing, repeated checking during course development and during instruction to determine the most effective instructional content, methods, aids, and testing techniques. Two ways to conduct formative evaluations are field testing and observation. Formative evaluation answers such questions as the following:

- Does the course provide the appropriate information and learning format?
- Is the instructor teaching the right content and using the most appropriate methods to facilitate learning?
- Have the participants learned in the most efficient way possible?

Field Testing

The course objectives serve as the primary criteria against which a judgment is made. Objectives are developed before the course is designed. Course design and how it works with the objectives are field tested in a *pilot program* before the course is finalized. A *field test* is the process of teaching the course on a trial basis to determine whether the following components are addressed:

- Sequence of material facilitates learning.
- Teaching methods and aids are appropriate to the objectives and effective in teaching the material.
- Objectives can be met within the designated time frame.
- Testing procedures are adequate and appropriate for the course objectives.
- Course can be revised before using it on a regular basis if necessary.

Observation

Evaluation is not a static, one-time event; it occurs at all times throughout development and instruction. Just as testing occurs throughout instruction, course evaluation occurs before, during, and after instruction. Instructors can gather different kinds of evidence, most of which is observational, during the course. The instructor wants to be attentive to the following factors:

- Learner interest in the subject
- Level of general participation
- Learner reaction to exercises and activities
- Level of learner questions and comments
- Level of learner frustration
- Level of learner sense of achievement
- Test results

With feedback received during instruction, instructors can change or modify instructional methods to meet the needs of the class. In some cases, it will be clear that a course needs to be changed on a permanent basis. The judgment of instructors is critical when making decisions about a program that is under development.

Summative Evaluation (Reaction Survey)

A *summative evaluation* is an end-of-the-course appraisal. This type of evaluation commonly measures learning by some form of objective or subjective test, but it is also used to evaluate the reactions of learners and instructors to each other and to the course. Some examples of how summative or reaction evaluations are gathered are by observing and reporting unexpected learning outcomes or by performing follow-up surveys of learners, their supervisors, and their instructors about the effectiveness of the program, its activities, and its job relevance. Summative evaluations answer the following questions:

- Have program participants learned what is needed to perform their work duties on their jobs?
- Was instruction effective so that it met performance objectives?
- Were instructional activities relevant to job tasks?

Instructors and program developers gather evidence from two critical sources: test results and course feedback from participants. *Test results* measure the degree of learning that has occurred (see Scoring Methods and Interpreting Test Data section). *Course feedback* describes the learners' opinions of and reactions to the success of the course materials and the instruction. Instructor assessment of the program content and relevance is also an important source of feedback. In addition, supervisor input on program relevance to job tasks is considered when evaluating a training program.

Course Feedback

The final course evaluation (feedback) is an important part of the entire evaluation procedure and determines whether the instructional process met the course objectives. As a course of study or period of training

so that the evaluation instrument determines if the student has achieved the learning objectives, the instrument evaluates performance in an objective, reliable, and verifiable manner, and the evaluation instrument is bias-free to any audience or group.

3-5.2.1 *Prerequisite Knowledge:* Evaluation methods, development of forms, effective instructional methods, and techniques.

3-5.2.2 *Prerequisite Skills:* Evaluation items construction and assembly of evaluation instruments.

3-5.3 Develop a class evaluation instrument, given agency policy and evaluation goals, so that students have the ability to provide feedback to the instructor on instructional methods, communication techniques, learning environment, course content, and student materials.

3-5.3.1 *Prerequisite Knowledge:* Evaluation methods, test validity.

3-5.3.2 *Prerequisite Skills:* Development of evaluation forms.

3-5.4 Analyze student evaluation instruments, given test data, objectives and agency policies, so that validity is determined and necessary changes are accomplished.

3-5.4.1 *Prerequisite Knowledge:* Test validity, reliability, and item analysis.

3-5.4.2 *Prerequisite Skills:* Item analysis techniques.

Chapter 4 Instructor III

4-2.5 Construct a performance based instructor evaluation plan, given agency policies and procedures and job requirements, so that instructors are evaluated at regular intervals, following agency policies.

4-2.5.1 *Prerequisite Knowledge:* Evaluation methods, agency policies, staff schedules, and job requirements.

4-2.7 Present evaluation findings, conclusions, and recommendations to agency administrator, given data summaries and target audience, so that recommendations are unbiased, supported, and reflect agency goals, policies, and procedures.

4-5.2 Develop a system for the acquisition, storage, and dissemination of evaluation results, given agency goals and policies, so that goals are supported and those impacted by the information receive feedback consistent with agency policies, federal, state, and local laws.

4-5.3 Develop course evaluation plan, given course objectives and agency policies, so that objectives are measured and agency policies are followed.

4-5.4 Create a program evaluation plan, given agency policies and procedures, so that instructors, course components, and facilities are evaluated and student input is obtained for course improvement.

Chapter 10

Instructional Media and Technology

No instructional medium or technology is a substitute for experienced, skilled instructors. Instructors who depend on instructional media and technology to deliver their presentations become redundant. Even when instructors use advanced instructional media or technology, all the same skills and tricks of the trade such as strategic use of questions, positioning within the room, body language, and vocal inflection still apply. Instructors must not make the instructional medium or technology a crutch or prop. They must learn to use it effectively by integrating each element with the traditional skills they already have. Instructors must practice with instructional media and technology, experiment with them, and have *FUN* with them. Using instructional media and technology can enhance existing instructional skills to new levels. Today's learners who attend fire and emergency services training programs have grown up in the age of television, video games, and multimedia computing. These people do not want to attend straight lectures or read words on a screen — but they do want to learn. Effective use of instructional media and technology allows instructors to meet many learning styles, tailor instruction to individual needs, and design instructional media to deliver the intended messages. Instruction is not about the technology or fancy tricks, it's about the instructor and the message.

This chapter describes the many types of instructional media and technology commonly used in training programs. It discusses the advantages and disadvantages of the various media and technology so that instructors may consider which media are appropriate for training situations and which can be easily integrated into training programs. While instructional media and technology often enhance and reinforce training, they must be used with careful planning to achieve the desired training results. References and supplemental readings and job performance requirements are given at the end of the chapter.

The Need for Instructional Media

[NFPA 1041: 2-2.1, 2-2.2, 2-2.2.1]

Fire and emergency service instructors are faced with the task of conveying complex information on many topics to people of varied backgrounds within tight time frames. Instructional media allow an instructor to maximize the transfer of knowledge and skills within the time allowed. The old saying *"A picture is worth a thousand words"* conveys the same message: The efficient and effective communication of detailed or subtle concepts can be facilitated through stimulation of multiple senses. Instructional media have the benefits of appealing to the senses and saving time.

Appeals to Senses

Using multiple senses is more effective in learning situations. Studies have shown an increase in the retention of learning when learners read along with what they are hearing as compared to simply listening to a lecture (See Chapter 4, The Psychology of Learning). This concept is amplified by the use of media that allow learners to hear, read, see still or motion pictures, manipulate a model, operate a simulator, and/or participate in a real-time scenario (Figure 10.1).

Saves Time

Although it may seem more efficient in terms of time, a pure lecture medium is not very effective for most fire and emergency services subjects. The benefits of reaching large numbers of people in a short period of time through lecture are offset by the lost opportunity to access other pathways into their minds. Action-oriented, hands-on learners — types who are prevalent in the fire and emergency services — are receptive

Figure 10.1 Effective instruction appeals to the senses, which stimulates learning and increases retention. Practice with real scenarios, such as the one pictured, allows learner participation using multiple senses: seeing (flames), hearing (instructions and sounds of fire), and feeling (heat, hose, and water spray). *Courtesy of Bill Partridge, Utah Fire and Rescue Academy.*

to the use of varied and multiple media. The appropriate use of instructional media increases time efficiency in terms of the learning (knowledge and skill abilities) that individuals transfer and retain.

How to Select Appropriate Media

[NFPA 1041: 2-2.2, 2-3.1, 2-3.2, 2-3.2.1, 2-3.2.2, 3-3.2, 3-3.2.1, 3-3.2.2, 3-3.3, 4-2.6, 4-2.6.1, 4-2.6.2, 4-3.2, 4-3.2.1]

Using instructional media in training programs means planning to make them work with the program content and objectives and ensuring that they have a purpose. Consider the following when selecting instructional media:

- Behavioral objectives and content
- Required learner performance
- Class size and interaction
- Media equipment flexibility
- Pace of learning
- Practice factors

Behavioral Objectives and Content

First and foremost, the content of any instructional media that an instructor uses in a class must be relevant to the desired behavioral objectives or learning outcomes. It is often tempting to make use of a new, professionally produced instructional medium, but it may only give marginal benefit to the class. If a preproduced package is known to have limited relevance, consider using only the portion that applies to the class. Alternatively, adapt the package to make it more pertinent or current. It may be more appropriate to not use the instructional medium at all or substitute one that more closely matches the class objectives.

Required Learner Performance

The standardization of training curricula in recent years has made it easier to select appropriate media. Standards for professional competency are developed, published, and periodically updated. Textbooks, instructional media, and curricula are developed to specifically meet those standards. In many cases, the standards development and curriculum development occur simultaneously or with at least some overlap. This timing allows current and compliant materials to be available as soon as possible after the publication of standards.

In some jurisdictions, compliance with standards is mandated by legislation. Regardless of any legislated requirements, the concept of *due diligence* dictates that a prudent organization base required learner performance on any and all existing standards. These standards form the basis of comparison in any civil or other legal action in which the training of fire and emergency services personnel is concerned. By using instructional media that have been specifically developed to meet standards, the employer or teaching organization can more easily demonstrate compliance with legislation or due diligence.

In areas of instruction where no external standards exist, it is still important to ensure that instructional media are demonstrating or illustrating required learner performance. An example would be instructing battalion chiefs in their responsibilities as managers under a new municipal policy. In this case, the use of an instructional medium, whether commercially produced or developed within the organization, best benefits the class if it outlines the manager's responsibilities exactly as dictated in the policy.

Class Size and Interaction

The size of a class and the expected level of interaction are key factors in selecting instructional media. For example, using an easel pad in front of an audience of 200 people may not be appropriate. An overhead projector with a large screen is a better choice. On the other hand, when debriefing a group of 6 trainees after a practical exercise, an easel pad or marker board may be ideal. In the first example, interaction with the class may be limited to a question-and-answer period at the end of the presentation. In the second example, a highly participative coaching session could be facilitated with diagrams on an easel pad or marker board.

Media Equipment/Materials Flexibility

When selecting items of instructional media equipment (for example, projectors, television sets, and easel pads), give consideration to the flexibility of their uses. Ideally, an expensive piece of equipment is going to be useful for a variety of training programs in multiple locations. It is important to make maximum benefit of the equipment dollar. Any equipment that is not used regularly represents money that could have been spent more effectively elsewhere.

The same principal cannot be applied as readily to training materials such as textbooks, overhead transparencies, or videotapes. These materials are usually applicable to a single training program. Ensure that any materials purchased are as current as possible. Though these materials may be applicable to only one area, their useful life is maximized if they are the newest available editions.

Pace of Learning

Different instructional media can be effective when applied in different instructional environments. One factor in the suitability of the medium to the environment is the intended pace of learning. When the expected pace of learning is high, media can be integrated into the lesson plan in such a way as to allow a class to move quickly through the information. For example, a videotape can quickly review the basics of an incident management system at the beginning of a lesson on the responsibilities of the water supply officer.

If the content of the lesson is primarily unfamiliar to the learners, media that allow the instructor to proceed at a slower pace are more appropriate. One strategy is to use precourse reading assignments from textbooks or self-study guides followed by guided discussions on marker boards or easel pads. Learners can then view a videotape for review and reinforcement.

If the learners represent a variety of knowledge levels or experiences, a learner-centered strategy can include a mixture of the two scenarios. Those learners who are not yet at the required learning level can have access to self-study media prior to joining the more advanced learners at the appropriate time.

Practice Factors

It is always important to give learners an opportunity to apply new knowledge and skills. To make the new application meaningful, practice factors should be built into the lesson plan along with the instructional media needed. In order to apply the knowledge and skills obtained from a lesson on handling a hazardous material leak or securing a container, for example, the learners need the equipment and props so they can execute the steps (Figure 10.2).

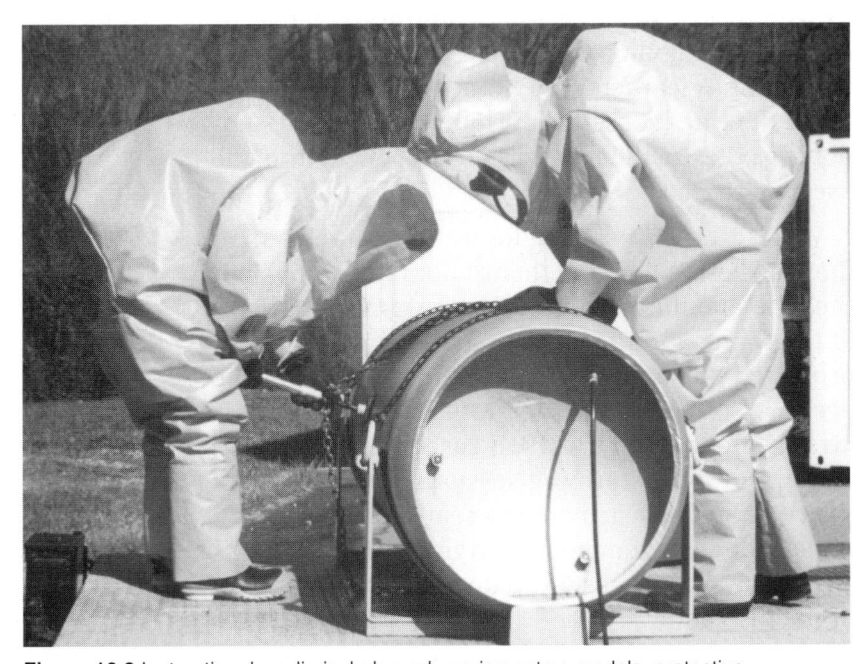

Figure 10.2 Instructional media include such equipment as models, protective equipment, and props. Learners use various equipment and props during practice sessions so they may execute the steps learned in classroom sessions. *Courtesy of Dan Gross, Maryland Fire and Rescue Institute.*

The media required to apply management or investigation skills such as incident management, supervision of personnel, problem solving, or accident investigation may be more difficult to determine. For prospective incident managers, scenarios can be conducted on paper with magnets and diagrams on a marker board, with tabletop models, or through interactive and virtual reality technology. Similarly, accident investigation skills can be applied by using case studies on paper, using computer-based scenarios, or staging an accident scene (complete with simulated casualties) for the learner to physically investigate. Each of these examples requires that the instructor carefully plan the use of instructional media into the lesson in order to give learners meaningful application of their new knowledge and skills.

Integration of Instructional Media into the Learning Environment

[NFPA 1041: 2-4.1, 2-4.2, 2-4.2.1, 2-4.2.2, 2-4.4, 2-4.6, 2-4.7, 3-3.3.1, 3-3.3.2, 3-4.2, 3-4.2.1]

Once instructors select the appropriate instructional media, they need to integrate them into the learning environment so that using them is a smooth and efficient process. Consider the arrangement of the classroom and distractions to the learning environment when using instructional media.

Classroom Arrangement

When planning the arrangement of learners in the classroom, instructors must take into account the visibility and audibility of any instructional media used. For example, learners who cannot see the screen because they are blocked by furniture, projectors, supporting poles, or other people do not benefit from the use of overhead transparencies.

When arranging seating for learners in relation to the instructional media, instructors must consider the type and limitations of the equipment, the size of the room, and the needs of the learners. Display projected or nonprojected visual media above the eye level of seated learners. An exception to this arrangement is the use of an easel pad during discussion or brainstorming sessions. In this type of instruction, the instructor can seat small groups in a horseshoe arrangement, which gives all participants a clear view of the easel pad down to the waist level of the instructor.

Instructors must ensure that the volume of any amplified sounds are at an audible but comfortable level for all participants. For an average-sized classroom, a television set has adequate volume to be heard at the back of the room, and most instructors do not need a microphone in order to be heard. In a large auditorium, however, participants in front may be deafened by the volume level of a videotape presentation that is set loud enough for those at the back to hear, and some instructors may need to shout in order to be heard if they are not using a public address system. It is important that a sound system provide an even volume level to all parts of the learning environment.

Learning Environment Distractions

Regardless of the instructional media used, instructors must take care to minimize distractions in the learning environment. By applying Maslow's Hierarchy of Needs (see Chapter 4, The Psychology of Learning), instructors can ensure that learners feel safe and comfortable. Eliminate or minimize any auditory or visual stimuli that are not related to the topic. Additional actions that help minimize distractions include keeping window blinds closed or coordinating with other personnel and operations to minimize pedestrian and/or vehicle traffic in the training area.

Beyond the basic and immediate considerations mentioned, instructors must also consider how they actually use the media. Inappropriate use of instructional media can actually create distractions. Techniques on how to avoid distractions when using specific types of instructional media are discussed as the media are introduced in later sections of this chapter. Some simple strategies for avoiding distractions are as follows:

- ***Introduce instructional media at the time they are to be viewed, heard, or manipulated*** — Withdraw them as soon as their immediate relevance is complete. For example, leaving a slide or transparency displayed on a screen while discussion continues on another point creates a distraction that may divide the attention of the class.
- ***Avoid simultaneous use of multiple media unless they are carefully and strategically written into the lesson plan*** — A common example is to pass a small item around the class for participants to examine. By the time it reaches the last person, it may no longer be relevant to the material the instructor is currently presenting. In the meantime, the participants' attentions have been divided between the instructor and the object in their hands or the wait for the object to get into their hands. Imagine the potential distractions when two or more objects are making their way around the classroom at the same

time. An alternative to passing items around the class is to use a visual presenter (see Projected Instructional Media section).

- ***Apply media in ways that emphasize the message, not the equipment*** — This technique is especially important when using sophisticated electronic media such as multimedia projectors. Avoid the temptation to use every feature of a new program, system, or simulator. When learners are paying more attention to flashing lights, overly showy slide transitions, and sound effects, the medium becomes the message.

Nonprojected Instructional Media

[NFPA 1041: 2-4.7.1, 2-4.7.2]

Nonprojected instructional media offer several major advantages over projected media (see Table 10.1). Nonprojected media are not dependant on high levels of technology or technical skill. As a result, they are easier to use and less likely to malfunction at inopportune moments. Nonprojected media also generally cost less to purchase, develop, and maintain. Some popular and easy-to-use nonprojected instructional media are as follows:

- Chalkboard, dry erase marker board, and easel pad
- Illustrations and display boards
- Duplicated materials
- Models
- Audiocassettes
- Casualty simulation media

Figure 10.3 A chalkboard is a commonly used instructional medium. It is used to list key ideas and show simple illustrations or detailed diagrams. Newer chalkboard models have magnetic capabilities that expand versatility. *Courtesy of Jarett Metheny, Midwest City (OK) Fire Department.*

Chalkboard, Dry Erase Marker Boar Easel Pad

The easiest, most frequently used, and most versa nonprojected media are the chalkboard, dry eras marker board, and easel pad. These items may be fixed or mounted to a wall, movable (on wheels or stands), or portable (folding and compact for travel). Instructors find these media useful for numerous instructional activities and rely on them as supplements and backup when more technical or complex media fail.

Chalkboard

The *chalkboard*, the traditional mainstay of classroom education, remains a versatile and effective training tool (Figure 10.3). The advantages of a chalkboard are low cost, low maintenance, high reliability, and high visibility. The wide availability of large-sized colored chalk (sidewalk chalk) makes it easy for instructors to vary their displays from the monochrome look of the usual white chalk on a black or green background. The drawbacks of a chalkboard are that the presentation, once it is produced, must either be left in place or erased. If left in place, that portion of the chalkboard is no longer available for other uses. If erased, the presentation must be built from the beginning if needed again. Chalkboards should be thoroughly erased between uses and periodically cleaned.

Dry Erase Marker Board

The *dry erase marker board* has many of the same uses as the chalkboard. It is versatile, reliable, and easy to use. Instructors who have developed skills and an affinity for either one of these media find it easy to transfer to the other. Dry erase marker boards allow instructors to write or draw on the board with markers of many colors (Figure 10.4). Dry erase markers leave colored marks that are erased with special erasers. If material is left on the board for more than a few minutes, a cleaning solution may be needed to completely remove all marks. It is important to use only dry erase markers on marker boards.

Table 10.1
Nonprojected Instructional Media

…tion	Advantages	Disadvantages
…Dry Erase Marker Boards	• Versatile and effective • Low cost • Low maintenance • High reliability • High visibility • Easy to vary display with colored chalk	• Once produced, presentation must be left in place or erased • *Dry erase boards require the use of dry erase markers and cleaning solutions*
Easel Pads	• Inexpensive and easily transported • Can prepare information in advance • Take little space • Allow flexibility of presentation order • Easels can be wall-mounted, freestanding, or placed on tabletop • Provide a permanent record	• Provide a limited writing space • Create limitations on room and audience size • Pads are not very durable although they can be reused
Illustrations/Display Boards	• Save time in showing a vast amount of information • Address a variety of learning styles	Distracting when passed around individually or when on display continuously
Duplicated Materials (handouts, study guides)	Provide a resource/reference copy of critical information not easily found in the text	Distracting when passed around individually *or before intended time of use*
Models	• Excellent for illustrating mechanical or spatial concepts • Allow learners to observe relationships of components and their functions • Allow manipulation • Some can be purchased or constructed at low cost	• Some are expensive to purchase or construct • Some take extensive time to construct
Audiocassettes	• Bring realism to training • Can be used to — Record and review lectures — Dictate and review notes — Listen to prerecorded books or messages	• Can be time-consuming to listen to recorded materials • Can be destroyed by heat or misuse
Simulation Aids	• Increase realism of training • Many simulations are inexpensive	• May be costly to train individuals in casualty simulation and to purchase materials • *May be* time-consuming to prepare simulations and to brief personnel to act as casualties

Figure 10.4 The dry erase marker board is becoming a favorite and more common instructional medium. By using colored markers, instructors make instructional information visually attractive and interesting. (left) *Courtesy of Toronto Fire Services.* (right) *Courtesy of Larry Ansted, Maryland Fire and Rescue Institute.*

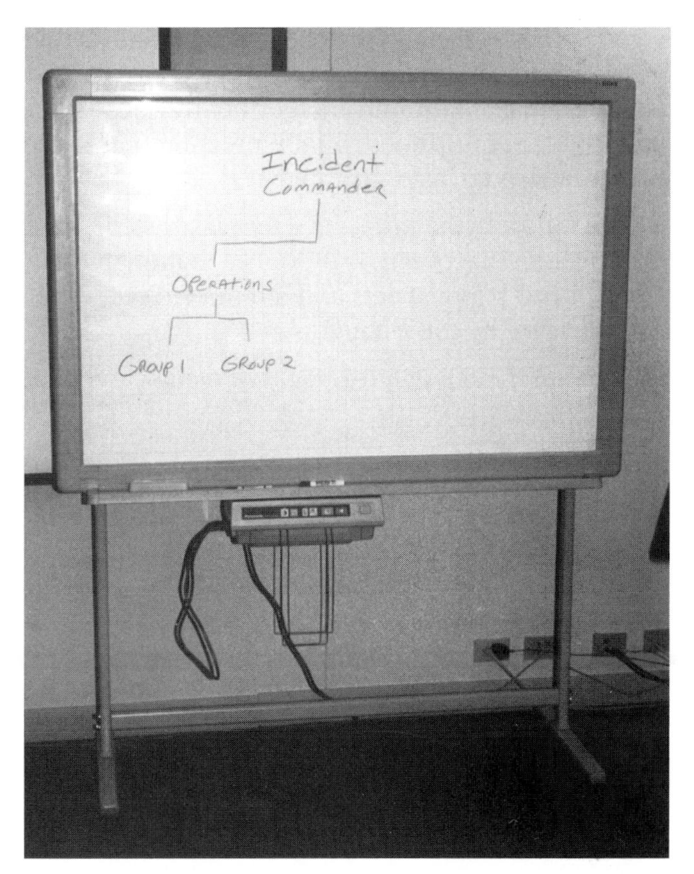

Figure 10.5 An electronic copyboard is a technical version of the dry erase marker board. The copyboard scans the information printed on the board and prints it to paper for handouts. *Courtesy of Jeff Tucker.*

Using other types of markers may result in lengthy cleanups or may even necessitate replacing the board. Dry erase markers are more expensive than regular markers intended for use on easel pads. However, using dry erase markers exclusively avoids the possibility of damaging a marker board.

Many marker boards are made of an enameled surface on a thin sheet of steel, allowing instructors to use magnets as part of the display. Instructors can draw diagrams using markers of multiple colors and add magnets to represent fire apparatus, ambulances, or other objects. Some chalkboards also accept magnets.

A versatile but expensive variation of the dry erase marker board is the electronic copyboard that scans and prints a reduced-size copy of material drawn or printed on the board (Figure 10.5). An electronic copyboard is expensive for most budgets, but its cost eventually may be reduced by increased demand and improved technology. This device is especially handy in cases where the instructor uses the board to develop a detailed plan or diagram for future use.

Figure 10.6 An easel pad readily supplements the chalkboard and dry erase marker board. It is also a useful medium for learner groups who use it for outlining tasks and recording activities. *Courtesy of Dan Gross, Maryland Fire and Rescue Institute.*

Easel Pad

An *easel pad* is a large pad of paper mounted on an easel for display (Figure 10.6). Paper may be lined, unlined, or ruled for graphs; various grades and colors are available. Easels can be wall-mounted, freestanding on the floor, or shortened for tabletop use. A wide variety of colored marker pens can produce attractive, interesting presentations. Instructors can easily transfer their chalkboard and marker board skills to the easel pad. Some professionally prepared easel-pad presentations are also available.

A variation on the easel pad is the plastic static-cling sheet. These sheets can be produced in the same way as easel-pad paper sheets and mounted directly on a wall or board by static cling.

Some of the many advantages of using easel pads are as follows:

- Pads and easels are easily transported.
- Information can be prepared in advance on pads lying flat on a table.
- Pages are turned one after the other, allowing the creation of a large presentation that takes up little space.
- Pages do not always need to be turned in sequential order, allowing flexibility in the order of the presentation.

Illustrations and Display Boards

Instructors can apply the concept of *"A picture is worth a thousand words"* by displaying illustrations or diagrams that they have purchased or prepared in advance. In addition to saving class time, illustrations contain vast amounts of information. Using graphical materials also addresses a variety of learning styles beyond those reached by written or spoken words. Some examples of the benefits of using illustrations and display boards are as follows:

- Technical diagrams such as mechanical or electrical schematics illustrate and help explain troubleshooting or repair procedures.
- Maps or plan diagrams illustrate routes or aid in debriefing incidents.
- Anatomical charts assist in explaining human anatomy and physiology in emergency medical services (EMS) classes.
- Flowcharts illustrate and help explain processes or procedures that are dependant on inputs or results of yes/no questions.
- Data charts such as pie charts or bar graphs illustrate common causes of injury, response types, or other statistical data.
- Photographs of incident scenes are invaluable in illustrating proper or improper procedures.

Instructor Tips

- *Write **lightly** any easily misspelled words or technical terms or **lightly** draw diagrams and other similar material on the easel-pad paper or chalkboard in advance with a lead pencil. The pencil marks are visible to the instructor standing at the front of the class but not to the learners. Quickly trace over the pencil marks with chalk or marker. This technique ensures accuracy or quick reproduction of material in class.*
- *Draw complex or detailed diagrams in advance. Mark preprepared material with instructions not to disturb if the classroom is to be used by other instructors in the interim. Keep diagrams covered until needed in order to avoid distracting the class. Use one of the following methods:*
  - *— Cover material with a sliding section of a chalkboard or marker board (Figure 10.7).*
  - *— Tape paper across material on boards.*
  - *— Cover material on an easel pad with blank sheets or title sheets.*

Figure 10.7 Some chalkboards are mounted on runners in several "layers." Instructors can prepare an outline or diagram on an inner layer, slide an outer layer in front of it, and expose the information when it is time for it to be introduced. Although the marker board shown is portable, some have the same sliding feature. *Courtesy of Toronto Fire Services.*

- ◆ *Avoid speaking when facing boards or easel pads. Develop the skill of pausing while speaking when turned away from the learners to write (Figure 10.8).*
- ◆ *Write only what is necessary. It is rarely necessary to transcribe verbatim every spoken word of the lesson. Using concise points minimizes the time needed to write while maximizing the space available.*
- ◆ *Use chalk or marker pen colors that contrast with the background. Contrasting colors means using light colors of chalk and dark colors of markers. For example, the use of yellow or light orange markers in a large classroom is not recommended because those colors are difficult to see.*
- ◆ *Write in letters large enough to be seen at the back of the room. Experiment with letter size in advance. Remember that some learners may be ten or twenty times further from the board or easel pad than the instructor.*
- ◆ *Bring chalk and marker pens of several colors when scheduled to teach in an unfamiliar location.*

Figure 10.8 Instructors are tempted to say what they are writing on the easel pad or board as they write. It is important and effective to write first and then face the class to emphasize the written points.

When displaying any of these types of graphical materials, it is important to remember the ability of such materials to distract the learner. Avoid leaving diagrams and illustrations in view when they are not immediately relevant. By integrating graphical material into easel-pad presentations, instructors can easily control these distractions.

Duplicated Materials

Duplicated materials include any printed matter that instructors distribute before, during, or at the end of a class. As with other instructional media, instructors must strategically plan their use in the lesson to gain maximum benefits. Some examples with their benefits and suggested uses are included in the following list:

- Distribute handouts of lecture material at the end of a presentation unless it is necessary for learners to refer to them during the class. This strategy avoids learners dividing their attentions between the handout and the instructor.
- Consider using handouts as precourse material if they contain the same material as the presentation. This method of early distribution has the additional advantage of advancing the achievable learning level of the class.
- Give self-study guides to participants to assist them in working through textbooks or other learning materials at their own paces. Guides with self-quizzes offer review mechanisms that indicate areas of study where learners can concentrate their efforts. The instructor can prepare the study guides or arrange to purchase them as part of the curriculum.
- Give assignments to learners for completion within the class. Participants can complete case studies, style inventories, or other in-class materials individually or in groups according to the lesson plan.
- Give activities such as research projects, practice applications of theory learned in class, or review of material presented by the instructor as take-home assignments.
- Provide note guides (outline or shell of the lesson) for participants to fill with notes during class. When instructors design note guides into the lesson effectively, they help focus the attentions of participants on the material presented. As participants write their notes, they are attentive to the major points of the lesson that are reinforced.

Instructor Tips

- *Ensure that handouts are legible and complete. A photocopy of a photocopy of a fax copy may have poor-quality text or be askew on the page.*
- *Staple multiple pages. Learners are less likely to lose or rearrange multiple page handouts that have been stapled together in advance.*
- *Provide handouts with pages that are three-hole punched in advance unless they are otherwise bound.*

Models

A *model* is an excellent medium for illustrating mechanical or spatial concepts. Learners can clearly observe the types of relationships between parts of a model as they watch it function or manipulate it. Instructors can obtain many types of models or construct them at low cost, but some types, depending on their size and complexity, require a heavy investment of time and money. Some examples of models that are used as instructional media include the following:

- *Tabletop miniatures* of fire and emergency apparatus are routinely used along with model buildings to practice strategy and tactics (Figure 10.9). Tabletop models allow participants to play out a simulated incident in compressed time (for example, 5 minutes of simulation can represent 1 hour). Instructors can monitor a model town by video camera for display in other rooms. Advanced versions of tabletops may include smoke and flame effects that are static (nonmoving) or that change continuously according to the inputs received by the instructor.
- *Cutaway models* are of great value when participants are learning the inner workings of mechanical systems such as valves or pumps. Instructors can often obtain cutaways at little or no cost by dismantling obsolete or surplus equipment.
- *Anatomical models* are available in three dimensions, some of which have cutaway or take-apart features. These models benefit learners who are developing their EMS skills and knowledge in such areas as the mechanisms of internal injuries. A life-size or even small-scale skeleton is also a good anatomical model.

Audiocassettes

The ability to record and play sounds on *audiocassettes* using a small, portable audiocassette machine allows instructors to bring realism to many areas of training. In addition to classroom uses, audiocassettes can be used by learners to record and review lectures, dictate and review their own notes, or listen to prerecorded books on tape. Some examples of sounds that can be recorded for use in the classroom include the following:

- Engine and pump sounds (identify problems such as pump cavitation)
- Dispatch radio traffic (telecommunicator training or post-incident review)
- Heart, breathing, or blood pressure sounds (EMS training)

Casualty Simulation Media

Any medium that increases the realism of a simulation increases its value. Simulated casualties give tremendous benefits in increasing realism for EMS training involving hands-on applications. Instructors can simulate injuries using commercially available moulage kits and prostheses or by applying Plasticine® modeling paste, wax, and makeup (Figure 10.10).

Moulage kits typically contain plastic "wounds" that instructors can apply to a simulated casualty. Prostheses such as simulated amputated limbs, devices to simulate arterial bleeding, and other lifelike injury effects are also available. Minimal training is required to use moulage kits. The plastic wounds also have the advantage of being relatively quick to prepare and apply.

If a higher quality of simulation is required, instructors can arrange to use personnel who are qualified as casualty simulators. Casualty simulators may be EMS instructors who have been trained to use Plasticine®, mortician's wax, makeup, prostheses, and simulated blood to produce *very* realistic wounds and other effects. The process is more time-consuming than moulage but results in a higher degree of realism. For practical examinations or if simulations are videotaped for later use, the time invested is worthwhile. For mass-casualty disaster scenarios when the hands-on treatment is secondary to the bigger picture of response capacity and

Figure 10.9 A tabletop model of buildings and apparatus allows learners to see structure relationships and manipulate equipment. Other models include cutaways of pumps and hydrants, sections of hose and rope, samples of protective gear and clothing, and pieces of actual equipment and apparatus. *Courtesy of Toronto Fire Services.*

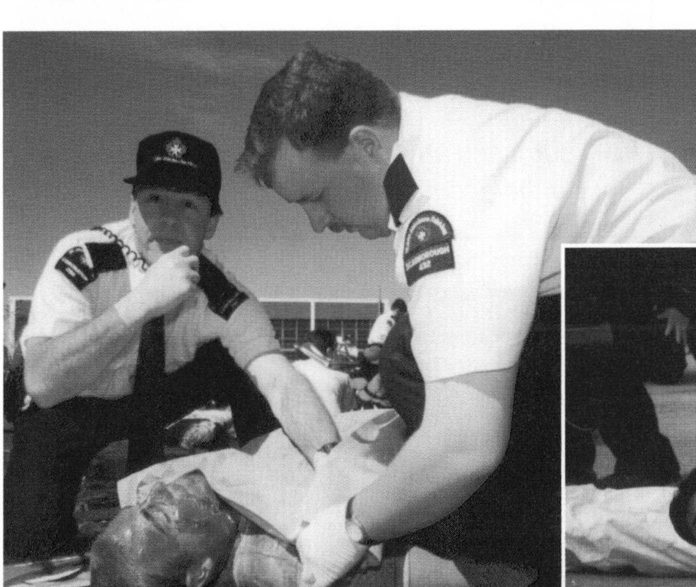

Figure 10.10 The injuries shown are not real, but they are convincing. Making individuals actually look injured adds realism to practice exercises, provides opportunities for learners and other personnel to portray different emergency roles, and enables instructors and supervisors to assess skill competencies under real-life conditions. *Courtesy of St. John Ambulance Brigade, Toronto.*

interorganization liaison, simple moulage simulations are sufficient.

For all cases of casualty simulation, it is important that instructors properly brief the individuals who are playing the roles of casualties on their injuries, the information they are to reveal to the rescuers, and any other pertinent scenario details. A realistic role-play helps enhance the lessons learned.

Projected Instructional Media

[NFPA 1041: 2-4.6, 2-4.6.1, 2-4.6.2, 3-5.1, 3-5.2, 3-5.2.1, 3-5.2.2]

Projected instructional media offer many advantages: Images are vivid, multicolored, and visible to a large audience. These media stimulate multiple senses simultaneously. There are some drawbacks, including an investment in audiovisual equipment and the costly purchase of projected media or extensive time spent in creating presentations (see Table 10.2). Despite these factors, a certain level of audiovisual support is considered the norm in today's instructional environment, and the expected level of support is increasing rapidly. Mixed-media presentations are also effective. Some projected instructional media include the following:

- Overhead projectors/transparencies
- Visual presenters
- Slide projectors/slides
- Scanners
- Digital still cameras
- Television
- Video and multimedia projectors/videotapes
- Video capture devices
- Liquid crystal display (LCD) projection panels
- Motion pictures/projectors

Overhead Projectors/Transparencies

The relatively low cost of *overhead projectors* along with the ability to produce high-quality *transparencies* (both black/white and color) on a computer printer make transparency presentations very popular and versatile media. Overhead projectors are common in most training environments. Newer models have the advantages of bright halogen bulbs and quiet operation (Figure 10.11). Most newer projectors can operate in a classroom without dimming the lights beyond the point where it becomes difficult to see the learners. Portable models are available that are very compact, but generally they do not project as brightly as regular models.

Commercially available transparency sets come mounted in cardboard frames. Such frames are a good idea for all transparencies because they keep the transparencies clean and neat. Frames also fill the space between the edge of the transparency and the edge of the projected image, a feature that minimizes the blinding effect of excess white space on the screen.

Transparencies can be produced on a printer or photocopied from a diagram, illustration, or any paper document. The type of transparency must be compatible with the output device — anyone who has dismantled a photocopier to remove the melted remnants of a transparency will not forget this point. When using paper documents to produce transparencies,

Instructor Tips

- *Dim the lights when using projected media even though the projection brightness of newer projectors is improving. If possible, leave lights at a level where the projected image is clearly visible but learners are still visible to the instructor. Create this level of visibility by dimming the light level only at the front where the screen or television is located.*
- *Maintain a backup set of training materials that can be used at a lower level of technology. For example, back up computer-generated presentations with overhead transparencies. Be prepared to present or demonstrate material if projection equipment fails.*
- *Keep a supply of spare projector bulbs, batteries for remote controls, an extension cord, and anything else that may be needed in the event of equipment problems.*

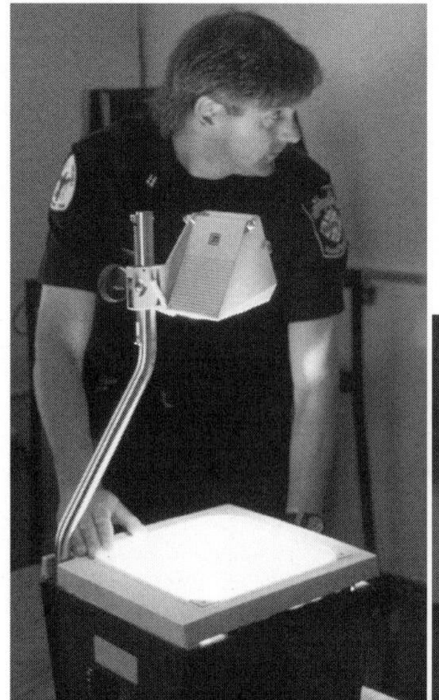

Figure 10.11 Many instructors favor the overhead projector because it is portable, reliable, and technically simple to use. Instructors and learners can create "on-the-spot" transparencies during lesson discussions and activities. *(right) Courtesy of Jeff Tucker.*

Test?

Table 10.2
Projected Instructional Media

Description	Advantages	Disadvantages
Overhead Projectors/Transparencies	• Transparencies are versatile and easy to produce • Allow easy change of presentation order • Easy to reveal individual points • Projectors are low in cost	• Transparencies may be overused because they are so easy to create • Transparencies are bulky and heavy when they are framed • Tendency to put too much information (clutter) on each transparency
Visual Presenters (Overhead Transparency Computers)	• Allow projection of actual images of objects or documents • *Extremely versatile*	• Cost • *Require the use of a television monitor or multimedia projector*
Slide Projectors/35 mm Slides	• High resolution of images • Project well to a large audience • Projector lenses less likely to keystone *due to long focal length*	• Stored slides may be bulky • Linear presentations • Slide development may be costly • Slide presentations often must be shown in a dark room, which restricts ability to take notes • *Multiple copies of slides may be expensive*
Scanners	Reproduce copy or images into computer for storage or alteration	Cost
Digital Still Cameras	• Produce high-quality photos and integrate them into documents without using photo shops • Photos easily edited with graphic or word processing software	Cost
Television	• Provides a variety of learning opportunities • Reaches large audiences over great distances	• May violate copyright laws to tape and show a program in class • Program may not air at a convenient time
Videotape Presentations	• Stimulate multiple senses • Usually affordable • Easy to produce high-quality videotapes in-house • Videotapes are available in a variety of formats	Requires extensive time to script, shoot, and edit *a professional-quality videotape*
Video Capture Devices	Digitize individual frames or short segments of videos for use in presentations	Cost

Continued on next page

consider that they have to be legible at the back of the room when displayed. Even if color options are not available, strive for a professional appearance and a concise, quality message. Ensure that transparencies support the lesson objectives and are clear and understandable to learners.

Table 10.2 *(Continued)*

Description	Advantages	Disadvantages
Video and Multimedia Projectors	• Combine the best aspects of motion picture projectors and television • Easily transported • Able to display computer-generated images • May be used with any size audience • Produce extremely professional images	• Cost • Time and ability to learn presentation software
LCD Projection Panels	• Smaller size • Greater portability • Lower cost than multimedia projectors	• Lower resolution • Need overhead projectors that project very brightly for best results • *Most do not have good video capability*
Motion Pictures/Projectors	Little, if any, with the advent of videotapes	• Films become obsolete quickly • Film may break and be costly to repair • May be difficult to feed film into projector and track correctly • Difficult to stop and discuss a point • *Noisy projectors*

Color transparencies are produced in several ways. Some of these ways are as follows:

- Blank transparencies can be purchased in many background colors; black images are then photocopied onto them.
- Color transparencies are available that have contrasting color images imprinted on them (for example, green images on a yellow background).
- Color printers can produce any combination of background and foreground colors on clear transparencies.
- Commercial printing services can take presentations that are saved on a computer disk and produce high-quality color transparencies mounted in cardboard frames.

The instructor is in total control of the order and method of presentation of each overhead transparency. If asked a question relating to a transparency already shown, the instructor can easily go back to it for review. Instructors can reveal bulleted lists one point at a time by placing an opaque sheet of paper between the transparency and the projection surface. The paper can be moved to reveal each point in turn. The static cling between the transparency and the projection surface keeps the paper in place. The instructor can read all points directly from the transparency. If there is a time gap between the displays of transparencies, turn off the projector to avoid a large bright, white screen.

Visual Presenters

Another instructional medium that can be used in many of the same ways as an overhead projector is the *visual presenter.* This device consists of a small video camera mounted vertically over a tabletop platform (Figure 10.12). Live images of objects or documents that have been placed on the platform are displayed on a television monitor or projection screen. An instructor who wants to project an item does not have to make a flat transparency when using a visual presenter. Some examples of how this medium could be used in the classroom include the following:

- Display a tool, material sample, or other item on the platform so that all learners may view it at the same time.
- Display paper copies of documents without the need to make transparencies first.
- Display original photographs or illustrations without the degeneration inherent in duplication.
- Display transparencies like an overhead projector.

Slide Projectors/Slides

Slide projectors and *35 millimeter (mm) slides* are capable of producing the same types of images as overhead projectors and transparencies, but they have some different advantages. The images produced on a slide are generally of higher resolution and project better to a large audience. A 35 mm slide is an excellent medium for presenting a photograph. Many publishers of educational materials offer 35 mm slide presentations in their catalogs. Slide projectors are less subject to *keystoning* (the effect of a rectangular image becoming trapezoidal on a screen) because they have a longer focal length than overhead projectors.

On the negative side, 35 mm slides are quite bulky when stored in a carousel. Slide presentations tend to be linear with the slides presented one after another in a predetermined order, and slides cannot be altered during the presentation. The room needs to be darker for a slide presentation as compared to one using transparencies, which makes it difficult for learners to take notes. While it is very easy to take photographs using slide film, multiple copies can be expensive. Producing data charts or bulleted lists on 35 mm slides requires either a commercial printing service or specialized equipment such as a desktop film recorder.

Scanners

A *scanner* is a peripheral computer device that allows photographs, diagrams, or illustrations to be digitized for integration into a presentation or document (Figure 10.13). The quality of the image depends on the resolution of the scanner. The higher the resolution, the larger the size of the file created. This factor must be considered when using scanned images. The desired end result, for example a projected image, may require very-high resolution. Other applications such as a black and white photo in a training document may not require such high resolution.

Another application of scanners is optical character recognition (OCR). This technology differentiates between text and graphics, creating a text file from a scanned document. Instructors can save a great deal of time by scanning existing documents through OCR as opposed to retyping them.

Digital Still Cameras

Digital photography has the advantage of not requiring a visit to the photo processor in order to produce a useful image. Instructors with access to a digital camera and a computer can take high-quality photographs and integrate them into documents within minutes. Pay close attention to resolution and memory specifications when purchasing a digital camera. The desired

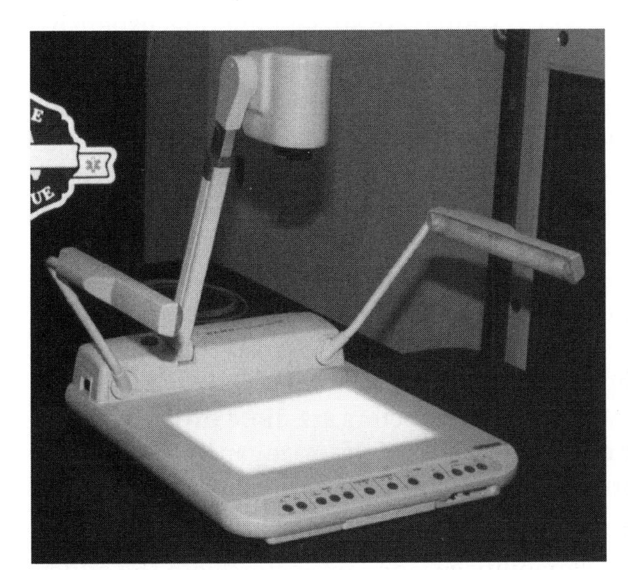

Figure 10.12 A visual presenter records the item on the platform by video camera and then projects the image on a screen. An entire class can easily see the details of an object or photo or the words on a printed document or text. *Courtesy of Jeff Tucker.*

Figure 10.13 Scanned information does not have to be retyped, redrawn, or recreated manually into another computer program. Scanners save time by digitizing information that can be saved on computer files and integrated into other documents. *Courtesy of Toronto Fire Services.*

Instructor Tip: *Regardless of the device used to create or import them, digital images can be edited using graphic or word processing software. The following options are available for editing digital images:*

- *Crop for size and shape*
- *Enlarge*
- *Reduce*
- *Label*
- *Enhance*
- *Highlight*
- *Transmit by electronic mail (e-mail) or other electronic means*

end result is a professional-quality, high-resolution image.

Television

Television has tremendous potential as an instructional medium. Distance learning programs broadcast on community-access channels by cable, satellite, or closed-circuit have the ability to reach large learner audiences over vast distances. State or provincial authorities, post-secondary institutions, or any organization wishing to reach a diverse group of learners can make effective use of televised lessons.

Videotape Presentations

A *videotape presentation* has the advantage of stimulating multiple senses, specifically sight and sound. Videotape becomes a powerful instructional medium with the added effects of motion, animation, diagrams, and text. Videotapes are commercially available on a tremendous variety of fire and emergency services subjects.

The quality of affordable videotape recording and editing equipment continues to improve dramatically. It is possible for many organizations to produce very high-quality videotapes in-house. Different videotape formats are available, including VHS, S-VHS, 8 mm, Hi-8™, and digital. Organizations must take into account, however, that the time required to properly script, shoot, and edit a videotape may be as much as one hundred times the length of the finished product.

When using a videotape in class, consider the following points:

- Start the videotape before turning on the television set, and turn it off before stopping the videotape. Some television sets automatically mute the sound and display a blue screen when not receiving input. Others display "snow" and a loud static noise that can be startling and uncomfortable.
- Cue the videotape to the desired spot in advance if showing only a portion of a tape.
- Rewind a videotape immediately if showing it in its entirety or rewind at least as soon as a class is finished.
- Pause the videotape while it is playing if adding to or illustrating a teaching point. If the videotape must continue to illustrate the point, mute the sound. Avoid shouting over the soundtrack of a videotape.
- Avoid leaving to attend to administrative or personal matters while a class views a videotape. Instructors who leave during the showing of a videotape are not aware of any technical problems, classroom disruptions, or other such issues that may occur in their absences.

Video and Multimedia Projectors

Projectors that are bright enough to display a quality videotape or computer-generated image in a classroom continue to decrease in price. The size and weight of these units are decreasing also, making them more portable and versatile. Video projectors designed to accept input only from a videocassette recorder (VCR) or other video source are smaller and less expensive than multimedia projectors.

Video projectors combine the best aspects of motion picture projectors (large, high-quality image) with those of television (near-silent operation with high-fidelity sound output). In addition, the portability of video projectors is similar to that of a VCR — they can easily be moved between classrooms. All features of the video source are available such as stop motion, slow motion, fast forward, or reverse. The only disadvantage in comparison to a television/VCR setup is cost. While video projectors are more expensive than the average television set, the price gap between large-screen televisions and video projectors is narrowing.

Multimedia projectors offer the same videotape projection capability as video projectors with the added capacity to display computer-generated images. The variety of uses of multimedia projectors is limited only by the capabilities of the presentation software. Presentations that contain sounds, video clips, data charts, animation, and photographs can be produced. Text can be introduced one point at a time with animated transitions between slides. The sophistication of presentation software increases with the release of each new edition. The key for an instructor is to learn to use this instructional medium effectively. Avoid the temptation to include every possible feature in the presentation. Making strategic and selective use of the software maximizes the benefits of multimedia presentations.

Multimedia projectors have uses beyond video and presentation software applications. For example, when a committee is meeting to develop or edit training documents such as a new policy or procedure, the document being considered can be projected for all to see. In this way, edits can be made as they are discussed, reducing the turnaround time between drafts of the document.

When shopping for a video or multimedia projector, consider the following factors:

- Images must be bright enough to be seen in the classroom where it will be used. Displays in a darkened sales showroom may not project as well in the instructional environment.
- Resolution of the video and computer-generated RGB (red-green-blue) images must be high enough to produce a quality image at the size desired in the classroom.
- Sound output of the projector must have sufficient volume and fidelity for the intended use.
- Projectors must be compatible with the video and computer devices that they are intended to support.

Video Capture Devices

An option for creating digital images is using a *video capture device.* This medium can take an individual frame or short segment of a videotape and digitize it for use in a document or presentation. High-resolution analog images from a reasonably priced Hi-8™ video camera can be incorporated as video clips or still images into presentations.

Liquid Crystal Display (LCD) Projection Panels

LCD projection panels function much the same way as multimedia projectors. They can receive RGB input from a computer. Some can receive video input from a VCR or video camera. An LCD panel produces a translucent image, and a light shines through to display the image on a screen when the panel is placed on an overhead projector.

The relative advantages of LCD panels are smaller size, greater portability, and lower cost when compared to multimedia projectors (although multimedia projectors are becoming much more affordable). The disadvantages are lower resolution (particularly video) and the need to have an overhead projector that projects very brightly for best results.

Motion Pictures/Projectors

The use of *motion pictures/projectors* as an instructional medium has declined significantly with the increased use of videotape. Any size advantage of a projected motion picture is being countered by the advent of video projection devices. Any organization using motion pictures/projectors should consider converting 8 mm or 16 mm motion pictures to videotape. The likelihood of any motion pictures having instructional relevance diminishes each year.

Multi-Image Delivery

Even without multimedia projectors, it is possible for a versatile instructor to project multiple images in a classroom. The combinations and arrangements are many. The key to any multi-image or mixed-media presentation is to have the transitions planned into the lesson. An instructor should practice any timings or other subtle details until they are mastered. Some examples of multi-image delivery are as follows:

- ***Two or more slide projectors displaying photographs on adjacent screens*** — The photographs can be compared to illustrate different views of an incident or technique.
- ***Slide projector and overhead projector displaying on adjacent screens or intermittently on the same screen*** — Photographs from a slide projector are integrated with text or data charts from an overhead projector.
- ***Video projector or television used in conjunction with an overhead projector*** — A technique or procedure is introduced on videotape and discussed in detail on overhead transparencies.
- ***Projected medium combined with a nonprojected medium*** — For example, a videotaped demonstration is discussed in detail while also using a marker board.
- ***Projection of an overhead transparency onto an easel pad*** — For example, this method can

accurately trace a diagram or chart. This use of two separate media results in high-quality paper copy charts in multiple colors that can be transported between training locations.

- ***Projection of an overhead transparency onto a marker board*** — For example, this method can show a diagram of an incident scene for a post-incident analysis. Markers and magnets can be used to show the movement of personnel and apparatus around the scene. When projecting an image onto a marker board or other smooth surface, avoid subjecting learners to a bright, reflected glare.

Simulators

[NFPA 1041: 2-4.7.1, 2-4.7.2, 3-5.1, 3-5.2, 3-5.2.1, 3-5.2.2]

A *simulator* includes any instructional medium intended to represent a system, process, or environment in which actual training would be unsafe, impractical, or prohibitively expensive. For example, it is unsafe to burn buildings in order to practice search and rescue techniques. It is impractical to defibrillate actual humans in an instructional environment. It is prohibitively expensive to fill high-rise apartment towers with hot smoke in order to study ventilation. Simulators bring elements of reality to the instructional environment. The greater the degree of reality, the more effective the learning. The following sections describe various types of simulators (also see Table 10.3).

Electronic Simulations

Electronic components are engineered to simulate a wide variety of situations and environments. Elements of communication systems, alarm/detection systems, ventilation fans, theatrical smoke machines, electric lights, miniature models, video cameras, and projection and recording media can all be integrated into a tabletop tactical simulator. For example, electronic simulations of a pump panel may include working gauges, realistic recorded sounds, and warning or trouble messages.

Mechanical Simulations

Mechanical simulators are built to simulate fire and emergency service functions such as roof ventilation or forcible entry (Figure 10.14). Often these simulators can be constructed at low cost from obsolete or surplus materials.

Display Boards

The component parts of a mechanical or electronic system can be mounted on a display board. In this way, the system can be manipulated by an instructor or participant without crawling under a chassis or into a machine. An example would be a display board mounted with the actual working parts of a vehicle air brake system. Participants can practice maintenance inspections quickly and safely. The parts can be scavenged from a vehicle that is being scrapped.

Smoke Simulators

The movement of smoke through a building or area is of critical importance to fire and emergency services personnel. Smoke movement can be simulated or illustrated by using small-scale mechanical models of high-rise stairwells. Developments in computer modeling allow realistic simulations of smoke movement through complex structures such as shopping malls, tunnels, and atria.

Computer Formats

Computerized simulations of burning buildings, injured casualties, hostage scenarios, and other fire and emergency services applications are becoming widely available. Due to the greater amounts of memory required for video and sound files, many of these simulations are being produced in compact disc-read only memory (CD-ROM) and digital video disc (DVD) formats. Some formats feature still photographs or illustrations with text. The more advanced DVD simulations

Figure 10.14 The locking devices shown are examples of mechanical simulators. By displaying the locking arrangements, instructors can demonstrate a simulation of the actual function of the locks. *Courtesy of Toronto Fire Services.*

feature high-quality, full-motion video and high-fidelity sound.

With the availability of compact disc-recordable (CD-R) and compact disc-rewritable (CD-RW) technology, instructors now have the ability to produce instructional programming that may include video, text, photographs, and interactive feedback from the participant. Other high-memory capacity storage media exist that may be used in the same manner.

Virtual Reality

A *virtual reality simulation* displays a field of view as if the participant is part of the simulated environment. The environment can be manipulated in response to inputs from the participant. Programmed changes occur based on the actions taken by the participant in relation to the simulated scenario. Virtual reality simulators are continually improving in realism. These simulators are tremendous tools for reinforcing procedure-based or protocol-based skill sets such as incident management or EMS skills.

Most of these simulations require an investment in equipment and software. This investment must be weighed against the instructional value offered by the increased realism of the product. Some simulations have built-in study materials and self-tests. In some cases, centralized training records and statistics are obtained directly from the simulation software when networked through an organization's computer network.

Table 10.3
Simulators

Description	Advantages	Disadvantages
Electronic	Can engineer components to simulate a wide variety of situations and environments	Purchase and setup costs
Mechanical	• Can simulate fire and emergency service functions/operations • Can build at low cost	Need storage place and ability to relocate to training site
Display Boards	• Allow convenient manipulation of an electronic *or mechanical* system • Low in cost if parts are obtained from scrapped vehicles	• Needed parts may not be accessible • Require individual knowledgeable in constructing simulators
Smoke	Allow realistic smoke movement and address visual learning	• Cost of models or computer programs • Layout of models may be extensive and a time factor in setting up
Computer Formats (CD-ROM/DVD)	Ability to produce instructional programming with video, text, photos, and interactive feedback from participant	• Cost • Technical ability required
Virtual Reality	Manipulates environment in response to participant input for lifelike scenarios	Investment in equipment and software
Anatomical/Physiological Manikins	Available to match any level of EMS training, often at reasonable costs	Some manikins are expensive, highly technical, and high-maintenance

Figure 10.16 Training demands increase as new equipment and systems come on-line. Today's fire service is witnessing an increase in technology, including the integration of GPS into dispatch/telecommunication functions. *Courtesy of Terry L. Heyns.*

of those residing in a home or apartment. Terminals placed in the emergency vehicle and linked to the computer at the telecommunication/dispatch center have the potential to provide fire and emergency responders with critical information.

New Technology Training

The key to the productive use of new technology systems is the training received by fire and emergency responders. In a sense, the use of instructional media becomes more important. Instructors will not only use instructional media as aids in the classroom, but will "train" fire and emergency responders to actually use new technology or types of media. The use of a computer simulation, such as a fireground incident command "game," may not be that different from using a computer terminal onboard the first-due pumper while en route to the scene of an emergency. As technology develops, individual firefighters may even have their own computers built into their SCBAs. The use of such technology is a challenge that requires special training and careful practice. Fire and emergency service instructors are sure to play important roles in training personnel in the use of future technology.

Access and Security

Access to computer programs and records can be managed with security measures. These measures can limit a user's access to certain programs, systems, and data sources so that training records, test documents, and course material content cannot be viewed or altered without authorization. System security and access are usually regulated by policy and maintained by a designated system manager.

References and Supplemental Readings

Brandt, Richard C. *Flipcharts: How to Draw Them and How to Use Them.* San Diego, CA: University Associates, Inc., 1986.

Burn, Bonne E. *Flip Chart Power.* Johannesburg: Pfeiffer & Company, 1996.

Hannum, Wallace. *The Application of Emerging Training Technology.* Alexandria, VA: American Society for Training and Development, 1990.

Kemp, Jerrold E. and Don C. Smellie. *Planning, Producing, and Using Instructional Media.* New York: Harper & Row, Publishers, 1989.

Mason, Robin. *Using Communications Media in Open and Flexible Learning.* London: Kogan Page Limited, 1994.

Job Performance Requirements

This chapter provides information that addresses the following job performance requirements of NFPA 1041, *Standard for Fire Service Instructor Professional Qualifications* (1996 edition). Colored portions of the standard are specifically covered in this chapter.

Chapter 2 Instructor I

2-2.1 Definition of Duty. The management of basic resources and the records and reports essential to the instructional process.

2-2.2 Assemble course materials, given a specific topic, so that the lesson plan, all materials, resources, and equipment needed to deliver the lesson are obtained.

2-2.2.1 *Prerequisite Knowledge:* Components of a lesson plan; policies and procedures for the procurement of materials and equipment and resource availability.

2-3.1 Definition of Duty. The review and adaptation of prepared instructional materials.

2-3.2 Review instructional materials, given the materials for a specific topic, target audience and learning environment, so that elements of the lesson plan, learning environment, and resources that need adaptation are identified.

2-3.2.1 *Prerequisite Knowledge:* Recognition of student limitations, methods of instruction, types of resource materials; organizing the learning environment; policies and procedures.

2-3.2.2 *Prerequisite Skills:* Analysis of resources, facilities, and materials.

2-4.1 Definition of Duty. The delivery of instructional sessions utilizing prepared course materials.

2-4.2 Organize the classroom, laboratory or outdoor learning environment, given a facility and an assignment, so that lighting, distractions, climate control or weather, noise control, seating, audiovisual equipment, teaching aids, and safety, are considered.

2-4.2.1 *Prerequisite Knowledge:* Classroom management and safety, advantages and limitations of audiovisual equipment and teaching aids, classroom arrangement, and methods and techniques of instruction.

2-4.2.2 *Prerequisite Skills:* Use of instructional media and materials.

2-4.4 Adjust presentation, given a lesson plan and changing circumstances in the class environment, so that class continuity and the objectives or learning outcomes are achieved.

2-4.6 Operate audiovisual equipment, and demonstration devices, given a learning environment and equipment, so that the equipment functions properly.

2-4.6.1 *Prerequisite Knowledge:* Components of audiovisual equipment.

2-4.6.2 *Prerequisite Skills:* Use of audiovisual equipment, cleaning, and field level maintenance.

2-4.7 Utilize audiovisual materials, given prepared topical media and equipment, so that the intended objectives are clearly presented, transitions between media and other parts of the presentation are smooth, and media is returned to storage.

2-4.7.1 *Prerequisite Knowledge:* Media types, limitations, and selection criteria.

2-4.7.2 *Prerequisite Skills:* Transition techniques within and between media.

2-5.3 Grade student oral, written, or performance tests, given class answer sheets or skills checklists and appropriate answer keys, so the examinations are accurately graded and properly secured.

2-5.3.1 *Prerequisite Knowledge:* Grading and maintaining confidentiality of scores.

2-5.4 Report test results, given a set of test answer sheets or skills checklists, a report form and policies and procedures for reporting, so that the results are accurately recorded, the forms are forwarded according to procedure, and unusual circumstances are reported.

2-5.4.1 *Prerequisite Knowledge:* Reporting procedures, the interpretation of test results.

2-5.4.2 *Prerequisite Skills:* Communication skills, basic coaching.

2-5.5 Provide evaluation feedback to students, given evaluation data, so that the feedback is timely, specific enough for the student to make efforts to modify behavior, objective, clear, and relevant; include suggestions based on the data.

2-5.5.1 *Prerequisite Knowledge:* Reporting procedures, the interpretation of test results.

Chapter 3 Instructor II

3-3.2 Create a lesson plan, given a topic, audience characteristics, and a standard lesson plan format, so that the job performance requirements for the topic are achieved, and the plan includes learning objectives, a lesson outline, course materials, instructional aids, and an evaluation plan.

3-3.2.1 *Prerequisite Knowledge:* Elements of a lesson plan, components of learning objectives, instructional methods and techniques, characteristics of adult learners, types and application of instructional media, evaluation techniques, and sources of references and materials.

3-3.2.2 *Prerequisite Skills:* Basic research, using job performance requirements to develop behavioral objectives, student needs assessment, development of instructional media, outlining techniques, evaluation techniques, and resource needs analysis.

3-3.3 Modify an existing lesson plan, given a topic, audience characteristics, and a lesson plan, so that the job performance requirements for the topic are achieved, and the plan includes learning objectives, a lesson outline, course materials, instructional aids, and an evaluation plan.

3-3.3.1 *Prerequisite Knowledge:* Elements of a lesson plan, components of learning objectives, instructional methods and techniques, characteristics of adult learners, types and application of instructional media, evaluation techniques, and sources of references and materials.

3-3.3.2 *Prerequisite Skills:* Basic research, using job performance requirements to develop behavioral objectives, student needs assessment, development of instructional media, outlining techniques, evaluation techniques, and resource needs analysis.

3-4.2 Conduct a class using a lesson plan that the instructor has prepared and that involves the utilization of multiple teaching methods and techniques, given a topic and a target audience, so that the lesson objectives are achieved.

3-4.2.1 *Prerequisite Knowledge:* Use and limitations of teaching methods and techniques.

3-5.1 Definition of Duty. The development of student evaluation instruments to support instruction and the evaluation of test results.

3-5.2 Develop student evaluation instruments, given learning objectives, audience characteristics, and training goals, so that the evaluation instrument determines if the student has achieved the learning objectives, the instrument evaluates performance in an objective, reliable, and verifiable manner, and the evaluation instrument is bias-free to any audience or group.

3-5.2.1 *Prerequisite Knowledge:* Evaluation methods, development of forms, effective instructional methods, and techniques.

3-5.2.2 *Prerequisite Skills:* Evaluation item construction and assembly of evaluation instruments.

Chapter 4 Instructor III

4-2.6 Write equipment purchasing specifications, given curriculum information, training goals, and agency guidelines, so that the equipment is appropriate and supports the curriculum.

4-2.6.1 *Prerequisite Knowledge:* Equipment purchasing procedures, available department resources and curriculum needs.

4-2.6.2 *Prerequisite Skills:* Evaluation methods to select the equipment that is most effective and preparation of procurement forms.

4-3.2 Conduct an agency needs analysis, given agency goals, so that instructional needs are identified.

4-3.2.1 *Prerequisite Knowledge:* Needs analysis, task analysis, development of job performance requirements, lesson planning, instructional methods, characteristics of adult learners, instructional media, curriculum development, and development of evaluation instruments.

4-5.2 Develop a system for the acquisition, storage, and dissemination of evaluation results, given agency goals and policies, so that the goals are supported and those impacted by the information receive feedback consistent with agency policies, federal, state, and local laws.

4-5.2.1 *Prerequisite Knowledge:* Record keeping systems, agency goals, data acquisition techniques, applicable laws, and methods of providing feedback.

4-5.2.2 *Prerequisite Skills:* The evaluation, development, and use of information systems.

Chapter 11

Management and Supervision of Training

One of the most important tasks facing any program manager responsible for organizing training is the effective management and supervision of training programs and personnel. Just as with any other organizational function, training must be effectively managed if it is to be successful in achieving its goals and objectives. This chapter addresses the management and supervision areas shown on the following list. References and supplemental readings and job performance requirements are given at the end of the chapter. The information given in the following sections suggests steps to follow as the training manager fulfills the many roles and responsibilities included in managing and supervising training programs.

- Designing training programs
- Developing training policies, records, and standards
- Determining organizational training needs
- Recruiting and selecting instructors
- Scheduling training programs
- Evaluating training programs and instructors
- Providing budget and resource management

Training Program Design

[NFPA 1041: 4-3.3, 4-3.3.1, 4-3.3.2]

Before an organization can effectively deliver any training program, it must design the overall training program. Designing the training program is a critical step toward its success. In fact, the design and structure of a training program can either contribute to its effectiveness or result in failure to meet the needs of the organization. Most organizations appoint an individual to perform the role of training manager who takes on the responsibility of designing or overseeing the design of a training program that meets organizational goals and objectives. The training manager follows two basic steps when designing a training program: (1) Identify its purpose and (2) consider program design factors. Several goals and components are identified and discussed for each design process step.

Identify Program Purpose

The first step identifies the *purpose* of the training program. A fire training manager, for example, responds to NFPA 1201, *Standard for Developing Fire Protection Services for the Public,* which states the purpose of fire department training as follows: *"The fire department shall have a training program and policy that ensures that personnel are trained and competency is maintained to effectively, efficiently, and safely execute all responsibilities. . . ."* This purpose can be applied to any fire and emergency services organization. The training goal is to meet the following requirements:

- Develop and maintain the skills needed by all personnel to do their specific jobs in the organization.
- Instill the organization's values and culture in every member — an often overlooked but implicit element in every training program.
- Ensure the program meets the multiple requirements of local, state/provincial, and federal agencies such as responding to hazardous materials incidents, using infectious disease control procedures, and using personal protective gear appropriately.
- Provide quality programs that challenge personnel. Training that is too easy tends to cause participants to discount its importance.
- Provide opportunities for professional growth that prepare personnel for future responsibilities at the next organizational level *prior* to being promoted to the position.

Consider Design Factors

The second step considers several factors in the actual *design* of the training program. Whether the design is for a new program or modification of an existing program, the following factors have a direct impact on the structure and scope of the training program:

- ***Organizational and personnel training needs*** — Ensure that the design of the training program is based on the needs of the organization. Every organization is different. The training manager must ensure that the program provides adequate and appropriate training to meet the needs of the different departments within the organization. Adequate and appropriate training include the necessary support resources such as supplies, equipment, and instructional personnel. It is also important to ensure that these programs are scheduled or designed around work schedules so all personnel can attend.
- ***Needs analysis*** — Have a method or policy in place for performing a needs analysis to identify the most critical training needs. Today, there are multiple training topics and requirements, all of which have high priorities. It may be necessary to implement policies to ensure that the training program addresses the topics identified by the needs analysis (see Organizational Training Needs section).
- ***Basic program philosophy*** — Develop the basic philosophy for the training program, and strive to meet it. The philosophy of the training program must provide a customer-service approach to the development and delivery of training. For example, determine what programs meet the training needs of personnel. Every training program has customers with specific needs. Customers are organizational personnel who need training in order to perform the skills required to do their jobs — their needs vary based on their jobs. For example, drivers of engines need different skills than drivers of ladder trucks, and all drivers need different skills than officers.
- ***Overall organizational strategic goals*** — Ensure that all personnel have the skills necessary to meet the strategic goals of the organization (Figure 11.1). For example, if a strategic goal of an organization is to keep all personnel in their districts as much as possible while training, the design of the training program must ensure that training can be delivered to personnel in appropriate locations. The focus of every function within an organization must be directed toward helping that organization meet its strategic or performance goals. A training program is no different in this regard. Goals are achieved by personnel who are well-trained and have the skills necessary to perform their jobs. These skills are developed as a result of a well-designed, well-managed training program.
- ***Program and administrative structures*** — (1) Provide a clear line of authority and accountability for achieving an organization's training goals. (2) Identify the individual who is in charge of managing and leading a training program. Examples:
  - — NFPA 1201 is a guide for developing a fire training program structure. A component of this standard, for example, requires that the fire chief appoint a department training manager who is responsible for developing training goals and performance objectives, ensuring training equipment operation and facility availability, coordinating training schedules, selecting instructors, and evaluating program success.
  - — In large organizations, the training program structure may also include other training staff who report to the training manager such as instructors, curriculum developers, media specialists, administrative clerks, and other support personnel.

Figure 11.1 Personnel must have the skills necessary to meet organizational goals.

- In small organizations, the structure may identify one person who is charged with performing all tasks associated with training.

- *Appropriate training policies and procedures* — Develop and adopt appropriate training policies and procedures that answer the *who, what, when,* and *where* questions about training (see Policies section). It may be necessary to develop and adopt other policies based on local needs. A training issue that requires consistency over time with different work groups is an appropriate issue for a policy. These policies/procedures can answer the following questions:
  - Who is responsible and accountable for developing, delivering, and evaluating training?
  - Who is required to attend training programs?
  - Who is responsible for ensuring that training goals are achieved?
  - Who is responsible for evaluating the training program, the instructors, and the participants?
  - Who is expected to deliver training sessions?
  - Who is responsible for acquiring resources?
  - What are the training priorities?
  - What is the mission of the training program?
  - What are the job requirements and prerequisites for instructors?
  - What are the minimum hours of training required for all positions?
  - What are the topics and skills to be taught?
  - When is training to be conducted?
  - What training reports are required to be kept, who keeps them, and when are they due?
  - Where is training to be held?

- *Program evaluation process (overall training program, instructors, and learners)* — Evaluate the courses, the instructors, the learners, and the entire program. Instructors evaluate individual courses and learners, and training managers evaluate instructors and the entire program (such as a recruit firefighter training program). By considering the program evaluation in the program design, the training manager may prevent problems and identify information that must be collected such as the following:
  - What elements of the program must be evaluated (for example, prerequisites, learner demographics, sequence of courses, use of time, and costs)?
  - When are these elements evaluated?
  - Who collects the data needed?
  - Who analyzes the data and formulates recommendations for changes?
  - How are the recommendations implemented?

Program Design Tips: *Regardless of the size or scope of a training program, the following design tips ensure an effective organizational training program:*

1. ***Focus on the needs of the customer*** *— For the training organization, the customer includes all organizational personnel or members who attend training.*
2. ***Design the program so that training is visible within the organization*** *— Training should be seen as well as heard! Publicize training activities so that all personnel are aware of training opportunities and the associated training goals and objectives of the organization.*
3. ***Design the program so that accountability for achieving training goals is clearly established and communicated*** *— Accountability goes well beyond the training manager. It extends to every supervisor who is responsible for ensuring that assigned personnel are trained. Further, all individuals must be held accountable for their own skills. The design of the training program must clarify responsibility and accountability through both policies and training classes.*
4. ***Deliver quality training classes and drills*** *— In fire and emergency services training, quality is more important than quantity. Personnel expect to actually participate in training that is challenging and enjoyable while enhancing their skills and knowledge. Training should not be scheduled or required simply to add hours to a ledger. Rather, training programs must ensure that personnel can do their jobs effectively, efficiently, and safely.*

- ***Resources required to meet goals and objectives*** — Consider necessary resources because they are essential parts of the delivery of any training program. Failure to adequately plan could result in failure to meet learning objectives, delays in the program, and an increase in costs. Consider the following questions:
  — Who is responsible for logistics?
  — What equipment or apparatus is needed?
  — How many instructors are needed?
  — What handouts, workbooks, or texts are needed?
  — What sites or locations can be used?
  — What scheduling conflicts are there for the same resources or like resources?
  — What outside organizations are needed?

Training Policies, Records, and Standards

[NFPA 1041: 4-2.1, 4-2.2, 4-2.2.1, 4-2.2.2, 4-2.3, 4-2.3.1, 4-2.3.2, 4-2.7, 4-2.7.1, 4-2.7.2]

Training managers have multiple responsibilities when managing training program operations. Among these responsibilities are those of developing recommendations for adoption of policies (including procedures and guidelines), administering training record systems, and assuring that standards are followed.

Policies

A *policy* is a guiding principle or rule that organizations develop, adopt, and use as a basis or foundation for decision-making. Policies help organizations address specific issues or problems. Training policies serve to guide an organization's training function on a day-to-day basis and are often accompanied by procedures for fulfilling the requirements of the policies. For policies to be effective, they must be well-written, adopted through a process that provides critical feedback, and explicitly supported by the organization's administration and training manager (Figure 11.2).

In some cases, policies are developed at the local level. In other cases, organizational training policies are based upon state, provincial, or federal laws or standards. For example, the Code of Federal Regulations (CFR) identifies specific requirements for training personnel who respond to hazardous materials incidents. These federal regulations often serve as the basis for locally adopted training requirements. Sometimes, these regulations, such as *Respiratory Protection Training,* CFR 1910.134, require the development of local policies and procedures. A resource for developing local policies are some of the NFPA standards, including NFPA 1201, *Standard for Developing Fire Protection Services for the Public,* and NFPA 1500, *Standard on Fire Department Occupational Safety and Health Program.*

Procedures and Guidelines

Training managers are responsible for assuring that instructors comply with training policies. Instructors also need directions or methods by which they can comply with policies. To facilitate this compliance, training managers develop procedures and guidelines that delineate steps to follow and outline what latitude can be used to make decisions. Procedures and guidelines may accompany policies, or they may stand alone. They are both essential management tools for a training program. Consider the following explanations:

- ***Procedure*** — A *procedure* identifies the steps that must be taken to fulfill the intent of a policy — it is written to support a policy. For example, a policy may state that *"all live fire training evolutions must be conducted in compliance with NFPA 1403, Standard on Live Fire Training Evolutions."* The procedures accompanying this policy list steps the instructor takes to ensure the policy is achieved, including such criteria as establishing the Incident Management System (IMS), inspecting the training structure, providing a safety officer for evolutions,

Figure 11.2 Organizational policies guide personnel in many tasks (from simple to complex), in performing procedures, and in making decisions. Policies may include guidelines on sick leave, canceling classes due to inclement weather, or selecting texts or videotapes for use in a training program. *Courtesy of Robert Wright, Maryland Fire and Rescue Institute.*

requiring use of appropriate protective equipment, and using qualified instructors who are knowledgeable in fire behavior. An advantage of adopting well-defined procedures is consistency in implementing the policy. Procedures are essential in programs where adherence to policy is a critical factor such as implementing safety precautions during training, hiring or evaluating personnel, and acquiring structures.

- *Guideline* — Unlike a policy or procedure that provides a clear rule or step-by-step process, a *guideline* identifies a general philosophy. Guidelines may be part of a policy, or they may stand alone. They provide direction with latitude for achieving the overall goal of the guideline or policy. For example, an organization may have a policy for conducting training in inclement or extreme weather conditions. Included in the policy are guidelines that give instructors information so they may make decisions on when it is appropriate to cancel or reschedule training. The information in the guidelines gives instructors the *parameters (leeway)* they need in considering all factors involved before making a decision.

When deciding whether an issue requires a policy, procedure, or guideline, consider the goal to be achieved. If a specific rule or philosophy is needed that allows no variance, a policy is the answer. If a step-by-step outline is required for an administrative task or training operation, a procedure is the answer. If a general philosophy is needed that provides direction, but allows the instructor to use some discretion, a guideline is most appropriate.

Adoption Process

Before a policy, procedure, or guideline can be used in the management of a training program, the training manager must develop it and follow organization policy for having it adopted. *Adoption* is the process by which the chief and/or designated administrators review, amend, and approve the policy, procedure, or guideline. If this document is to have credibility and be effective in day-to-day operations, it is essential that it go through a formal adoption process.

The adoption process begins well before a draft is put on paper. Consider the following process steps:

Step 1. Identify a need.

First, it must be evident that there is a need for a policy, procedure, or guideline. A need becomes apparent when a challenge arises or there is a change in the scope or structure of a training program. Needs can be identified by the following methods:

- ✔ ***Record and maintain documents on issues, challenges, and incidents that occur during training, and refer to them when researching information for a needs analysis*** — For example, if during summer drills there are frequent incidents of heatstroke, a policy, procedure, and/or guideline may be needed for addressing weather conditions, hydrating learners, taking breaks during training, or rescheduling classes.
- ✔ ***Communicate with other training organizations and share information regarding training issues, challenges, and incidents*** — Learn from other organizations, and contribute to the safety of all training activities. Sometimes the need for a policy, procedure, or guideline becomes evident through training challenges that other organizations experience. For example, during the 1980s, several fatal training fires in acquired structures were well-publicized (see the case study on the Milford, Michigan, incident in Chapter 8, Practical Training Evolutions). Becoming aware of these incidents had positive results: Organizations developed and adopted strict policies on live fire training in acquired structures and burn buildings.

Step 2. Develop a draft document.

After a need is identified, an individual or committee develops a draft of the document. Use the following process:

- ✔ Determine whether a policy, procedure, guideline, or all three are the most appropriate for the issue or incident. Always select the least restrictive type of document. When possible, use similar existing policies, procedures, or guidelines as references.
- ✔ Make the wording in the document clear, concise, and easy to follow. Use appropriate terms, and follow proper organizational format.
- ✔ Include the date a document becomes effective if adopted.

- ✔ List the person or unit responsible for managing the new rule and any other policy, procedure, or guideline to which it is related.
  - — A policy must also state to whom or what it applies and the specific rules of application.
  - — A procedure must contain the steps to follow and the policy to which it applies.
  - — A guideline must include to whom or what it applies and the guideline statement.
- ✔ Conduct research on the legalities of the rule because the documented issue or incident likely has potential legal implications.
  - — Determine whether the issue or incident is addressed in existing organizational policies.
  - — Review NFPA standards that address the issue or incident.
  - — Seek similar issue or incident resolutions from other organizations.
- ✔ Submit the draft document to the organization's legal counsel for review. It is the training manager's responsibility to ensure that any document is legally and administratively sound prior to leaving the draft step in the adoption process.

Step 3. Submit the draft for organizational review.

Next, submit the draft document for organizational review and comment. Review and comment opportunities are especially important for those documents that are controversial or affect multiple groups in an organization. For example, a proposed policy on attendance at training sessions certainly affects participants more than instructors or program administrators. A critical consideration about feedback on draft documents is the fact that those affected by a new rule may not always support the change. Their lack of support does not necessarily mean that the rule is not worthwhile or needed. Ultimately, it is the training manager and the organization's chief executive officer who are held accountable to the obligation of providing safe and effective operations in the training program. If a rule is required to meet those obligations, then it is appropriate to adopt one, even if it is not popular with organization members. Use the following review process:

- ✔ Provide personnel an opportunity to respond with feedback and input on any document that affects them. An important point is that the draft does not need to be reviewed by *all* personnel, but it should be reviewed by the chief officers responsible for managing the fire and emergency service responders affected by the rule and representative personnel who will be affected.
- ✔ Provide a comment period for the draft document, and review all comments.
- ✔ Evaluate comments, and amend the draft document as necessary.

Step 4. Adopt the document.

Once feedback has been evaluated and the document has been amended if necessary, the policy, procedure, or guideline is ready for adoption. The appropriate manager or administrator endorses the document. If it is a policy for the training program, for example, the training manager endorses it. If it is a broader policy affecting the entire organization, the endorsement must come from the organization's chief executive officer. An endorsement demonstrates to organizational personnel that the rule has official sanction.

Step 5. Publish the document.

Once the document has all the necessary signatures, it is time to publish the policy, procedure, or guideline. Anyone potentially affected by the rule must be informed of the change. Informing personnel is often done by memos, but if the document implements a substantial change or addresses a critical issue, the best method for communicating the new rule is a face-to-face meeting with personnel and supervisors. The face-to-face method provides an opportunity for personnel to ask questions, gain clarification, and ensure understanding. Regardless of the method used, it is essential to inform everyone with an interest in the rule. Without this communication, it is likely that neither supervisors nor subordinates will follow the content of the document.

Standards

Organizations adopt *standards* to provide the basis for performance or operational requirements. Every day, training managers make decisions based on standards. They are key elements to any training program.

The most common standards used by fire and emergency service organizations are the NFPA standards. These standards address many issues including professional qualifications, firefighter health and safety programs, and organizational structure. Some of the more commonly used NFPA standards are in the following list.

Fire and Emergency Services NFPA Standards

- NFPA 472, *Standard on Professional Competence of Responders to Hazardous Materials Incidents*
- NFPA 473, *Standard for Competencies for EMS Personnel Responding to Hazardous Materials Incidents*
- NFPA 600, *Standard on Industrial Fire Brigades*
- NFPA 1001, *Standard for Fire Fighter Professional Qualifications*
- NFPA 1002, *Standard for Fire Department Vehicle Driver/Operator Professional Qualifications*
- NFPA 1003, *Standard for Airport Fire Fighter Professional Qualifications*
- NFPA 1021, *Standard for Fire Officer Professional Qualifications*
- NFPA 1031, *Standard for Professional Qualifications for Fire Inspector*
- NFPA 1033, *Standard for Professional Qualifications for Fire Investigator*
- NFPA 1035, *Standard for Professional Qualifications for Public Fire and Life Safety Educator*
- NFPA 1041, *Standard for Fire Service Instructor Professional Qualifications*
- NFPA 1051, *Standard for Wildland Fire Fighter Professional Qualifications*
- NFPA 1061, *Standard for Professional Qualifications for Public Safety Telecommunicators*
- NFPA 1201, *Standard for Developing Fire Protection Services for the Public*
- NFPA 1401, *Recommended Practice for Fire Service Training Reports and Records*
- NFPA 1403, *Standard on Live Fire Training Evolutions*
- NFPA 1404, *Standard on a Fire Department Self-Contained Breathing Apparatus Program*
- NFPA 1500, *Standard on Fire Department Occupational Safety and Health Program*
- NFPA 1521, *Standard for Fire Department Safety Officer*

Other standards used in the fire and emergency services include governmental standards and regulations. Many of the policies and requirements for hazardous materials programs are found in federal laws. Also, most states and provinces have specific requirements for emergency medical services (EMS) training and certification. An organization's training program must comply with the standards that apply to the subjects taught.

Most standards reflect state, provincial, or national norms for fire and emergency services. For example, professionals from appropriate fire and emergency services organizations develop NFPA standards through a consensus process. NFPA members then have the opportunity to either ratify or reject the proposed standards. This process ensures that standards reflect the needs and current practices of the fire and emergency services.

When a standard is accepted and followed, it becomes the accepted "standard of care" by the group publishing it. Standards promulgated by NFPA are considered minimum standards against which an organization could possibly be judged, regardless of whether or not that organization adopts the standard. In essence, the consensus of organizations considers that NFPA standards carry the weight of law.

Many standards can affect an organization's training program, so it is advisable for training managers to learn about those standards that affect their particular organizations. If an organization participates in a state or provincial fire and life safety education certification program, for example, the training manager should review NFPA 1035, *Standard for Professional Qualifications for Public Fire and Life Safety Educator.*

If there is an NFPA or other standard used for identifying either organizational training requirements or operations, the organization should formally adopt it. Adopting the standard gives it formal "power" in the organization and allows the training manager to enforce the requirements set forth in the standard. Of course, adoption also allows the training manager to hold organizational personnel accountable for meeting the requirements in the standard. The adoption process would be the same as that used for adopting policies, procedures, or guidelines.

Records

Training records are essential components of a successful training program. Not only do accurate records give an organization long-term inventories of its training activities, records may also be important and necessary in legal proceedings and in management reviews such as those conducted by the Insurance Services Office, Inc. (ISO).

In addition to the factors of inventories, legal records, and management reviews, the benefits of keeping accurate training records also include the following:

- Training documents may be reviewed by appropriate or authorized parties when necessary.
- Records identify training areas that are emphasized as well as areas that require more attention.
- Records provide documentation of required training completion.
- Information in records can be used for planning training programs and for scheduling training.

In addition, some training records that organizations must maintain are required by law or by Occupational Safety and Health Administration (OSHA) or Environmental Protection Agency (EPA) requirements such as training records of personnel who are exposed to hazardous chemicals and infectious disease and records for respiratory protection programs. These requirements, in effect, make training records legal documents.

In developing or improving a records system for a training program, training managers must consider what information to gather for the records. Important information to be gathered for any training records system include the following:

- Participant attendance rosters
- Topics taught at each session
- Lesson plans, workbooks and texts, tests, attendance sheets, videotapes, and other program event documentation and processes
- Evaluation/testing scores of participants when applicable
- Dates of each training session
- Names of instructors for each training session
- Descriptions of instructors' qualifications
- Any other information deemed appropriate by the organization such as employee identification numbers and locations of the training sessions

The type and format of training records may vary widely depending upon the specific needs of the organization. NFPA 1401, *Recommended Practice for Fire Service Training Reports and Records,* provides examples of different training forms as well as other helpful information on the design and procedures for effective management of training records. The format of records may also depend on the type of training for which information is being gathered (Figure 11.3).

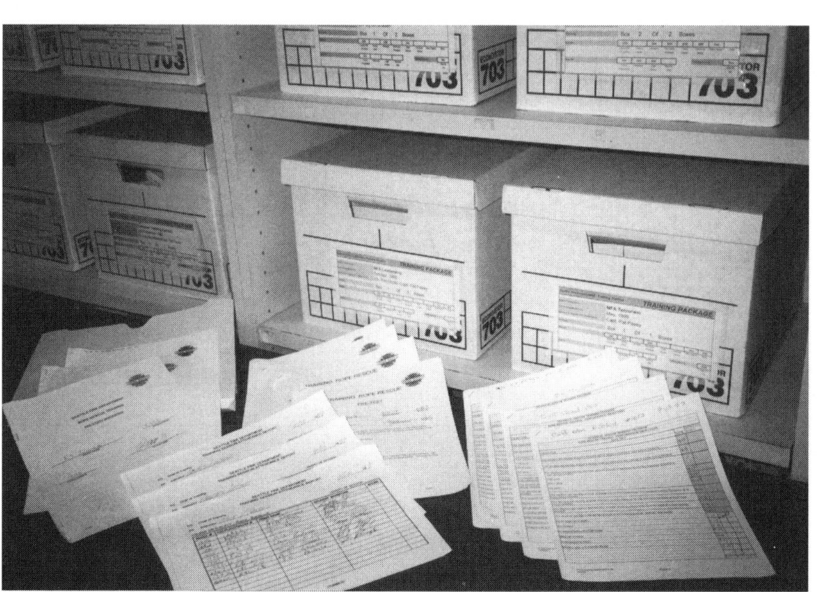

Figure 11.3 Collecting, organizing, and archiving training course materials and documents completed by learners are essential in the management of training programs. Examples of the different types of training records and documents are shown. *Courtesy of Ron Hiraki, Seattle (WA) Fire Department.*

Training managers will likely collect information for some of the following types of training:

- Daily training delivered by a training division
- Company training delivered by a company officer or a member of the company
- Organizational training delivered to all members of an organization
- Self-study by an individual
- Individual training provided by an organization
- Special training such as that received outside the organization

State, provincial, and federal governments generally have specific laws that direct record maintenance and in some cases the information that is gathered and stored in these records. For example, many organizations no longer use an employee's Social Security Number for records identification. The practice of using other identification methods reduces the opportunity for improper use. Training records may also be considered part of an individual's personal, private employment file — a fact that requires an organization to limit access to training records. And even if this practice is not required by law, organizations should develop and adopt policies that limit access to training records only to those personnel with a legal need to know.

Other information that is protected by law is an individual's evaluation/testing scores. In the United States, the Family Education and Privacy Act prohibits the release of this type of information. Similarly, the Canadian province of Ontario has the Municipal Freedom of Information and Protection of Privacy Act (MFIPPA), which places the responsibility on instructors to know their duties and responsibilities under the applicable legislation within their jurisdictions. What do these protections mean for training record management? The old practice of posting test scores and other personal information on bulletin boards is no longer allowed. Scores and personal data are considered privileged information and are available only to management and a few other designated personnel with authorization and a specific need to know.

While many organizations develop their own training records systems, there are many commercial software programs available for keeping training records. Computer software programs improve the efficiency of records management and make it easy to record, access, and process training information. The cost of such software may limit its availability to small

Training Records Examined After Rescue

On December 30, 1997, a worker was installing fireworks on Seattle's Space Needle for a New Year's Eve show. Near the 200-foot (60 m) level, the worker's safety line became caught in a descending elevator. The worker was pulled into the elevator shaft and fell approximately 45 feet (14 m). The worker landed on top of an elevator car, climbed onto a crossbeam, and waited for help.

Seattle Fire Department firefighters responded to the scene. The first-arriving firefighters used the elevator to reach the injured worker. The firefighters carefully helped the worker into the elevator and returned to the ground. The worker was treated by paramedics.

The Department of Labor and Industries, the state agency responsible for worker safety, reviewed the accident. In that review, the agency questioned the level of training that firefighters received in rope-rescue procedures.

The Seattle Fire Department had initiated operations level, rope-rescue training for all firefighters and officers. The training program developed by the Seattle Fire Department's Technical Rescue Team and the Training Division included a workbook, test, and specific performance objectives sign-off sheet in addition to the lesson plan. The Seattle Fire Department produced these items for each of the firefighters directly involved in the rescue. The amount and scope of rope-rescue training for these firefighters was appropriate and well-documented. The Department of Labor and Industries took no action against the Seattle Fire Department.

In addition to providing the training and documenting attendance, the Seattle Fire Department demonstrated a desired level of competence through the workbook, test, and specific performance objectives sign-off sheet.

Courtesy of Deputy Chief Ron Hiraki, Seattle (WA) Fire Department.

organizations, but they can still develop an effective records system by using organized files. Regardless of the system that an organization uses, appropriate personnel must be able to easily retrieve the information contained in the records and protect it from general access. Laws may require organizations to maintain training records for a specified period. Training managers or personnel responsible for maintaining training records must be aware of and follow local and state or provincial requirements.

Organizational Training Needs

[NFPA 1041: 4-3.2, 4-3.2.1, 4-3.2.2, 4-3.5, 4-3.5.1, 4-3.5.2]

Determining training needs is a key role for a training manager. For example, NFPA 1201 clearly defines and assigns the management function of determining the fire training needs of an organization to the fire training manager. In most organizations, this role is a challenging responsibility, typically because there are more training requirements than available training time (Figure 11.4).

It is important to realize that training is not always the answer to performance problems in an organization. Many times the training manager or instructor is charged with using training to correct problems and issues that are simply not training issues. Personnel performance problems may be due to a lack of skills or knowledge, in which case training is the appropriate solution. In other cases, performance problems may actually be caused by a lack of appropriate or usable equipment. Providing the equipment is the simple remedy. Sometimes performance problems are the result of "system" problems caused by organizational structures, policies, or resources. These and other limiting factors require changes to the organization that are outside the duties of the training manager and scope of the training program.

Figure 11.4 Training managers must manage and respond to multiple challenging tasks: planning training budgets and schedules, prioritizing training to meet safety and certification/standard requirements, selecting and supervising qualified instructors, and coordinating training dates with personnel schedules and duties. *Courtesy of Terry L. Heyns.*

Finally, the training manager must consider that performance problems may be due to willful actions of personnel. The solutions to these types of performance problems are found in supervisory and disciplinary actions rather than training. Carefully review the cause of performance problems before scheduling a training session to correct them. Training is not always the answer.

When determining the training needs of an organization, a training manager considers the following issues:

- ***What training is essential to achieve the organization's mission?*** Every organization has a specific mission, which may include fire suppression, hazardous materials mitigation, EMS, search and rescue, and fire and life safety education. All training must ensure that members of the organization are able to learn and maintain knowledge and skill in order to effectively and efficiently perform their jobs and achieve the mission of the organization.
- ***What training is essential to ensure personnel safety?*** A critical function of the organization — and a priority in any training program — is to provide the skills necessary to work *safely*.

- ***What training is mandated by law or organizational policy?*** Specific training topics and program hours may be mandated by law, including such programs as hazardous materials mitigation, infectious disease control, and self-contained breathing apparatus use. Training managers must be aware of the laws that mandate training and schedule programs accordingly.
- ***What training is listed in applicable NFPA standards?*** NFPA standards not only identify skills for different jobs within the fire and emergency services but in some cases also mandate specific training. For example, NFPA 1404, *Standard for a Fire Department Self-Contained Breathing Apparatus Program,* requires annual ongoing training and evaluation of personnel using breathing apparatus. In addition, NFPA 1500 requires that training based upon NFPA standards be provided before personnel engage in assigned duties. Similarly, NFPA 1201 requires compliance with the NFPA professional qualifications standards such as NFPA 1001 (*Standard for Fire Fighter Professional Qualifications*), NFPA 1021 (*Standard for Fire Officer Professional Qualifications*), and NFPA 1035 (*Standard for Professional Qualifications for Public Fire and Life Safety Educator*).
- ***What areas of training do current personnel desire and/or need?*** Often, personnel know what skills they need to improve and what new skills they require to do their jobs. Training managers should consider and include personnel desires when determining training needs.

Identifying the training needed, prioritizing those needs appropriately, and identifying resources required to deliver a program are essential components of a successful training program. The process for determining these components is called a *needs analysis.* The benefits gained from conducting a needs analysis include identifying who needs the training, determining skills and topics with the most immediate need, determining the resources required, and maximizing the effectiveness of limited training time.

Selecting Training Topics

A needs analysis helps a training manager determine the type of training needed, the topics with the most need, and the desires of personnel. There are several methods used during a needs analysis that help determine training needs. Some examples are as follows:

- ***Survey*** — A *survey* is a written questionnaire designed to ask personnel for feedback about training and input on training plans. The purpose of a survey is to identify what training is needed the most. It requires time to develop a valid survey. Processing the information is simple if questions are clear and the design enables the processor to select the best answers quickly. Survey information may not be valid if personnel misinterpret questions or provide information they *think* is desired rather than true information.
- ***Interview*** — An *interview* is a discussion with a person. An advantage of an interview is that it allows a training manager to gather very specific information in either a face-to-face meeting or a person-to-person telephone call. An interview enables a training manager to establish rapport with a person, but it is time-consuming. In addition, the information may be subjective or based upon the bias of the person and the interviewer.
- ***Skills evaluation*** — Personnel can be *evaluated* based upon the skills set forth in a standard such as NFPA 1001 and NFPA 1002. Evaluations can be performed at a training site, on a drill ground, in a classroom, or in a fire station. The advantage of a skills evaluation is that evaluators or training managers actually see personnel perform the skills and can identify deficiencies. A skills evaluation is time-intensive, and the evaluation process may be intimidating to some personnel, which may negatively effect performance.
- ***Observation*** — Performance can be *observed* during emergency, routine, or other situations that require personnel to apply their skills. The advantage of observation is that this method allows a training manager, officer, or evaluator to view personnel in the actual work environment performing many skills. In fact, the observation method generally gives a real-life picture of personnel skills. In small organizations it may be necessary to view personnel during multiple situations. Training managers should not normally set training priorities based on observations of a single situation. However, a manager may observe a single incident in which it becomes immediately obvious that training has been overlooked for a situation or incident and must be designed and implemented as soon as possible.

Prioritizing Training Topics

Once training topics have been identified by the needs analysis process, the training manager must then prioritize them (Figure 11.5). A consideration in prioritizing training topics is to ensure that instructors attend programs to maintain their technical skills and professional certifications. While there is no specific process or set of steps for prioritizing topics, the training manager can start by categorizing them into the following three areas:

- ***Must-know topics*** — Allow personnel to perform their jobs effectively, efficiently, and safely based on standards and policies.
- ***Need-to-know topics***—Allow personnel to perform their jobs based on job descriptions and responsibilities.
- ***Nice-to-know topics***—Supplement or complement personnel jobs.

For example, training on the proper use of self-contained breathing apparatus (SCBA) is a *must-know* topic. SCBA training is essential for effective performance and the safety of personnel. It is required by NFPA 1500, NFPA 1001, and NFPA 1404. This type of training has a higher priority than training that covers the proper use of a fire hose washer. While the use of a fire hose washer may be a part of a firefighter's job, it is not a component of fire fighting effectiveness or safety. Yet cleaning hose is a necessary component of equipment maintenance and readiness, so it is considered a *need-to-know* topic. Finally, a topic such as the history of the fire service is a *nice-to-know* topic and receives a low priority because such knowledge neither contributes to fire fighting effectiveness or safety nor enhances equipment effectiveness.

Figure 11.5 Specialized equipment calls for special training, especially by instructors who are qualified. For example, fire or EMS training managers have to prioritize training and recertification on the operation of various types of fire or medical equipment. Here firefighters review procedures and a checklist for a specific piece of medical equipment. *Courtesy of Terry L. Heyns.*

Certain jobs that personnel perform may affect the prioritization of training topics. For example, driver/operators have different training priorities than firefighters or officers. When prioritizing training topics, training managers need to ask the following questions:

- What topics are required for personnel to achieve the organizational mission or program goals?
- What topics are required for personnel to perform jobs effectively and safely?
- What topics are mandated by laws or standards?
- What topics are perceived as most important by the customer?
- What topics are a priority of the organization's administration?

Based on the answers to these questions, training managers must establish training priorities and program goals that meet the needs of the organization. There is no set formula for establishing priorities and goals. Every organization has different needs that require different priorities and goals.

Determining Resources

The final component in a needs analysis is determining the resources required to deliver the training that has been identified and prioritized. A list of resources is also necessary when the training manager is developing the training budget. In order to determine what resources are necessary for program delivery, training managers can include the following points in their considerations:

- Instructional requirements for developing, delivering, and evaluating training
- Instructional materials such as curriculum or instructor guides, student texts, and handouts
- Other program delivery supplements, support materials, equipment, and facilities such as classrooms, audiovisual equipment, training equipment, burn sites, driving ranges, and vehicles

Recruiting and Selecting Instructors

[NFPA 1041: 3-2.6, 3-2.6.1, 3-2.6.2, 3-4.3, 4-2.4, 4-2.4.1, 4-2.4.2]

Effective instructors are the foundations upon which effective training programs are built. Effective instructors can teach, lead, motivate, inspire, and change the attitudes of an organization. However, organizations often invest too little effort and consideration in the selection of competent instructors. Selecting competent instructors is a key management function of training managers. In addition, determining whether instructor roles are long-term or short-term aids in the selection process. Is this instructor going to be a member of the training staff for several years? Is this instructor simply being used to teach one topic during a short-term period of time? Establish these factors before going any further.

Next, determine the instructors' roles in the organization and the qualifications they need to teach the training programs. Establishing their positions in the organizational structure gives importance to the instructor role. Instructors, whether long- or short-term, are critical parts of program planning. They also act as intermediaries between administration and personnel in training. Instructors not only teach job knowledge and skills, but they are active in applying knowledge and skills on the job at various ranks or positions in the organization. Their experiences are critical components to their qualifications.

Few instructors are qualified to "teach it all." Rather, instructors tend to specialize in certain areas and become proficient in teaching knowledge and skills specific to their areas of experience and expertise. Base instructor selection upon the topics to be taught and the designated instructor roles in the organization and training program.

In addition to topic expertise, instructors must have credibility with the personnel being trained. Personnel perceive that instructors have credibility when they display technical proficiency and evidence of formal training and education and demonstrate instructional experience. Credibility is also indicated by rank, reputation, and respect among members of the organization.

When determining instructor qualifications, establishing credibility in technical proficiency and professional experience in the subject to be taught are important factors. If instructors have neither skill nor credibility in the desired teaching areas, they will not be effective. And instructors must be effective communicators along with having technical credibility. Topic expertise and credibility are only effective in training if instructors are able to transmit knowledge and demonstrate skills in methods that program participants can understand and apply (see Chapter 1, Challenges of Fire and Emergency Services Instruction).

Along with the technical qualifications, a person must also be a qualified instructor. For example, fire instructors should meet the Instructor I requirements of NFPA 1041, *Standard for Fire Service Instructor Professional Qualifications.* If no personnel meet these requirements, training managers need to provide instructor training programs. Training, program resources, and information are generally available from local, state, or provincial training organizations.

After instructor roles and qualifications are determined, the training manager must advertise or market the position to people both inside and outside the organization. Too often instructors are selected from a small cadre because internal personnel who may be eligible for an instructor position are unaware of the open position. All personnel must be aware of the opening. An organization may have policies or other requirements (such as required rank) that limit the number of applicants, but these requirements must also be advertised so that interested personnel may choose an appropriate course of action.

While a position is being advertised, interested people submit applications. The training manager or a selection committee then chooses the most qualified to interview. Ideally and in order to observe and evaluate an applicant's qualifications, the interview process includes an instructional scenario in which the candidate demonstrates the desired abilities and qualifications. This activity gives the training manager or selection committee an opportunity to see all candidates simulating the job they would perform if selected. Many potential instructors have great résumés but cannot effectively communicate or demonstrate effective teaching methods. Training managers should select instructors based on organizational needs.

If there is a shortage of qualified applicants, the organization should provide instructor training so personnel who wish to teach can improve their skills. Training managers are responsible for ensuring that the instructors they use are competent. To ensure competence, provide opportunities for instructor training and the development of current and potential instructors.

Once the selection has been made after the interview process, the training manager orients the new instructor. Orientation includes, in particular, defining the instructor's teaching role. Instructors must have a clear understanding of the job before they enter the classroom or arrive at the drill ground. The new instructor needs to know the following information:

- Topics to teach
- Policies of the training program to promote and enforce
- Instructional support to expect
- Limitations to expect

Training Program Scheduling

[NFPA 1041: 3-2.2, 3-2.2.1]

In addition to selecting instructional personnel, the training manager also schedules the training. Providing a training schedule enables supervisors to assign personnel to training with minimal interference with or impact on duties (Figure 11.6). For volunteer fire personnel, for example, training schedules allow them to integrate required training with job responsibilities and personal activities. Posted training schedules provide training topics, dates, times, and locations. The organization may post schedules also to notify potential instructors of upcoming classes that are available for them to teach.

When developing training schedules, the training manager considers several factors that have an impact on supervisors, personnel, and instructors:

- ***How will the training schedule impact on other job activities?*** The training manager must remember that personnel have responsibilities other than attending training. Some examples are inspection workload, expected alarm load, special events, and annual leaves. Some responsibilities may have a higher priority to the organization *at the moment* than training, and the training manager must be aware of these other activities when creating the training schedule.
- ***When are instructional resources (including instructors) most readily available?*** If training resources such as classrooms, burn building, or driving courses are shared with other organizations, the training manager must make cooperative arrangements, consider weather that may cause cancellations and rescheduling, and coordinate timely arrival and departure of classes. Training managers also have to consider instructor schedules and conflicting assignments. Instructors may not always be available for assignments at certain times of the day or week, and available instructors may not be qualified to teach certain topics in some programs.
- ***What is the most appropriate time for the training?*** If a training program must be taught outdoors, it is best to schedule training when the weather is suitable to the topic. Evenings and weekends may be the best time to schedule some training.
- ***How quickly must the training be completed?*** In some cases, an organization's administration may want a specific training program completed immediately in order to fulfill a regulatory requirement. To meet the need for immediate training, the training manager can plan several program offerings in several time frames; that is, maximize the times and opportunities for personnel to complete the training. In addition, a training manager can reprioritize other organizational activities and training programs where possible. For most training, programs may be scheduled over longer time periods and more frequently throughout the year so personnel

Figure 11.6 In large organizations, training schedules must be planned well in advance. Not only do the needs of the organization enter into consideration but also vacations, holidays, sick leaves, promotions, and the process of hiring and training new recruits. Seasonal and other unforeseen events can also affect the training schedule. *Courtesy of Terry L. Heyns.*

can take advantage of required programs at their convenience.

The process of developing a training schedule is not always a simple one. Responding to all the factors mentioned is taxing. To simplify the task, training managers can establish guidelines that outline steps that provide consistency to the planning or scheduling process. Steps that assist in developing a training schedule are as follows:

Step 1. Develop a consistent procedure for creating and publicizing a training schedule.

- ✔ Consider the time period needed for training. The training schedule may cover one month, six months, or a year. The longer the time period covered by the training schedule, the easier it is for personnel to prepare for training, but the greater the likelihood the schedule will have to be modified because of the following events:
  - — Unforeseen activities
  - — Extreme weather
  - — Annual leaves
  - — Other circumstances.
- ✔ Ensure that all personnel are aware of the schedule and that all personnel have access to it. This is especially important in a volunteer organization, for example, where personnel must plan their personal lives around the organization's activities.
- ✔ Create the schedule for time periods most convenient to all personnel.

Step 2. Assign one individual to create and keep record of the training schedule.

Whether the schedule is planned by the training manager or other assigned individual, one person should be responsible for the following activities:

- ✔ Tracking schedules
- ✔ Handling cancellations
- ✔ Rescheduling
- ✔ Sharing of time slots or facilities with other organizations

Step 3. Schedule a reasonable amount of training programs within a given time period.

- ✔ Do not make the common mistake of scheduling too much training in any given period and not allowing for unforeseen circumstances. There should be enough time left open in the training schedule so that it can be modified without requiring a major change in the overall training plan for the year.
- ✔ Schedule multiple sessions to ensure that all personnel have access to the training.
- ✔ Allow for travel to and from the training site, setup time for instructors and evaluators, and cleaning or restocking time for classes that follow one another.

Step 4. Schedule all resources in a timely manner, and confirm that these resources are available.

- ✔ Confirm that instructors are available and aware of the classes or topics they are to teach.
- ✔ Confirm that equipment or vehicles are available when needed.
- ✔ Reserve facilities so that use does not conflict with other training.

Step 5. Publish and distribute the schedule.

- ✔ Distribute a schedule to all personnel who are required to attend the training, will deliver the training, or are affected by the training.
- ✔ Post the training schedule in areas that are accessible to all members of the organization.

Step 6. Develop an alternate plan.

Consider events that may occur that may prevent the scheduled training from being delivered as planned such as the following:

- ✔ Instructor illness or injury
- ✔ Equipment failure or required emergency use
- ✔ Other organizational priorities
- ✔ Weather conditions

Training Program/Instructor Evaluations

[NFPA 1041: 3-2.5, 3-2.5.1, 3-2.5.2, 4-2.5, 4-2.5.1, 4-5.1, 4-5.2, 4-5.2.1, 4-5.2.2, 4-5.3, 4-5.3.1, 4-5.3.2, 4-5.4, 4-5.4.1, 4-5.4.2]

The term *evaluation* often brings to mind testing in a classroom or assessing skills on the drill ground. Yet, evaluation also applies to and is necessary for the training program and the instructor. In fact, evaluation is essential to the success of any organization's training program. The training program's goals and objectives form the bases for evaluation. Goals and objectives identify what the training program is to accomplish. The success of the program is determined by how well the training program and the instructor achieve their goals and objectives (see Chapter 9, Testing and Evaluation).

Evaluating the training program includes the step of determining the satisfaction of program customers. Personnel participating in an *effective* training program appreciate the time they spend there and the information they learn. Likewise, participants are the first to recognize a training program that is ineffective or inappropriate and will make their displeasures known.

The training program evaluation process also includes surveys of supervisory personnel such as incident commanders, supervisors, and program managers. These supervisory personnel work with the individuals who complete the training programs. They are aware of the skills and attitudes these individuals now display on the job. Supervisory personnel provide invaluable information about the successes and results of training programs.

In addition to evaluating training programs, training managers must also evaluate instructors. Instructor evaluation requires the actual observation of instructor performance in the classroom and on the training ground (Figure 11.7). Training managers perform instructor evaluations for the following purposes:

- Ensure that instructors achieve training objectives.
- Improve overall instructor performance and quality.
- Demonstrate that quality instruction is important to the organization.

Before evaluating instructor performance, develop an appropriate evaluation form. Refer to Appendix I, Evaluation Forms, for examples. An evaluation form ensures that all evaluations are performed in the same manner and according to the same criteria. The form should identify criteria for behaviors that are necessary for instructor success and include the following points:

- Interaction between the instructor and the learners
- Proper use of verbal and nonverbal communication techniques
- Proper use of audiovisual equipment and materials

Figure 11.7 A training manager evaluating an instructor in a classroom situation. *Courtesy of Larry Ansted, Maryland Fire and Rescue Institute.*

- Communication that includes discussion or demonstration of all objectives in the lesson
- Frequent and appropriate use of questions
- Use of appropriate instructional methods
- Appropriate application of evaluation instruments
- Rates of student success/failure

When planning an evaluation, schedule a time that is convenient for both instructor and evaluator and does not interfere with the particular lesson or the instructional process. Scheduling evaluation time is not considered giving a warning to instructors so they can prepare and perform their best for the evaluation. Scheduling time is a fair planning process for both persons. The purpose of the evaluation is to give feedback for improvement not to surprise or catch an instructor off guard. An instructor should understand the evaluation process and the role of the evaluator and welcome an opportunity for feedback.

When performing an evaluation during instruction, observe the learning and instructional process rather than participate — do not intrude on the process. By observing, the evaluator experiences the *actions* of the instructor and the *reactions* of the learners. If possible, videotape the session and use it during the evaluation review and discussion with the instructor to highlight and critique behaviors.

When critiquing instructor performance, first provide positive comments, and then point out areas that need improvement using constructive suggestions that guide or direct the instructor on how to improve. Comments should refer to specific behaviors. Assign instructors who are having difficulty adapting appropriate teaching methods to observe or work with other more experienced instructors or mentors. Working with other instructors gives inexperienced instructors opportunities to observe "models" of the desired instructional behavior. Carefully select instructors to act as mentors so that appropriate teaching methods are modeled.

The training manager's evaluation of the instructor is not the only source of instructor effectiveness. Participants spend the entire program with their instructors and can provide a wealth of information. Many organizations have standard program and instructor evaluation forms, but there are other methods of gathering information from participants. One method for gathering participant reaction is by distributing surveys following a training session. A survey gives the program participant an opportunity to give feedback, not only on instructor performance but on other issues such as the following:

- Overall program
- Lesson content and activities
- Program's equipment
- Lesson handouts and other materials
- Learning environment

After discussing the evaluation results with the instructor, the training manager must hold the instructor accountable for any areas that require improved performance. Holding instructors accountable is an important responsibility of a training manager just as it is for supervisors of other personnel. With no accountability, there is no change in behavior or performance. Follow up on the evaluation, and ensure accountability by returning to the classroom or training ground to determine if critical instruction points have improved. If an instructor's performance hasn't improved, hold further discussions with the instructor and consider further training. If the quality of instruction is not acceptable, it may be necessary to remove the instructor from the program and assign another instructor who is more qualified.

An instructor who is having difficulty should not simply be "thrown away." It is a challenge finding personnel who have an interest in becoming instructors. Once interested individuals are found, orient them with the appropriate training, education, and supervised experience they need to be successful. The orientation process not only benefits these individuals personally, it increases the resources available to the training manager as well.

A final purpose of the information gathered in the instructor evaluation process is to modify instructor training programs. The feedback often provides insight into improving the content of the instructor training program or modifying the selection process for new instructors.

Budget and Resource Management

[NFPA 1041: 3-2.3, 3-2.3.1, 3-2.3.2, 3-2.4, 3-2.4.1, 3-2.4.2, 4-2.6, 4-2.6.1, 4-2.6.2]

Training managers often find that there is never enough funding to support every training need. Budgeting and using funds to fulfill the needs of the organization and its personnel require careful planning and managing of the resources allocated for training. The training manager is typically responsible for the following budget and resource management tasks:

- Tracking expenditures and revenues associated with the training program
- Identifying and communicating to administrators the resources required to meet the training goals and objectives
- Seeking alternative sources for training resources
- Ensuring that legal and organizational requirements for managing resources are met

In order to fulfill the role of managing the organization's training resources, the training manager must understand and apply the following five principles for budget management:

Principle 1. Base the training budget on program needs.

Many times, the training program is designed around the possible funding for training. The training needs of the organization *should be driving the requests in the budget process, not the other way around.* The budgeting process is a planning process based upon the organization's training goals and objectives. The goals and objectives identify what training is required and thus also identify the resources required to accomplish the training. From these needs, the training manager determines the funding required to accomplish the goals and objectives. Financial growth of the training program can only occur if the budget process is used to achieve the organization's training goals and objectives.

Principle 2. Plan ahead.

Budget requests should identify both short-term needs such as curriculum and textbooks as well as long-term needs such as additional staff, new facilities, and up-to-date equipment. When the training manager projects budgetary needs into the future, it is easier to find the needed funds in the organization's overall budget.

Principle 3. Create partnerships and share resources whenever possible.

Most organizations find it difficult to fully fund all the needs of a training program. However, when several organizations form partnerships and share training resources, they can often meet the needs of each organization's training program.

Principle 4. Know and follow the organization's rules and policies for budgeting and managing funds.

Organization administrators expect the training manager to appropriately manage the organization's resources. The more closely the training manager follows the rules, the greater the chances of being successful in the budget process.

Principle 5. Create and maintain accurate budget records.

The training manager must account for all the resources and funds provided by the organization, including grants and financial gifts. Failure to keep accurate and timely records is not only poor management but may be a legal violation.

Budget Categories

When developing budget requests, the training manager categorizes the requests into one of the following three areas:

- ***Salaries and benefits*** — Expenses associated with human resources; include the following:
  - — Wages
  - — Benefits
  - — Worker's compensation costs
  - — Insurance
  - — Contract labor
- ***Operations*** — Expenses required for the normal, day-to-day operation of the training program; include items such as the following:
  - — Office supplies
  - — Heating and air-conditioning
  - — Telephones
- ***Capital goods*** — Items having a long-term useful life (usually 5 years or more, but the figure varies from organization to organization) and a high cost; include such items as the following:
  - — Office machines
  - — Apparatus
  - — Real estate (land)
  - — Buildings

Some organizations also include a budget category for grants and gifts. The organization closely monitors this type of category, which may have special requirements and restrictions. The training manager must be knowledgeable of and carefully follow organizational and local requirements for managing grants and gifts

Budget Development

Once management principles and budget categories are established, the training manager is ready to develop the budget (Figure 11.8). Considerations for developing a budget are as follows:

- ***Review and understand organizational policies and guidelines for budget requests*** — Understand the organization's rules.
- ***Identify specific needs*** — Base training needs on organizational training goals and objectives. Needs are not "pulled out of thin air" but are based on short- and long-term training plans.
- ***Identify a cost for all the listed needs*** — Identify multiple human resource needs, operating expenses, and capital expenses for each specific need (Figure 11.9). For example, if a training objective is to provide training for a specific curriculum, the cost of every component of that curriculum must be listed in the budget. In addition, if the need is for a capital item, it may be necessary to get a bid from more than one vendor or source. Bids must identify specific needs and whether it is acceptable to make variations or substitutions. Most organizations have policies guiding the development, advertisement, and selection of bids. Plan for the following factors when identifying costs:
  - *Delivery costs* of a program (instructor salary, personnel time, and facility use)
  - *Development costs* (personnel such as program developers, secretarial staff, graphic artists, and photographers)
  - *Preparation costs* (purchasing texts and audiovisuals and copying handouts)
- ***Justify budget requests*** — Develop the rationale for budget requests based upon the organization's training plan. The days of justifying budgets simply by saying *"Training is important"* are over. Provide sound reasons for budget requests:
  - Correlate the budget request to the support of the organization's mission.
  - Show that budget requests enable the organization to more effectively meet its mission and provide service to its customers.
  - Identify legally mandated training.
  - List benefits that training provides to the organization.
- ***Manage the budget once it has been determined*** — Track expenditures, report expenditures and account balances on a regular basis, and ensure that spending is within the budget. Closely monitor expenditures and revenues. Budget management is an ongoing responsibility, not simply a task to be done at the end of the budget year.

References and Supplemental Readings

Colorado Springs Fire Department. *Curriculum Development.* Colorado Springs: Colorado Springs Fire Department, 1996.

Dubois, David D. *Competency-Based Performance Improvement: A Strategy for Organizational Change.* Amherst: HRD Press, 1993.

Gill, Stephen J. *Linking Training to Performance Goals,* Alexandria: American Society for Training and Development, 1996.

Mager, Robert F. and Peter Pipe. *Analyzing Performance Problems* (3rd ed.). Atlanta, GA: The Center for Effective Performance, Inc., 1997.

Figure 11.8 Managing the training budget is a challenge that requires diligent tracking of records, careful research of needs, and precise justification of purchase requests. *Courtesy of Seattle (WA) Fire Department.*

Seattle Fire Department
In Service Training Program

Cost Estimate Worksheet For
Confined Space Rescue

Course Development Costs (Creating The Course)				
Staff Hours				
Position	*Rank*	*OT Rate/Hr.*	*Hours*	*Total*
Instructor II	Captain	$50	80	$4,000
Admin. Support Spec.	N/A	$25	16	$400
Initial Course Materials				
Item	*Unit*	*Cost*	*Quantity*	*Total*
Student Manuals Or Handouts	N/A	$0	0	$0
Instructor Manuals	Each	$10	1	$10
Transparencies	Each	$25	1	$25
Slides	Each	$1	120	$120
Videos	N/A	$0	0	$0
Props	N/A	$0	0	$0
Dedicated Equipment	N/A	$0	0	$0
TOTAL COURSE DEVELOPMENT COSTS				$4,555
Course Preparation Costs (Instructor Training)				
Instructor Training				
Position	*Rank*	*OT Rate/Hr.*	*Hours*	*Total*
Instructor I (Three)	Captain	$50	6	$900
Instructor I (Two)	Lieutenant	$45	6	$540
Instructor I (Four)	Firefighter	$40	6	$960
Initial Course Materials				
Item	*Unit*	*Cost*	*Quantity*	*Total*
Student Manuals Or Handouts	N/A	$0	0	$0
Instructor Manuals	Each	$10	9	$90
TOTAL COURSE PREPARATION COSTS				$2,490
Course Delivery Costs (Teaching Students) Per Course				
Course Specifications	*Course Hours*	*Classes @ Hours*	*No. Of Students*	
	12	1 x 8 Hrs.	15	
	4 Hrs. Self Study			
Delivery				
Position	*Rank*	*OT Rate/Hr.*	*Hours*	*Total*
Instructor I	Lieutenant	$45	8	$360
Instructor I	Lieutenant	$45	8	$360
Consumable Course Materials				
Item	*Unit*	*Cost*	*Quantity*	*Total*
Student Manuals Or Handouts	Each	$2	15	$30
Rental Of Prop	Per Day	$125	1	$125
Other Materials		$0	0	$0
TOTAL COURSE DELIVERY COSTS				$875
Total Cost Per Student				$44

11/29/98 Page 1

Figure 11.9 Example of a training program cost estimate worksheet. *Courtesy of Seattle (WA) Fire Department.*

Mager, Robert F. *Measuring Instructional Results* (3rd ed.). Atlanta, GA: The Center for Effective Performance, Inc., 1997.

Maryland Fire and Rescue Institute. "Respiratory Protection Training." College Park, Maryland, Maryland Fire and Rescue Institute, University of Maryland, 1998.

National Fire Protection Association. *NFPA 1401 Recommended Practice for Fire Service Training Reports and Records* (1996 ed.). Quincy, MA: National Fire Protection Association, 1996.

National Fire Protection Association. *NFPA 1404 Standard for a Fire Department Self-Contained Breathing Apparatus Program* (1996 ed.). Quincy, MA: National Fire Protection Association, 1996.

Pike, Bob. *Managing the Front End of Training.* Minneapolis: Lakewood Publications, 1994.

Seymour, Thomas H. "A Comprehensive Analysis of the OSHA Respiratory Protection Standard, 29 CFR 1910.134." Fairfax, VA: International Association of Fire Chiefs, 1998.

Zemke, Ron and Allison Rossett. *Be A Better Needs Analyst,* Alexandria: American Society for Training and Development, 1985.

Job Performance Requirements

This chapter provides information that addresses the following job performance requirements of NFPA 1041, *Standard for Fire Service Instructor Professional Qualifications* (1996 edition). Colored portions of the standard are specifically covered in this chapter.

Chapter 3 Instructor II

3-2.2 Schedule instructional sessions, given department scheduling policy, instructional resources, staff, facilities and timeline for delivery, so that the specified sessions are delivered according to department policy.

3-2.2.1 *Prerequisite Knowledge:* Departmental policy, scheduling processes, supervision techniques, and resource management.

3-2.3 Formulate budget needs, given training goals, agency budget policy, and current resources, so that the resources required to meet training goals are identified and documented.

3-2.3.1 *Prerequisite Knowledge:* Agency budget policy, resource management, needs analysis, sources of instructional materials, and equipment.

3-2.3.2 *Prerequisite Skills:* Resource analysis and forms completion.

3-2.4 Acquire training resources, given an identified need, so that the resources are obtained within established timelines, budget constraints, and according to agency policy.

3-2.4.1 *Prerequisite Knowledge:* Agency policies, purchasing procedures, budget management.

3-2.4.2 *Prerequisite Skills:* Forms completion.

3-2.5 Coordinate training record keeping, given training forms, department policy, and training activity, so that all agency and legal requirements are met.

3-2.5.1 *Prerequisite Knowledge:* Record keeping processes, departmental policies, laws affecting records and disclosure of training information, professional standards applicable to training records, databases used for record keeping.

3-2.5.2 *Prerequisite Skills:* Record auditing procedures.

3-2.6 Evaluate instructors, given an evaluation form, department policy, and job performance requirements, so that the evaluation identifies areas of strengths and weaknesses, recommends, changes in instructional style and communication methods, and provides opportunity for instructor feedback to the evaluator.

3-2.6.1 *Prerequisite Knowledge:* Personnel evaluation methods, supervision techniques, department policy, effective instructional methods and techniques.

3-2.6.2 *Prerequisite Skills:* Coaching, observation techniques, completion of evaluation forms.

3-4.3 Supervise other instructors and students during high hazard training, given a training scenario with increased hazard exposure, so that applicable safety standards and practices are followed, and instructional goals are met.

Chapter 4 Instructor III

4-2.1 The administration of agency policies and procedures for the management of instructional resources, staff, facilities, records, and reports.

4-2.2 Administer a training record system, given agency policy and type of training activity to be documented, so that the information captured is concise, meets all agency and legal requirements, and can be readily accessed.

4-2.2.1 *Prerequisite Knowledge:* Agency policy, record keeping systems, professional standards addressing training records, legal requirements affecting record keeping, and disclosure of information.

4-2.2.2 *Prerequisite Skills:* Development of forms, report generation.

4-2.3 Develop recommendations for policies to support the training program, given agency policies and procedures and the training program goals, so that the training and agency goals are achieved.

4-2.3.1 *Prerequisite Knowledge:* Agency procedures and training program goals, format for agency policies.

4-2.3.2 *Prerequisite Skills:* Technical writing.

4-2.4 Select instructional staff, given personnel qualifications, instructional requirements, and agency policies and procedures, so that staff selection meets agency policies and achievement of agency and instructional goals.

4-2.4.1 *Prerequisite Knowledge:* Agency policies regarding staff selection, instructional requirements, selection methods, the capabilities of instructional staff and agency goals.

4-2.4.2 *Prerequisite Skills:* Evaluation techniques.

4-2.5 Construct a performance based instructor evaluation plan, given agency policies and procedures and job requirements, so that instructors are evaluated at regular intervals, following agency policies.

4-2.5.1 *Prerequisite Knowledge:* Evaluation methods, agency policies, staff schedules, and job requirements.

4-2.6 Write equipment purchasing specifications, given curriculum information, training goals, and agency guidelines, so that the equipment is appropriate and supports the curriculum.

4-2.6.1 *Prerequisite Knowledge:* Equipment purchasing procedures, available department resources and curriculum needs.

4-2.6.2 *Prerequisite Skills:* Evaluation methods to select the equipment that is most effective and preparation of procurement forms.

4-2.7 Present evaluation findings, conclusions, and recommendations to agency administrator, given data summaries and target audience, so that recommendations are unbiased, supported, and reflect agency goals, policies, and procedures.

4-2.7.1 *Prerequisite Knowledge:* Statistical evaluation procedures and agency goals.

4-2.7.2 *Prerequisite Skills:* Presentation skills and report preparation following agency guidelines.

4-3.2 Conduct an agency needs analysis, given agency goals, so that instructional needs are identified.

4-3.2.1 *Prerequisite Knowledge:* Needs analysis, task analysis, development of job performance requirements, lesson planning, instructional methods, characteristics of adult learners, instructional media, curriculum development, and development of evaluation instruments.

4-3.2.2 *Prerequisite Skills:* Conducting research, committee meetings, and needs and task analysis; organizing information into functional groupings; and interpreting data.

4-3.3 Design programs or curriculums, given needs analysis and agency goals, so that the agency goals are supported, the knowledge and skills are job related, the design is performance based, adult learning principles are utilized, and the program meets time and budget constraints.

4-3.3.1 *Prerequisite Knowledge:* Instructional design, adult learning principles, principles of performance based education, research, and fire service terminology.

4-3.3.2 *Prerequisite Skills:* Technical writing, selecting course reference materials.

4-3.5 Write program and course goals, given job performance requirements (JPRs) and needs analysis information, so that the goals are clear, concise, measurable, and correlate to agency goals.

4-3.5.1 *Prerequisite Knowledge:* Components and characteristics of goals, and correlation of JPRS to program and course goals.

4-3.5.2 *Prerequisite Skills:* Writing goal statements.

4-5.1 Develops an evaluation plan, collects, analyses and reports data and utilizes data for program validation and student feedback.

4-5.2 Develop a system for the acquisition, storage, and dissemination of evaluation results, given agency goals and policies, so that the goals are supported and those impacted by the information receive feedback consistent with agency policies, federal, state, and local laws.

4-5.2.1 *Prerequisite Knowledge:* Record keeping systems, agency goals, data acquisition techniques, applicable laws, and methods of providing feedback.

4-5.2.2 *Prerequisite Skills:* The evaluation, development, and use of information systems.

4-5.3 Develop course evaluation plan, given course objectives and agency policies, so that objectives are measured and agency policies are followed.

4-5.3.1 *Prerequisite Knowledge:* Evaluation techniques, agency constraints, and resources.

4-5.3.2 *Prerequisite Skills:* Decision-making.

4-5.4 Create a program evaluation plan, given agency policies and procedures, so that instructors, course components, and facilities are evaluated and student input is obtained for course improvement.

4-5.4.1 *Prerequisite Knowledge:* Evaluation methods, agency goals.

4-5.4.2 *Prerequisite Skills:* Construction of evaluation instruments.

Code of Federal Regulations (CFR)
Superintendent of Documents
U.S. Government Printing Office
Washington, DC 20402
http://law.house.gov/cfr96.htm

Employment Standards Administration (ESA)
for information on mine safety and regulations implementing the Federal Coal Mine Health and Safety Act of 1969
http://www.dol.gov/dol/e...eg/proposed/97-13166.htm

Federal Register
Office of the Federal Register
National Archives and Records Administration
Washington, DC 20408

Legal Services of Northern California (LSNC)
for Code of Federal Regulations, Federal Regulations, and Federal Register
http://www.lsnc.net/fedreg.html

National Fire Protection Association
#1 Batterymarch Park
P. O. Box 9101
Quincy, MA. 02269-9101
1-800-344-3555

New York Firefighter and Code Enforcement Standards and Education Committee
Department of State OFPC
41 State Street
Albany, NY 12231
www.dos.state.ny.us/fire/firewww.html

Appendix B

OSHA Plan States

OSHA State-Plan States and Nonstate-Plan States

State-Plan States	Nonstate-Plan States
Alaska	Alabama
Arizona	Arkansas
California	Colorado
Connecticut (state and local government employees only)	Delaware
Hawaii	District of Columbia
Indiana	Florida
Iowa	Georgia
Kentucky	Guam
Maryland	Idaho
Michigan	Illinois
Minnesota	Kansas
Nevada	Louisiana
New Mexico	Maine
New York (state and local government employees only)	Massachusetts
North Carolina	Mississippi
Oregon	Missouri
Puerto Rico	Montana
South Carolina	Nebraska
Tennessee	New Hampshire
Utah	New Jersey
Vermont	North Dakota
Virginia	Ohio
Virgin Islands	Oklahoma
Washington	Pennsylvania
Wyoming	Rhode Island
	South Dakota
	Texas
	West Virginia
	Wisconsin

Appendix C

Lesson Plans

Lesson Plan Format

(suggested by Dr. Robert Kizlik, "Five Common Mistakes in Writing Lesson Plans." Http://www.adprima.com/mistakes.htm, 1998.)

Teacher ______________________ **Subject** ______________________

Grade Level ______________________ **Date** ______________________

I. **Content:** Indicate what you intend to teach, and identify which forms of knowledge will be included in this lesson.

II. **Prerequisites:** Indicate what the student must already know or be able to do in order to be successful with this lesson. (List one or two specific behaviors necessary to begin this lesson.)

III. **Instructional Objective:** Indicate what is to be learned. Write a *complete* objective (behavior, given, standard).

IV. **Instructional Procedures:** Describe what you will do in teaching the lesson, and include a description of how you will introduce the lesson to the students, what actual instructional techniques you will use, and how you will bring closure to the lesson. Include what specific things students will actually do during the lesson.

V. **Materials and Equipment:** List all materials and equipment to be used by both the teacher and learner and how they will be used.

VI. **Assessment/Evaluation:** Describe how you will determine the extent to which students have attained the instructional objective. Be sure this part is directly connected to the behavior called for in the instructional objective.

VII. **Follow-up Activities:** Indicate how other activities/materials will be used to reinforce and extend this lesson. Include homework, assignments, and projects.

VIII. **Self-Assessment:** (To be completed after the lesson is presented.) Address the major components of the lesson plan, focusing on both the students' strengths and their areas of needed improvement. A good idea is to analyze the difference between what the objective requires and what the student attained.

Lesson Plans

Regardless of the format or various names of components, a basic lesson plan typically includes the following components. All lesson plan formats and their many components will fit into the structure below:

I. Basic Lesson Plan Model

Lesson Topic:

Time Frame:

Level of Instruction:

Behavioral Objectives:

Materials Needed:

References:

Course Text (Lesson Chapter Assignment):

Preparation
Administration:
Introduction:
Motivation: (Why am I here? Why do I need to know this?)
Overview (of key ideas):

Presentation and ***Application***

Content Outline	Teaching Strategies
I.	Visuals/handouts/activities Teaching methods
II.	
III.	
IV. (Etc.)	

Summary

Evaluation
Cognitive evaluations (written quizzes or exams, oral questions or exams)
Performance evaluations (lab practices, skills mastery assessments)
Assignment

II. National Fire Academy Model

The National Fire Academy, with extensive experience and success at developing easy-to-follow lesson plans, uses a two-page format. The right-hand page is in two columns: the left column for instructor notes and the right column for the presentation outline. The left-hand page also uses two columns: the left column for instructor instructions or directives, usually at the beginning of a unit, and the right column for thumbnail photos of visuals, which are aligned with the matching section of the outline.

Page 1 of the Unit: (Preparation)

Unit Number and Title

Terminal Objective

Enabling Objectives

The left-hand pages of the unit:

Preparation Instructions, information, or directivfes to the instructor: What is included in the lesson — the lesson focus and an overview. How to prepare: Estimated time: Number of visuals:	Thumbnail photo Thumbnail photo

The right-hand pages of the unit:

Instructor Notes: Segment Time: 10 min. Visual number: Slide 9-6 Slied 9-7 Background information on topic: Questions to ask: Activities to do:	***Presentation and Application*** Presentation outline: I. Topic A. B. II. Topic A. B. C. (Inserted pages with Activities, Quizzes.) ***Evaluation*** III. Summary

Inserted pages: Activity pages and Quizzes are in a one-column format.

- Activity pages are inserted at the end of sections where participation/application will reinforce learning.
- Quiz questions (diagnostic tests) are inserted for review and reinforcement at the end of sections where evaluation of understanding and learning can be measured before proceeding to the next section.

III. The Hunter Model

The Madeline Hunter lesson plan model emphasizes the practice of new skills to be learned. This model includes components of the psychomotor domain.

- **Anticipatory Set:** Convince learners of the need for the information *(Why am I here? Why do I need to know this?)*. This step develops "ownership" between learners and the behavior. It emphasizes benefits of information or skills and allows learners to make comparisons between performing a skill or not performing it for safety, efficiency, or accomplishment. (Preparation)

- ***Objective/Purpose:*** Overview the objectives or the lesson purpose so that learners understand the reason for the lesson and what to expect in the presentation. Objectives must specify a behavior, what will be given in order to perform the behavior, and to what standard or degree learners will perform so they may meet intended outcome or purpose of the presentation. (Preparation)

- ***Input:*** The instructor presents information that learners need in order to achieve the intended outcome or lesson objective(s). The instructor presents information in such a way that learners do not have to search for meaning in order to understand. The instructor does this by matching appropriate instructional methods with the needs and background of the audience. Methods may include demonstration and practice activities; information presentation through discussion, visuals, models, and props; and field excursions to job-related locations. (Presentation and Application)

- ***Modeling:*** After presenting the information, the instructor models the skill through actual demonstration of the behavioral objective. Then every learner imitates the modeled skill slowly as the instructor provides input and feedback. Finally, learners perform the behavior independently. (Presentation and Application)

- ***Check for Understanding:*** Throughout the lesson, the instructor checks for understanding — a critical step. If learners do not understand the behavior or the information and cannot perform the skill, they cannot move on to the next bit of information or skill. The instructor must go back to the previous steps of *input* and *modeling.* (Presentation and Application)

- ***Guided Practice:*** After checking for understanding, the instructor allows learners to practice the behavior or apply the information. Provide enough time for practice so that everyone has the opportunity to master the behavior. The instructor spends time with individuals to ensure that they can perform the desired behaviors. Encourage learners to assist and coach each other, which leads to the next step. (Presentation and Application)

- ***Independent Practice:*** This is similar to guided practice except that learners perform without guidance or assistance from the instructor. This step provides learners with opportunity to practice behaviors independently. (Presentation and Application)

- ***Closure:*** In the final step, learners prove that they have achieved or mastered the behavioral objectives. Instructors use a variety of methods in the closure step, including oral, written, or performance tests, learner skill demonstration, and discussion and questioning. (Evaluation)

IV. IFSTA Model: Fire Protections Publications Essentials 4 Curriculum

Chapter 8 — Forcible Entry
Lesson 8A — Forcible Entry Tools

Objectives

LESSON PREREQUISITES

None

OBJECTIVES

Course

After completing this course, the candidate will have met the standards for Firefighter I as outlined in Chapter 3 of NFPA 1001, 1997, with the exception of that related to response to hazardous materials *(NFPA 3-1.1).*

Lesson

After completing this lesson, the candidate will be able to identify and know appropriate applications and maintenance procedures for forcible entry tools.

Enabling

After completing this lesson, candidates will be able to —

1. Identify cutting tools. *(NFPA 1001, 3-3.3)*
2. Identify prying tools. *(NFPA 1001, 3-3.3)*
3. Identify pushing/pulling tools. *(NFPA 1001, 3-3.3)*
4. Identify striking tools. *(NFPA 1001, 3-3.3)*
5. Match selected forcible entry tools to their basic applications. *(NFPA 1001, 3-3.3)*
6. Identify tools used for through-the-lock forcible entry. *(NFPA 1001, 3-3.3)*
7. **Break a door lock. *(NFPA 1001, 3-3.3b, 3-3.3b; Job Sheet 8A-1)***
8. Identify tools for breaking padlocks. *(NFPA 1001, 3-3.3)*
9. **Break a padlock. *(NFPA 1001, 3-3.3b, 3-3.3b; Job Sheet 8A-2)***
10. List forcible entry tool safety rules. *(NFPA 1001, 3-3.3)*
11. Describe correct methods for carrying forcible entry tools. *(NFPA 1001, 3-3.3)*
12. List general care and maintenance practices for forcible entry tools. *(NFPA 1001, 3-5.3a)*

LESSON OUTLINE

Subject Block	Obj. No.	Est. Time
Administration		5 min.
Introduction/Motivation		5 min.
General Forcible Entry Tools	1 – 5	60 min.
Tools for Locks & Locking Devices	6 – 9	30 min.
Teaching Option 8A-1: Forcible Entry Tool I.D.		15 min.
Tool Safety	10	15 min.
Carrying Tools	11	10 min.
Care and Maintenance of Forcible Entry Tools	12	10 min.
Teaching Option 8A-2: *Forcible Entry* Video		20 min.
Lesson Summary: Forcible Entry Tool Basics		10 min.
Written Test		30 min.
TOTAL		2 hr., 55 min.
TOTAL WITH TEACHING OPTIONS		3 hr., 30 min.

Chapter 8 — Forcible Entry
Lesson 8A — Forcible Entry Tools

Planning Pages

- Read Chapter 8, ***Essentials of Fire Fighting***, Fourth Edition, pages 234 through 243.
- Read and familiarize yourself with this lesson. Prepare your own notes and examples.
- Refer to transparency summary at end of lesson. Review transparencies and stack in teaching order.

  Hints: — Slip a sheet of paper between the transparencies to enable you to see them better and to prevent them from clinging together during use.

  — Tape a piece of cardboard over the overhead projector lens. When showing a transparency, flip the cardboard up. When finished with the transparency, flip the cardboard down. This process eliminates the distractions of turning the projector on and off and leaving the transparency up or the blank screen lighted.

- Write flipcharts by referring to flipchart summary at end of lesson. Be sure to leave blank pages after those that require student input.

  Hints: — Before making flipcharts, test to see whether the marker will bleed through the paper. If it does, place a blank sheet under the page you are creating.

  — Lightly, and in small print, pencil in information or desired responses that will be added during the classroom session.

  — Leave a blank sheet after each chart so that the writing on the page does not show on the page on top of it.

  — If you cannot accustom yourself to making flipcharts, and if your department or jurisdiction has a poster printer, you may type your flipchart information on letter-size paper and run it through the printer. This technique is easier than handwriting the flipcharts and creates professional looking flipchart pages.

  — If you have no access to a poster printer and are having difficulty creating flipcharts, your best bet is to use a computer to create transparencies or slides of flipchart material. A computer allows you to add clip art and color. Make sure that the lettering on the transparency/slide is large enough to be read from the back of the classroom.

- Decide which student applications you will teach and assign. Revise the lesson objectives as applicable.
- Determine what teaching options to use. If you plan on using the video teaching option, order the video from the address on the next page.
- Duplicate the appropriate number of Lesson 8A Written Tests.

  ✔ Note: Duplicate one extra written test so that you can fill in the answers and use it as a master copy for grading and critiquing exercises.

- Gather and set up materials needed for completion of the job sheets. Schedule a block of time for job sheet performances.

  ✔ Note: You may want to schedule performance of the two job sheets in this lesson with those in Lesson 8B. There is no Performance Test with this lesson because the 8A job sheet skills are incorporated into the Chapter 8 Performance Test in Lesson 8B.

- Arrange for equipment and materials:

— Variety of forcible entry tools for display and examples

— Television/VCR

— Video

Forcible Entry

The first part of this 20-minute videotape identifies tools for forcible entry, explains their uses, and presents procedures for through-the-lock entry on doors. The second part demonstrates forced entry through different types of doors and windows.

Order videotape from Fire Protection Publications, Oklahoma State University, 930 N. Willis, Stillwater, OK 74078-8045, phone 1-800-654-4055, or fax 1-405-744-8204.

— Overhead projector/screen

— Transparencies

T 8A-1 — Cutting Tools: Axes

T 8A-2 — Cutting Tools: Handsaws

T 8A-3 — Cutting Tools: Power Saws

T 8A-4 — Metal-Cutting Tools

T 8A-5 — Manual Prying Tools

T 8A-6 — Hydraulic Prying Tools

T 8A-7 — Manual Pushing/Pulling Tools

T 8A-8 — Striking Tools

T 8A-9 — Tools for Forcing Locks

T 8A-10 — Tools for Breaking Padlocks

— Prepared flipcharts/easel

F 8A-1 — Saw Blade Maintenance

F 8A-2 — Tool Cautions & Procedures

— Colored marking pens for writing on flipcharts

— Masking tape for posting flipchart pages

— ***Lesson 8A Firefighter I Student Applications*** workbook activities

Job Sheets

JS 8A-1 — Break a Door Lock

JS 8A-2 — Break a Padlock

— Written Test

— Other (list)

____

____

____

REFERENCES CONSULTED

Essentials of Fire Fighting, 4th ed. Stillwater, Oklahoma: IFSTA/Fire Protection Publications, 1998.

Forcible Entry, 7th ed. Stillwater, Oklahoma: IFSTA/Fire Protection Publications, 1987.

NFPA 1001 Standard for Fire Fighter Professional Qualifications, Quincy, Massachusetts: National Fire Protection Association, 1997.

Chapter 8 — Forcible Entry
Lesson 8A — Forcible Entry Tools

Presentation

ADMINISTRATION		5 minutes
Take	roll.	
Refer	candidates to ***Lesson 8B Firefighter I Student Applications*** workbooks, page SA 8B-65, and ask them to complete the top portion of the Competency Profile.	
Collect	the completed profiles so that you have a standardized system for recording each candidate's lesson competency levels.	
Return	Lesson 7 Written Test and critique in class as a review.	
Inform	candidates of performance test time, place, requirements, and criteria.	

INTRODUCTION/MOTIVATION		5 minutes
Explain	that forcible entry tools are placed into the categories of cutting, prying, pushing/pulling, and striking tools. Each has its own specialized forcible entry purpose, safety guidelines, and maintenance procedures.	
Tell	candidates that today's lesson is based on the material about tools in Chapter 8 of ***Essentials.*** By the end of the day's lesson, each candidate should be able to identify and know appropriate applications and maintenance procedures for forcible entry tools.	
Refer	candidates to the Study Objectives, page SA 8A-1, in their ***Lesson 8A Firefighter I Student Applications*** workbooks. Discuss lesson and enabling objectives. Review those enabling objectives for which the students will be responsible (those you will teach). Point out that boldface objectives are performance objectives that will be evaluated on their respective job sheets, and that those in regular type are cognitive objectives that will be evaluated on the lesson's written test. Explain also the relationship between the action verbs that start each objective and the written test questions — e.g., "*Select facts . . .*" indicates a multiple-choice format.	

GENERAL FORCIBLE ENTRY TOOLS	Obj. 1 – 5	60 minutes
Present Objective 1	***Identify cutting tools.*** *(Transparencies 8A-1 through 8A-4)*	

Display Transparency 8A-1

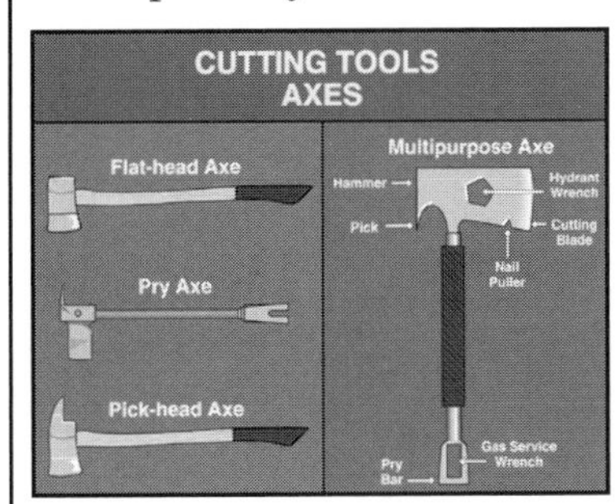

Discuss the different applications of the various axes illustrated in Transparency 8A-1. Point out design features that make specific axes appropriate for puncturing, picking, prying, or striking in addition to cutting.

Show candidates examples of as many actual axes as possible.

Display Transparency 8A-2.

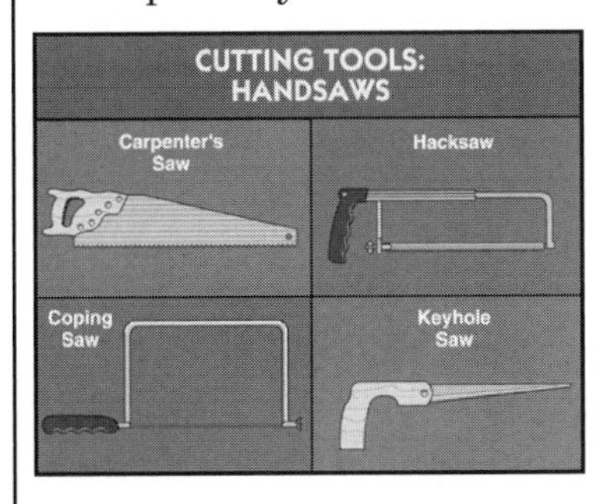

Explain the applications and advantages and disadvantages of handsaws:

- Are generally used for rescue, while power saws are more commonly used for forcible entry
- Are useful on objects that require a controlled cut but that are too big for shear jaws or unsuitable for a power saw
- Are more time-consuming to use than power saws or shears
- Are safer than shears or power saws when working close to a victim

Explain and provide examples of forcible entry uses for each of the saws illustrated.

Show candidates examples of as many actual handsaws as possible, pointing out the differences in the blades and teeth, and the different fire service uses for each.

Demonstrate the increased efficiency of the hacksaw when a second blade is installed in the opposite direction of the first, which allows cutting in both directions on the stroke.

Display Transparency 8A-3.

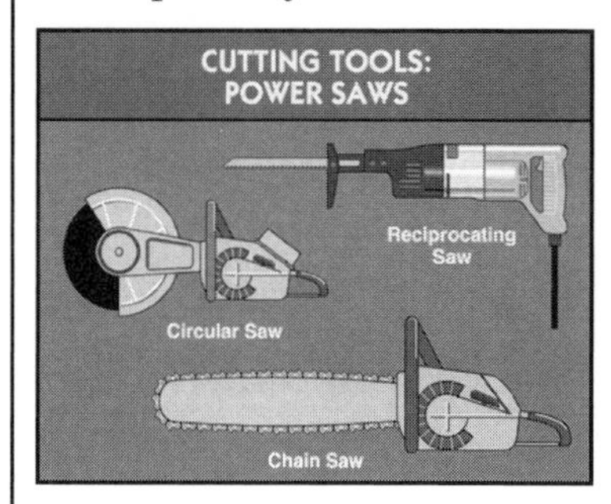

Explain forcible entry uses for the three power saws illustrated.

Ask the following questions:

- "What kind of cut is produced by a large-toothed as compared to a fine-toothed saw blade?" *(A large-toothed blade cuts faster but produces a less precise cut than a fine-toothed blade.)*
- "Why are carbide-tipped blades superior to standard blades?" *(They are less prone than other blades to dulling with heavy use.)*

Display Transparency 8A-4.

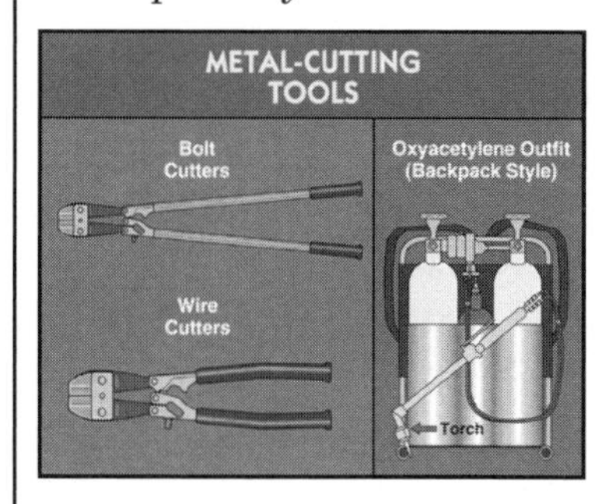

Explain forcible entry uses for the metal-cutting tools illustrated on the transparency.

Show candidates examples of actual bolt cutters, wire cutters, and oxyacetylene equipment, particularly backpack cutting equipment designed for rescue and forcible entry uses. Point out safety and design features.

Demonstrate using bolt cutters and wire cutters, and let each student return the demonstration.

Present Objective 2 ***Identify prying tools.*** *(Transparencies 8A-5 and 8A-6)*

Display Transparency 8A-5.

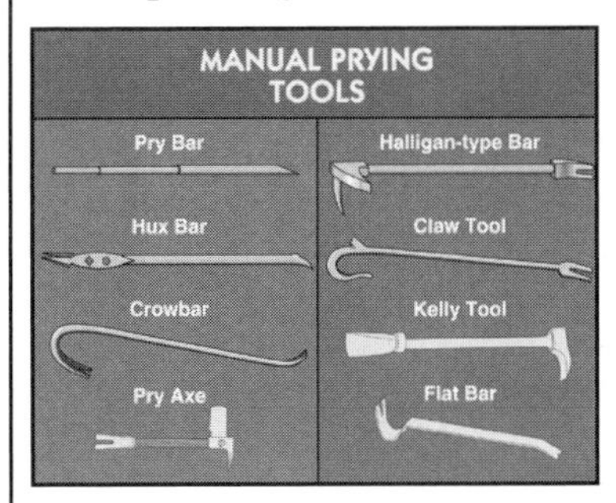

Explain forcible entry uses for each of the manual prying tools illustrated on the transparency.

Show candidates examples of as many actual prying tools as possible, pointing out design features that make the tool adaptable for specific uses.

Display Transparency 8A-6.

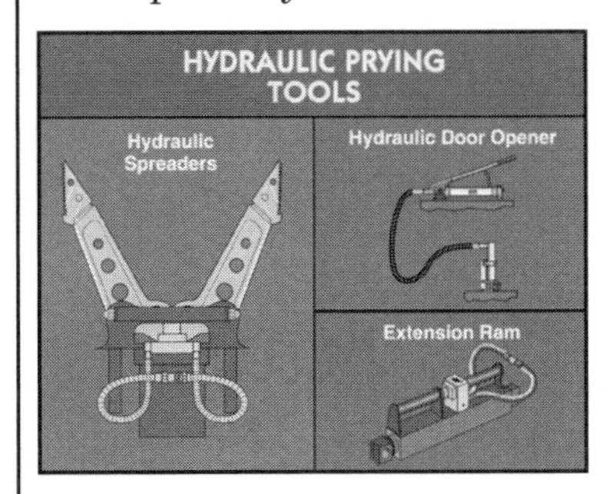

Explain the forcible entry uses of the hydraulic prying tools illustrated on the transparency.

Show candidates examples of an actual extension ram, hydraulic spreaders, and hydraulic door opener, pointing out safety and design features.

Demonstrate the safe operation of the extension ram, hydraulic spreaders, and hydraulic door opener.

Present Objective 3 ***Identify pushing/pulling tools***. *(Transparency 8A-7)*

Display Transparency 8A-7.

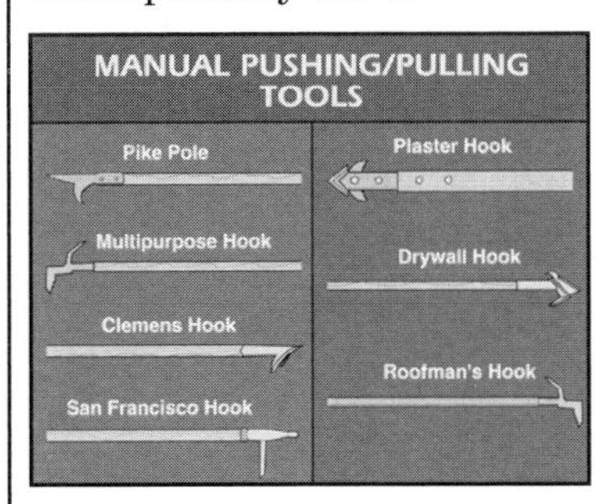

Explain the forcible entry uses of each of the manual pushing/pulling tools illustrated on the transparency.

Show candidates examples of as many manual pushing/pulling tools as possible and demonstrate their proper use.

Present Objective 4 ***Identify Striking tools.*** *(Transparency 8A-8)*

Display Transparency 8A-8.

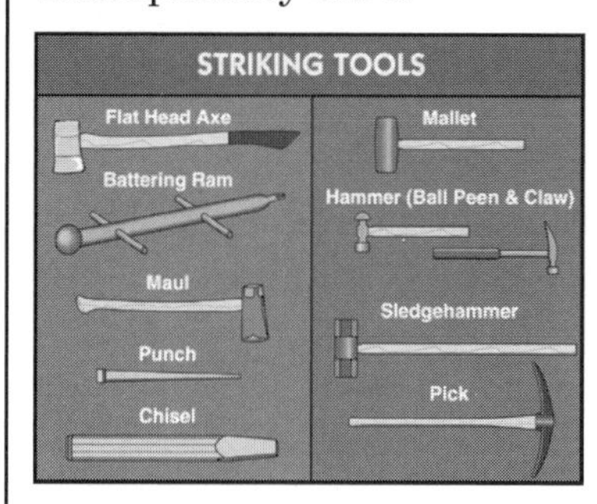

Explain forcible entry uses of each of the striking tools illustrated on the transparency.

Show candidates examples of as many striking tools as possible.

Demonstrate proper handling and striking techniques for each tool. Stress the hazards associated with using these tools. Have the students return your demonstration.

Present Objective 5 ***Match selected forcible entry tools with their basic applications.***

Review the uses of cutting, prying, pushing/pulling, and striking tools for forcible entry.

TOOLS FOR LOCKS & LOCKING DEVICES Obj. 6 – 9 30 minutes

Present Objective 6 ***Identify tools used for through-the-lock forcible entry.*** *(Transparency 8A-9)*

Display Transparency 8A-9.

TOOLS FOR FORCING LOCKS
K-Tool
J-Tool
A-Tool
Shove Knife

Explain the through-the-lock forcible entry tools.

Show candidates examples of as many actual through-the-lock forcible entry tools as possible.

Discuss methods used to pull or break locks.

Present Objective 7 ***Break a door lock.*** *(Job Sheet 8A-1)*

Refer candidates to Job Sheet 8A-1, page SA 8A-11 of their ***Lesson 8A Firefighter I Student Applications*** workbooks.

Demonstrate job steps and key points for breaking a door lock.

Announce job sheet performance and performance test dates.

Present Objective 8 ***Identify tools for breaking padlocks.*** *(Transparency 8A-10)*

Display Transparency 8A-10.

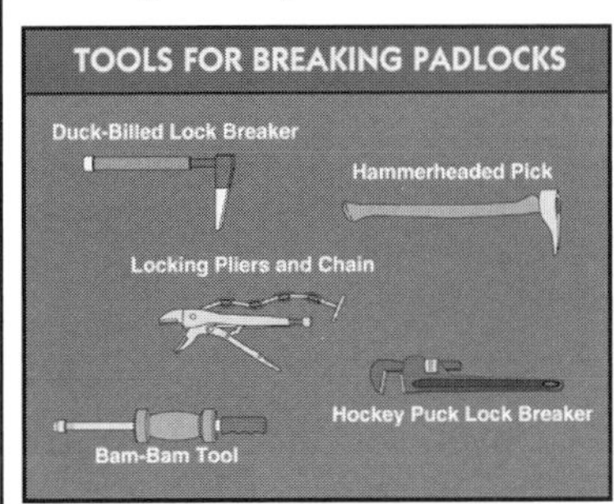

Present Objective 9 ***Break a padlock.*** *(Job Sheet 8A-2)*

Refer candidates to Job Sheet 8A-2, page SA 8A-15 of their ***Lesson 8A Firefighter I Student Applications*** workbooks.

Demonstrate job steps and key points for breaking a padlock.

Announce job sheet performance and performance test dates.

TEACHING OPTION 8A-1: FORCIBLE ENTRY TOOL I.D. — 15 minutes

Display on tables or the ground as many forcible entry tools as possible. Number each tool, and then make an "answer key" for yourself that lists the name of each tool next to its number.

Tell candidates to write their names at the top of a sheet of plain paper and to number the paper from 1 to __ (the number of tools displayed).

Allow students 10 minutes to walk through the display and identify the tools.

Collect the papers and redistribute them so that no one has his or her original paper. Ask the students to check each other's identifications.

Call on volunteers to name the tools as you and the class tour the display again.

Collect the papers and treat the exercise as a quiz, or return the papers to their owners for study.

TOOL SAFETY — Obj. 10 — 15 minutes

Present Objective 10 ***List forcible entry tool safety rules.*** *(Flipchart 8A-1)*

Ask "What safety rules should firefighters follow when using prying tools?" *(Never use a cheater bar; do not use a prying tool as a striking tool unless it is designed for that purpose.)*

Post Flipchart 8A-1.

SAW BLADE MAINTENANCE

- Keep clean.
- Keep sharp.
- Keep lightly oiled.
- Do not interchange different manufacturers' blades (power saws).
- Store in a clean, dry place.
- Do not store where gasoline fumes accumulate (composite blades).

Explain that maintaining saw blades properly is an essential element of power saw safety; review Flipchart 8A-1.

Refer students to ***Essentials***, page 240.

Review and explain each power saw safety rule.

CARRYING TOOLS — Obj. 11 — 10 minutes

Present Objective 11 ***Describe correct methods for carrying forcible entry tools.***

Demonstrate how to safely carry each of the following tools or types of tools. Allow candidates to practice by returning your demonstrations.

- Fire axe
- Pike poles and hooks
- Prying tools
- Striking tools
- Power tools
- Combinations of tools

CARE AND MAINTENANCE OF FORCIBLE ENTRY TOOLS — Obj. 12 — 10 minutes

Present Objective 12 ***List general care and maintenance practices for forcible entry tools.***

Refer candidates to pages 241 through 243 in ***Essentials***. Review the general care and maintenance principles for the following tool types and parts:

- Tool handles
- Cutting edges
- Plated surfaces
- Unprotected metal surfaces
- Axe heads
- Power equipment

Post and discuss Flipchart 8A-2.

TOOL CAUTIONS & PROCEDURES

- Try before you pry.
- Carry tools safely.
- Use tools safely.
- Use the right tool for the job.
- Keep tools clean.
- Maintain and store tools properly.
- Do not remove power tool safety guards.

TEACHING OPTION 8A-2: *FORCIBLE ENTRY* VIDEO — 20 minutes

Show video *Forcible Entry*.

Ask for and answer any questions students have on content.
✔ **Note:** If time does not allow for classroom viewing, consider assigning the videotape to be viewed before the next class session.

LESSON SUMMARY: FORCIBLE ENTRY TOOL BASICS — 10 minutes

Summarize lesson by asking for the following information:

- "Name three types of axes and describe their specific forcible entry uses."
- "In what types of situations are handsaws used for forcible entry?"

- "In what types of situations are power saws used for forcible entry?"
- "What would be an appropriate forcible entry application for an oxyacetylene cutting torch?"
- "Name four manual prying tools and describe their uses in forcible entry."
- "Name two power prying tools used in forcible entry."
- "Name four manual pushing/pulling tools used in forcible entry."
- "Name four striking tools used in forcible entry."
- "Describe tools for pulling or breaking a door lock and a padlock."

Review posted Flipcharts 8A-1 and 8A-2 to summarize saw maintenance and tool use safety precautions and procedures for forcible entry.

Assign Tell students to complete Study Sheet 8B and the Chapter 8 Review Test (if you are not planning on assigning it as a pretest or posttest) before the next class session.

WRITTEN TEST — 30 minutes

Distribute Lesson 8A Written Test, and administer to evaluate student mastery of the lesson's cognitive objectives.

Collect completed written tests.

✔ **Note:** Before the next class session, grade tests and record competency ratings on candidates' competency profiles, or redistribute tests and have students grade each other's if classroom time allows. This will save you grading time later.

Chapter 8 — Forcible Entry
Lesson 8A — Forcible Entry Tools

Written Test

Name ______________________________ **Date** ____________________

> → **Directions:** Each test item has its own criterion standard. To show mastery of each tested objective, you must achieve a required number of points. The points you must achieve are listed first, followed by the points possible. For example, in a test item designated *(9/12)*, you must achieve 9 of the 12 points possible.

1. Identify the cutting tools illustrated below. Write the correct name below each tool. *(1 pt. each, 11/14)*

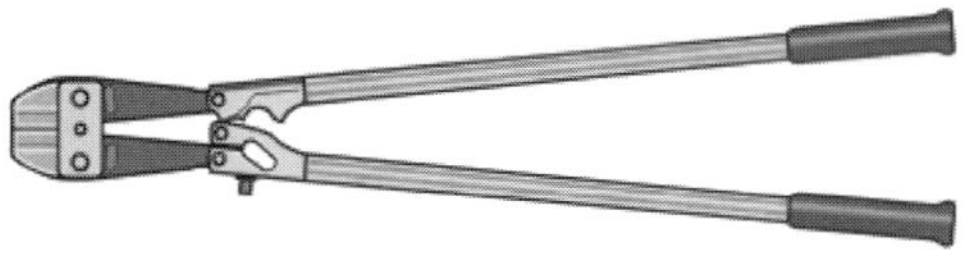

a. ______________________________

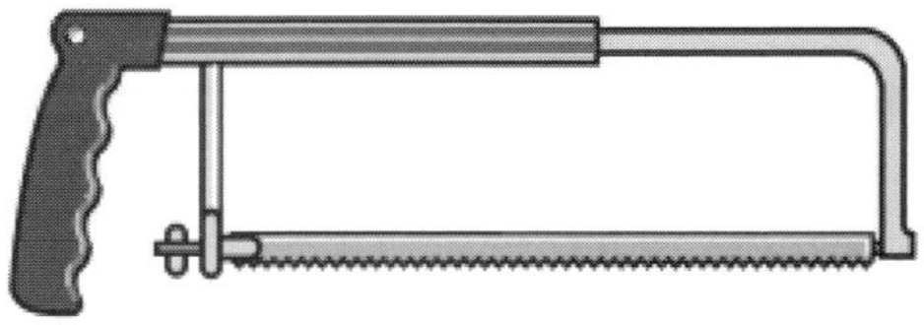

b. ______________________________

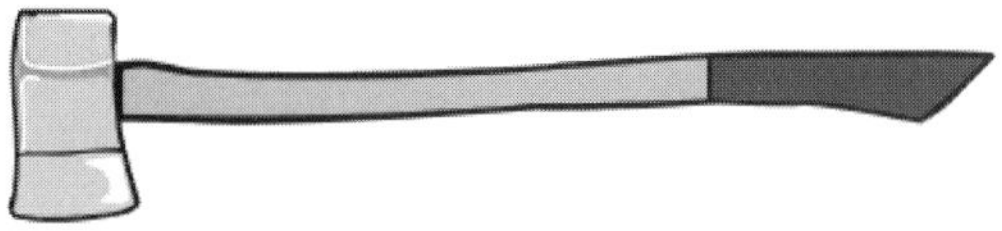

c. ______________________________

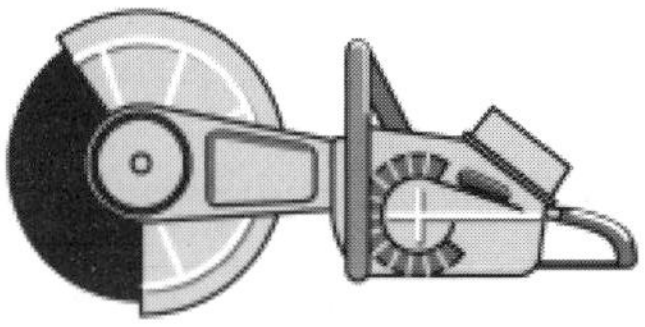

d. ______________________________

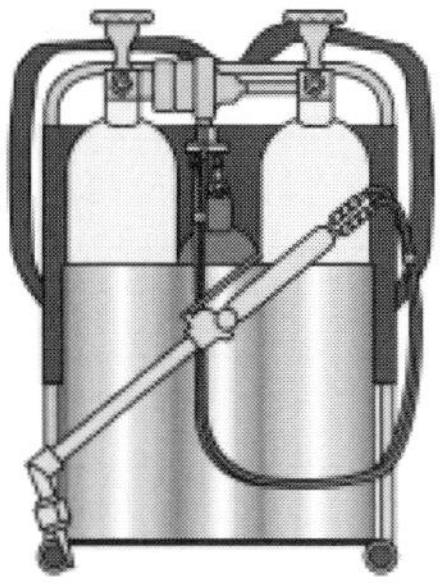

e. ______________________________

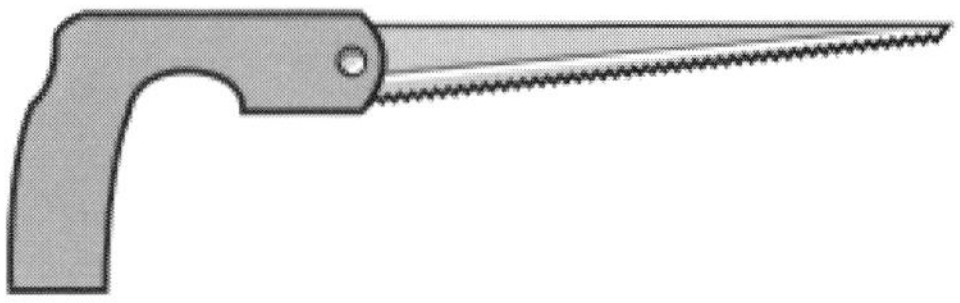

f. ______________________________

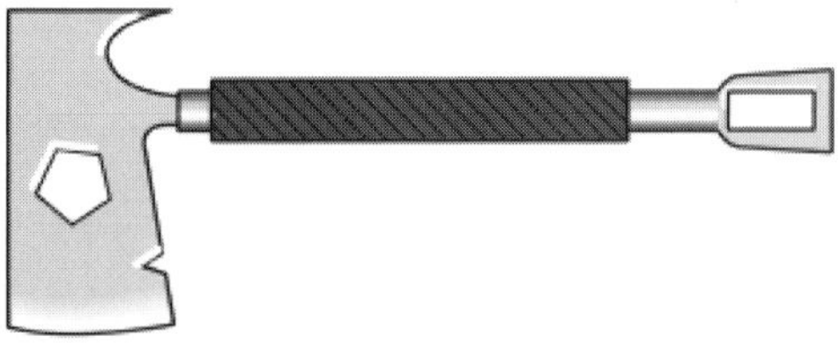

g. ______________________________

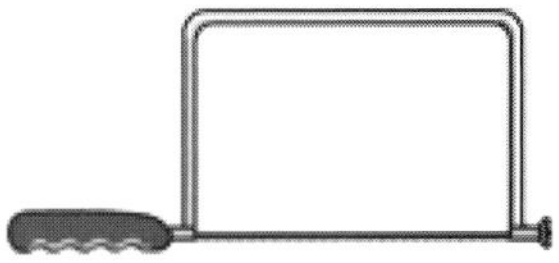

h. ______________________________

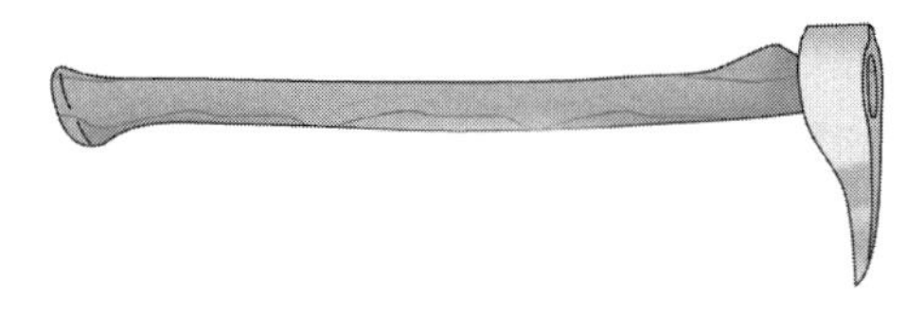

e. ______________________________

9. Break a padlock.

   ✓ **Note:** This objective is evaluated on Job Sheet 8A-2. If you have not attained competency for this objective, see your instructor.

10. List forcible entry tool safety rules for the tools named below. *(2 pts. each, 20/20)*

    Prying tools

    a. ______________________________

    b. ______________________________

    Circular saws

    a. ______________________________

    b. ______________________________

    Power saws

    a. ______________________________

    b. ______________________________

    c. ______________________________

    d. ______________________________

    e. ______________________________

    f. ______________________________

Plated surfaces

a. ______________________________

b. ______________________________

Unprotected metal surfaces

a. ______________________________

b. ______________________________

Power equipment

a. ______________________________

b. ______________________________

Scoring

Objective Number	Points Needed/Possible	Points Achieved	Additional Study Needed Yes	No
1	11/14	________	☐	☐
2	6/9	________	☐	☐
3	6/8	________	☐	☐
4	6/9	________	☐	☐
5	6/8	________	☐	☐
6	3/4	________	☐	☐
7	(JS)	________	☐	☐
8	4/5	________	☐	☐
9	(JS)	________	☐	☐
10	20/20	________	☐	☐
11	12/12	________	☐	☐
12	12/12	________	☐	☐

Written Test Answers

1. a. Bolt cutters
   b. Hacksaw
   c. Flat-head axe
   d. Circular saw
   e. Oxyacetylene cutting equipment
   f. Keyhole saw
   g. Multipurpose axe
   h. Coping saw
   i. Wire cutters
   j. Pick-head axe
   k. Pry axe
   l. Chain saw
   m. Reciprocating saw
   n. Carpenter's saw

2. a. Claw tool
   b. Flat bar
   c. Crowbar
   d. Halligan-type bar
   e. Hydraulic door opener
   f. Pry bar
   g. Kelly tool
   h. Hux bar
   i. Hydraulic spreader

3. a. Multipurpose tool
   b. Clemens hook
   c. Roofman's hook
   d. Extension ram
   e. Plaster hook
   f. Pike pole
   g. San Francisco hook
   h. Drywall hook

4. a. Chisel
   b. Sledgehammer
   c. Battering ram
   d. Flat-head axe
   e. Punch
   f. Pick
   g. Mallet
   h. Ball peen hammer
   i. Maul

5. a. 1
   b. 2
   c. 3
   d. 2
   e. 1
   f. 4
   g. 2
   h. 1

5. a. A-tool
   b. Shove knife
   c. K-tool
   d. J-tool

7. Evaluated on Job Sheet 8A-1

8. a. Bam-bam tool
   b. Locking pliers and chain
   c. Duck-billed lock breaker
   d. Hockey puck lock breaker
   e. Hammerheaded pick

9. Evaluated on Job Sheet 8A-2

10. *Prying tools*
   a. Never use a cheater bar.
   b. Do not use a prying tool as a striking tool unless it has been designed for that purpose.

Appendix D

How to Research a Topic

Research is a systematic activity in which an individual looks for, discovers, and studies information about an issue, problem, or topic in order to learn more about it and to apply discoveries to situations or use them in tasks. The information can be used to develop solutions for solving problems, making adjustments or corrections, and curing ills. Researchers may spend years looking for small bits of information that will provide big answers to a specific problem. But where does a researcher, or an ordinary person, begin? How do instructors research a topic to get more information about a lesson they will teach? The usual steps in researching a topic are in the following sections (Figure D.1).

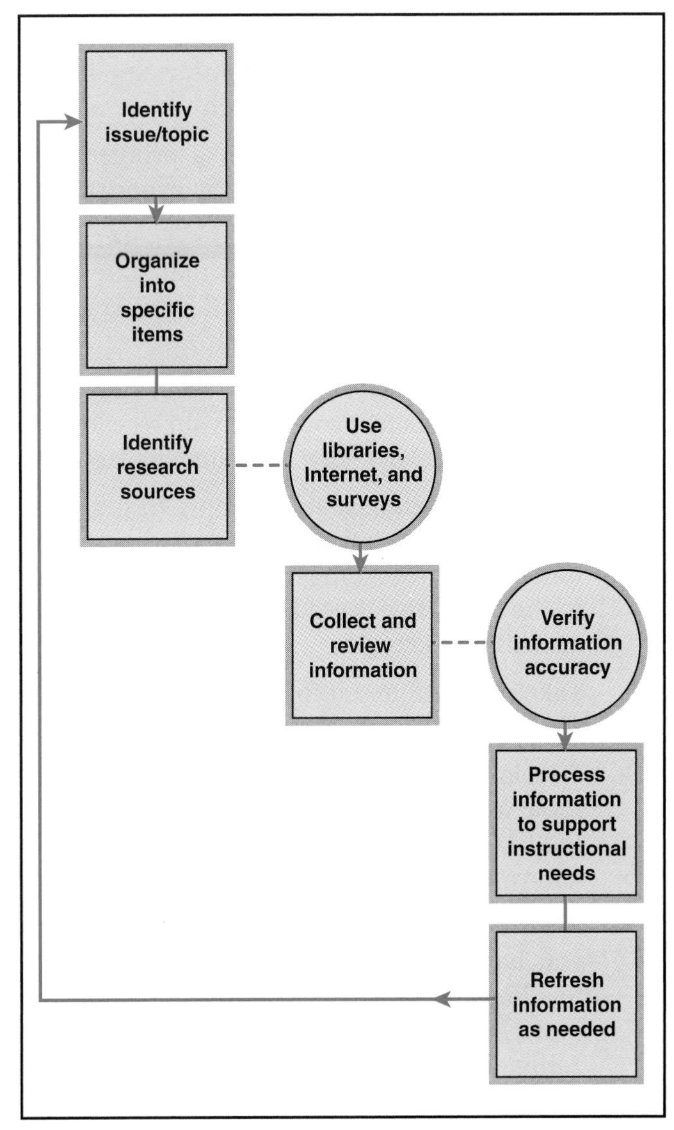

Figure D.1 Basic research model

Identify or Choose the Topic

- ***What do you want to do?*** This step may involve answering a question, solving a problem, or improving a skill or behavior. Your research examines the details of the topic you select — details that you look for and find in a variety of sources (professional journals, annual and investigative reports, personal and personnel interviews, textbooks, and other researchers' papers and reports). These details help you define the topic or narrow it down to the specific area in which you are interested.
- ***How much do you want to know (or share with others)?*** You may want more information on an area in a particular lesson or for an issue or problem on the job. The area you research should not be so broad it branches into too many areas that do not respond to your question or problem. It should also not be so narrow that it does not fully find the answers to the question or solutions to the problem. If you are trying to reinforce your own opinion of or belief in an issue, you want validation from other reliable, credible sources. A topic you share with others must not be too personal or too technical. It is too personal if it expresses your own feelings, opinions, and experiences without validation; it is too technical if the research requires knowledge, education, experience, or equipment that you do not have or if the result is beyond the understanding of those who read it, hear it, and use it.

Organize the Topic into Specific Items

- ***Decide what organizational pattern you will use.*** Selecting an organizational pattern guides your research to a specific area and helps you search for specific items. Decide which of the following organizational patterns most interests you, answers your question, or solves your problem:
  - Chronological or historical organization that examines events over time
  - Comparison/contrast organization that examines similarities and differences
  - Topical organization that examines small units or subtopics and analyzes each
  - Problem/solution that states a problem and analyzes solutions proposed by experts in the field
  - Opinion/reason organization that states your own opinion and shows reasons for it, which are well supported by the research
- ***Develop an idea sheet.*** An idea sheet helps you explore the topic before you begin your research. To prepare an idea sheet, take the following steps:
  - Do general reading on the topic to get background information and understanding. Determine key words from your readings that relate to the topic. Use these key words to find other articles on the topic.
  - Take notes on any information that relates to the organizational pattern you have selected.
  - Develop questions about the relevancy of the topic using *who, what, where, when, why,* and *how.* Look for answers that relate to the problem you want to solve.
  - Write down all ideas, questions, and reactions you have during your preliminary research. Reading and writing help define and limit your topic, find new possibilities for different organizational patterns, and recognize the efforts of your research.

Identify Research Sources

Where can you find information? Public libraries may have information, but the more technical or job-specific your research area, the less you will find in public libraries. Try college or university libraries, medical or legal libraries, and technical libraries (see Appendix A, Instructor Resources).

When using journals for formal research, be sure they are learned and refereed. A *learned journal,* under the opinion of an editorial board, publishes papers of the latest thinking or research findings of leading scholars or professionals in a particular academic or career field. A *refereed journal* lists the editorial board. An *editorial board* is a selected group of content experts who provide an opinion on the quality of the article submitted for publication. Though journals without editorial boards have credible articles, they are not acceptable resources for formal research, but they may certainly be used for informal research and varied opinions from experienced professionals on a variety of topics.

The Internet and computer databases are good sources. More and more organizations and libraries have web pages, and abstracts of research papers can be found by searching key words. Libraries with computers often have literature search software that can locate topics by key words, authors, and article titles.

Surveys are a source for information, but will you use information from a survey that someone else has done or will you have time to develop your own? If you develop your own, you need time to distribute it, wait for responses, and analyze the results before you can finish your research.

Collect and Review Information

- ***Copy or record your information.*** Download or print out computer or Internet information; copy or write out printed information. Be sure to include the citation for the source for each item including author(s), publisher or Internet source, publication date, and chapter and page numbers.
- ***Review and organize your information.*** Sort your information and place it in order based on progression of topic or date and sequence of events. Use information that has been researched and published within the past 5 years or less. Unless you are organizing your research for a chronological or historical document or your resources have historical impact on current practices or philosophies, information over 5 years old is considered outdated.

Process or Prepare Information to Support Instructional Needs

What do I do with this information now? Make a notation in your lesson plan that references your information, and introduce it at the appropriate time in class. Be sure that the information complies with organizational policy and if necessary, get approval from your supervisor before sharing information or implementing practices.

To get approval, you may have to submit a report. Your report should clearly state the problem, develop your reasoning and solution, and draw a conclusion. Reports that state validated facts based on credible research that show how an organization can save money, use personnel or equipment more efficiently, reduce injuries, or save lives are more effective in gaining support and acceptance.

Refresh Information as Needed

The fire and emergency services are dynamic professions that change constantly and rapidly with new or revised procedures, protocols, and standards. Other changes include new equipment and methods of using it and different actions and reactions to issues and problems. Instructors must continually participate in research and professional development to remain abreast of the latest innovation and determine whether it is a valid one.

Resources

Cash, Phyllis. *How to Develop and Write a Research Paper.* New York: Simon & Schuster, Inc. 1988.

Menegazzi, James J. *Research: The Who, What, Why, When & How.* Wilmington, Ohio: Ferno-Washington, Inc. 1994.

Support and Application Components

Chapter 6 — Ropes and Knots
Lesson 6 — Ropes & Knots

Information Sheet 6-1
Rope Materials

NATURAL FIBERS

Material	Identification	Use	Advantages	Disadvantages
Manila	Strong, hard fiber from the abaca plant; type #1 best for rope and identified by a colored string twisted into the fibers	Utility only	• Inexpensive • Biodegradable • Increases tensile strength by 15% when wet • Fairly good abrasion resistance • Good resistance to damage by sunlight	• Short, noncontinuous strands provide poor tensile strength • Poor shock load ability • Does not float • Low melting point (350°F *[177°C]*) • Good conductor of electricity • Rots, decays, and deteriorates easily • Easily damaged by chemicals • Must be stored dry
Cotton	Soft, pliable natural fiber; generally white or light colored	Utility only	• Inexpensive • Biodegradable • Increases tensile strength by 15% when wet • Good resistance to damage by sunlight	• Poorer tensile strength than manila • Poorer shock load ability than manila • Does not float • Low melting point (300°F *[149°C]*) • Good conductor of electricity • Very poor abrasion resistance • Rots, decays, and deteriorates easily • Easily damaged by chemicals • Must be stored dry

Chapter 1 — Firefighter Orientation and Safety
Lesson 1A — Orientation
Lesson 1B — Safety

Information Sheet 1B-1
Accidents and Injuries

Accidents

Accidents are unplanned events that may result in bodily injury, illness, or physical or property loss.

H. W. Heinrich of the Travelers Insurance Company devoted the greater portion of his life to the study of industrial accidents and their prevention. Heinrich proposed that there were five factors involved in an accident sequence:

1. Social environment
2. Human factors
3. Unsafe acts and conditions
4. Accident
5. Injury

He found that the last factor in the sequence, an injury, was always preceded by an accident. But, in order for an accident to occur, some unsafe act had to be committed or some unsafe condition had to exist. The unsafe act or condition was invariably caused by the human factor. In turn, the human factors responsible resulted from inherited characteristics or social and environmental conditioning. Heinrich called his findings the Domino Theory because like a row of end-standing dominoes, the activation of one factor precipitated the activation of the next, and the next, eventually resulting in an accident or injury.

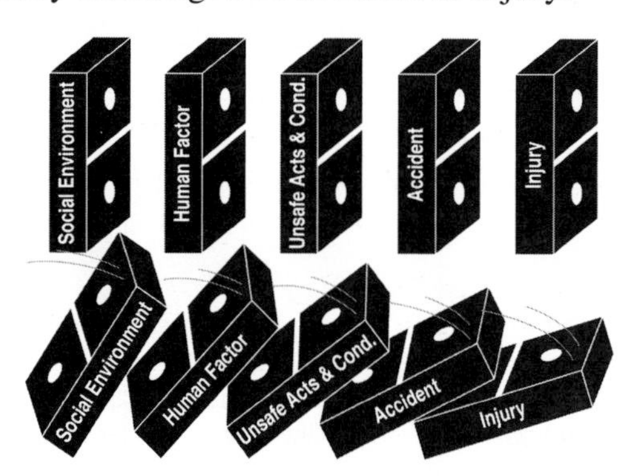

In Heinrich's Domino Theory, the emphasis on accident (and thus injury) control is in the middle of the sequence — at the unsafe act or mechanical or physical hazard. If this act or hazard is removed, the sequence is interrupted and the injury cannot occur.

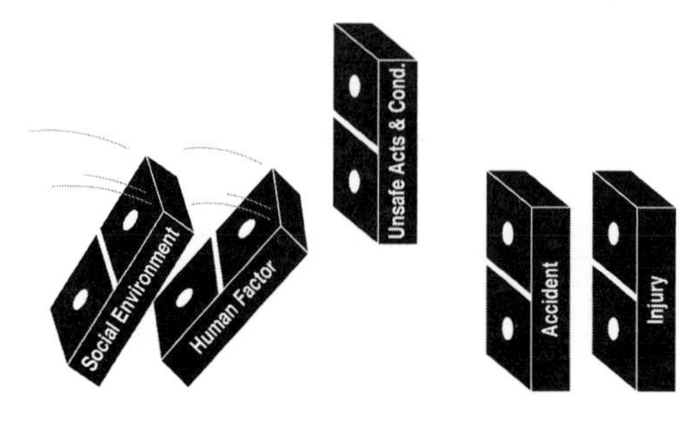

Job Sheet 6-4
Clean Rope

Name ____ **Date** _______________

Evaluator __________________________ **Overall Competency Rating** _______________

References

NFPA 1001, Prevention, Preparedness, and Maintenance 3-5.3
Essentials, pages 152, 153

Prerequisites

Job Sheet 6-2 — Coil and Uncoil Rope
Job Sheet 6-3 — Bag or Bird's-Nest Coil Rope for Machine Washing or Storage

Student's Instructions

To meet evaluation standards, you must perform this job within _____ *[amount of time, if applicable];* you may have _____ attempts. When you are ready to perform this job, ask your instructor to observe the procedure and complete this form. To show mastery of this job, you must perform all steps to receive an overall competency rating of at least 2.

Competency Rating Scale

3 — Skilled — Meets all evaluation criteria and standards; performs task independently on first attempt; requires no additional practice or training.

2 — Moderately skilled — Meets all evaluation criteria and standards; performs task independently; additional practice is recommended.

1 — Unskilled — Is unable to perform the task; additional training required.

☒ **— Unassigned** — Job sheet task is not required or has not been performed.

✔ **Evaluator's Note:** Formulate and inform the candidate of the standards for this task (time allowed and number of attempts). Observe the candidate perform the task, check the step/key point under the appropriate attempt number as accomplished, record total time (if appropriate), and then use the rating scale above to assign an overall competency rating. If the candidate is unable to perform any step of this job, have the candidate review the materials and try again.

Introduction

Rope can be cleaned by hand, with a rope washer, or with a clothes washer. When rope is cleaned with a clothes washer, a front-loading, tumbling-action machine should be used. This type of clothes washer causes less tangling of the rope and prevents the rope from being abraded by the agitator. In addition, the washer should not contain a plastic window that could damage the rope by heat fusion during the spin cycle.

The following procedures are the general steps for cleaning rope. Methods of washing vary with each rope manufacturer, so it is always advisable to contact them for specific cleaning and drying instructions for the type of rope being cleaned.

Equipment and Personnel

- One firefighter
- Rope to be cleaned
- Mild soap solution or cleaning agent recommended by rope manufacturer
- Front-loading, glass-windowed, tumbling-action washing machine
- Rope washer
- Cloth bag
- Nail brush or similar small, stiff-bristled brush
- Cleaning cloth
- Water source

Job Steps	Key Points	Attempt No. 1	2	3
NATURAL FIBERS BY HAND				
1. Position the rope.	1. a. Fully extended	___	___	___
	b. On clean, level surface	___	___	___
	c. Without dragging on ground or stepping on	___	___	___
2. Clean the rope.	2. a. Dry cloth or nail brush	___	___	___
	b. Systematically from one end to other	___	___	___
3. Return the rope to proper storage.	3. Per Job Sheet 6-2 or 6-3	___	___	___
	Time (Total)	___	___	___

Evaluator's Comments ____

____

____

Chapter 6 Competency Profile

Student Name ______________________________ Soc. Sec. No. __________
Last First Middle
Fire Department ______________________________
Address ______________________________
Phone ______________________________
Home Address ______________________________
Phone ______________________________
Date of Enrollment _____ - _____ - _____ Total Class Hours __________
Date of Withdrawal _____ - _____ - _____ Total Hours Absent __________
Date of Completion _____ - _____ - _____
Instructor's Name ______________________ Session Dates __________

Instructor's Directions

1. Check the student's competency rating (3, 2, 1, ☒) for each performance test task and psychomotor lesson objective (practical activity and job sheets) listed below.
2. List any additional performance tasks or psychomotor objectives (job sheets or practical activity sheets) under "Other," and check the candidate's competency rating.
3. Record the candidate's cognitive scores (written lesson tests and *administered* chapter review tests) in the spaces provided.

Level 3	Level 2	Level 1	☒	Psychomotor Competencies
				Practical Activity Sheets
☐	☐	☐	☐	PAS 6-1 — Tie an Overhand Safety Knot
☐	☐	☐	☐	PAS 6-2 — Tie a Half Hitch
☐	☐	☐	☐	PAS 6-3 — Tie a Bowline Knot
☐	☐	☐	☐	PAS 6-4 — Tie a Clove Hitch in the Open
☐	☐	☐	☐	PAS 6-5 — Tie a Clove Hitch Around an Object with No Free End
☐	☐	☐	☐	PAS 6-6 — Tie a Figure-Eight Knot
☐	☐	☐	☐	PAS 6-7 — Tie a Figure-Eight Follow Through Knot
☐	☐	☐	☐	PAS 6-8 — Tie a Figure-Eight Knot on a Bight
☐	☐	☐	☐	PAS 6-9 — Tie a Double-Loop Figure-Eight Knot
☐	☐	☐	☐	PAS 6-10 — Tie a Becket Bend or Sheet Bend Knot
☐	☐	☐	☐	Other ______________________________
☐	☐	☐	☐	______________________________

Points Achieved	Points Needed/ Total	Cognitive Competencies
		Written Test
______	6/6	1. Distinguish between life safety and utility rope applications.
______	5/5	2. List criteria for reusing life safety rope.
______	12/16	3. Match rope materials to their descriptions.
______	11/14	4. Select facts about rope construction.
______	4/6	5. List basic guidelines for rope care and maintenance.
______	4/4	6. List reasons for removing rope from service.
		7. Evaluated on Job Sheet 6-1.
______	10/12	8. Select facts about rope cleaning and storage.
		9. Evaluated on Job Sheet 6-2.
		10. Evaluated on Job Sheet 6-3.
		11. Evaluated on Job Sheet 6-4.
______	14/16	12. Label knot elements.
______	16/18	13. Match knots to their primary applications.
		14. Evaluated on Practical Activity Sheet 6-1 through 6-10.
______	3/3	15. List hoisting safety considerations.
		16. Evaluated on Job Sheets 6-5 through 6-10
		Review Test
______		Chapter 6 Review Test

Instructor's Signature ______________________________ **Date** ______________

Student's Signature ______________________________ **Date** ______________

Chapter 6 — Ropes and Knots
Lesson 6 — Ropes & Knots

Practical Activity Sheet 6-3
Tie a Bowline Knot

Name ______________________________ **Date** ______________

Evaluator ______________________ **Overall Competency Rating** ______________

References

NFPA 1001, General 3-1.1.2
Essentials, pages 155, 163

Prerequisite

Practical Activity Sheet 6-1 — Tie an Overhand Safety Knot

Introduction

The bowline is an important knot in the fire service, sharing a degree of acceptance in both life safety and utility applications. The bowline is easily untied and is a good knot for forming a single loop that will not constrict the object it is placed around.

An "outside" bowline may be tied with the working end outside the loop. The outside bowline is just as strong as the inside bowline, but the inside bowline is preferred because of its safety feature of having the working end locked against the object to which it is tied.

Firefighters should be able to tie a bowline in the open as well as around an object. The method below is one way of tying a bowline. Other methods may be just as effective.

Directions

Seal the ends of a 6-foot to 20-foot *(1.8 m to 6 m)* length of 1/2-inch *(13 mm)* rope with heat or by tying with a string and dipping in varnish or white glue.

Follow the instructions below for tying a bowline knot. When you are proficient at tying this knot, tie the knot in front of your instructor for evaluation.

To show competency in this activity, you must master starred criteria with a minimum rating of 2.

Activity

1. Measure off sufficient rope to form the size knot desired.
2. Form an overhand loop in the standing part.

PAS 6-3 — Tie a Bowline Knot

Competency Rating Scale

3 — Skilled — On the first attempt, product meets all criteria (correct knot; snug dressing); student requires no additional practice.

2 — Moderately skilled — On the first attempt, product meets critical criterion (correct knot) but is loosely dressed, *or* student may require more than one attempt; student may benefit from additional practice.

1 — Unskilled — Product does not meet critical criterion; student requires additional practice and reevaluation.

☒ — **Unassigned** — Task is not required or has not been performed.

✔ **Evaluator's Note:** Score the product as indicated below. Use the rating scale above to assign an overall competency rating. Note that critical criteria are marked with an asterisk and must be mastered (correct) to show competency. Record the overall competency rating on both the student's practical activity sheet and competency profile.

Criteria	Yes	No
1. Knot dressing snug	☐	☐
*2. Correct knot configuration	☐	☐
3. Secured with overhand safety	☐	☐

*Critical criterion

Chapter 6 — Ropes and Knots
Lesson 6 — Ropes & Knots
Study Sheet

Introduction

This study sheet is intended to help you learn the material in Chapter 6 of ***Essentials of Fire Fighting***, Fourth Edition. You may use it for self-study, or you may use it to review material that will be covered in the lesson and content review tests. The numbers in parentheses are the pages in ***Essentials*** on which the answers or terms can be found.

Chapter Vocabulary

Be sure that you know the chapter-related meanings of the following terms and abbreviations. Use a dictionary or the glossary in ***Fire Service Orientation and Terminology*** if you cannot determine the meaning of the term from its context.

- Bight *(154, 155)*
- Bird's-nest coil *(153)*
- Block creel *(148)*
- Core *(150)*
- Fiber *(148)*
- Hitch *(154)*
- Jacketed rope *(150)*
- Kernmantle rope *(150)*
- Knot *(154)*
- Loop *(155)*
- Manila fiber *(148)*
- Round turn *(155)*
- Running end *(154)*
- Sheath *(150)*
- Standing part *(154)*
- Static line *vs.* dynamic line *(148–150)*
- Strand *(148)*
- Tactilely *(151)*
- Working end *(154)*
- Yarn *(150)*

Study Questions & Activities

1. List some valuable applications of rope in the fire service. *(147)*

   ____

   ____

   ____

2. What ability is crucial to the safety of rope maneuvers? *(147)*

   ____

3. What are the two use classifications of fire service rope? *(147)*

   a. ______________________________

   ____

   b. ______________________________

   ____

4. What is the definition of life safety rope as stated by NFPA 1983? *(147, 148)*

   ____

   ____

f. Derived unit *(34)* ______________________________

____

3. Define the following terms related to energy and work.
   a. Energy *(34)* ______________________________

   ____

   b. Work *(34)* ______________________________

   ____

   c. Joule *(34)* ______________________________

   ____

   d. Foot-pound *(34)* ______________________________

   ____

   e. Chemical energy *(35)* ______________________________

   ____

   f. Mechanical energy *(35)* ______________________________

   ____

   g. Electrical energy *(35)* ______________________________

   ____

   h. Heat energy *(35)* ______________________________

   ____

   i. Light energy *(35)* ______________________________

   ____

   j. Fission *(35)* ______________________________

   ____

   k. Fusion *(35)* ______________________________

   ____

4. What is "power" and how does it apply to the study of fire behavior? *(35)*

   ____

   ____

   ____

5. Define the following terms related to heat and temperature.
   a. Heat *(36)* ______________________________

   ____

   b. Temperature *(36)* ______________________________

   ____

   c. Degrees Celsius *(36)* ______________________________

   ____

   d. Degrees Fahrenheit *(36)* ______________________________

Appendix F

Organizational or Program Rules, Policies, and Regulations

Course Completion Checklist

I, the undersigned, acknowledge, that in order to complete and receive credit for the instructor training course, I will complete the following items to the satisfaction of the instructor:

- ☐ Attend all class sessions. Makeups allowed only with permission of the instructor.

- ☐ Obtain a minimum average score of 70% on all quizzes and the midterm exam.

- ☐ Complete three 5-minute, one 15-minute, and one 20-minute presentation with a minimum score of 70%.

- ☐ Any requests for fire training leave or overtime for this class must be approved by the student's Deputy Chief. The only involvement the Training Division has in this matter is to verify attendance.

- ☐ Return textbook # _____________.

Name (print): ______________________________ **Date:** ______________

Student Signature: ______________________________

Instructor Signature: ______________________________

Section I 503

EMERGENCY MEDICAL SERVICE TRAINING

1.0 **REFERENCES**:

1.1 S.F.D. Operating Instructions:

1.1.1 I 204 - Discipline.

1.1.2 I 502 - Operation of Aid and Medic Units

1.1.3 I 504 - Infection Control Plan

1.1.4 I 601 - Training Standards

1.1.5 WAC 246-796, Emergency Medical Services and Trauma Care Systems, January 1993

1.1.6 WAC 296-305-01501, First-Aid Training and Certification

2.0 **POLICY**:

3.1 All personnel entering the Seattle Fire Department, as a Recruit Fire Fighter, shall show proof of a current Washington State Emergency Medical Technician Certification along with up-to-date continuing education hours/units.

3.2 All members Captain and below shall complete the bimonthly EMS exams, including the twice annual Competency Based Continuing Medical Education, with a passing score of 70%.

3.2.1 All company members shall be listed on a Form 31 with their test scores. Those members on vacation or disability shall be listed as such. The Form 31 shall be sent to Training Division within 10 days of the end of the testing period. Those members on vacation or disability during the testing period have 30 days upon return to duty to take the test and the results sent to Training Division.

3.2.2 Individuals not attaining a passing score of shall be re-tested within 30 days. Battalion Chiefs shall retain enough exams to perform the re-test procedure and to incorporate one exam in the company's "EMS-6, Tests and Training Memos " file.

3.2.3 All company members shall demonstrate competency regarding protocols, operation, and maintenance of the Automatic External Defibrillator. All company members shall drill monthly on the use of the Automatic External Defibrillator and record each drill separately on the Individual Training Record (F44).

3.3 An Individual Training Record (F44) shall be maintained for all members in the Operations Division through the rank of Captain, verifying all Emergency Medical Services Continuing Education.

3.4 Guidelines for Paramedic training will be established by the Seattle Fire Department/University of Washington School of Medicine in accordance with local and State laws WAC 248-15-060.

Rev. Date: February, 1997

Courtesy of Seattle (WA) Fire Department

3.0 **DEFINITIONS**:

3.1 Competency Based EMS Continuing Education: A program that allows for each student to prepare ahead of time to demonstrate their competence in performing pre-selected objectives with practical skills evaluation and written skills examination.

3.2 Paramedic: An individual who has completed the Seattle Fire Department/University of Washington Paramedic Training Program or equivalent, and is been certified by the Washington State Department of Social and Health Services or the University of Washington to provide mobile intensive care services as defined in RCW 18.71.200.

3.3 Physician Coordinator: The Medical Program Director for Medic One (Medic 55).

3.4 Seattle Fire Department Ongoing Emergency Medical Program: A program for non-EMT Fire Fighters to maintain NFPA and WAC 246-976 requirements of a minimum of Industrial First Aid training for Fire Fighters.

3.5 Washington State Emergency Medical Technician: A person who is authorized by the Secretary of the Department of Health to render emergency medical care pursuant to RCW 18.73.081.

4.0 **RESPONSIBILITY**:

4.1 Operations Division Personnel:

4.1.1 All members of the Operations Division through the rank of Captain shall be responsible for maintaining either the Washington State Emergency Medical Technician Certification or the Seattle Fire Department Ongoing Emergency Medical Certification.

4.2 Company Officers

4.2.1 Ensure that all members under their supervision meet the minimum training requirements of this article within the established Department standards and time frames.

4.2.2 Instruct acting officers on the standards of training requirements.

4.2.3 Maintain and review all records and forward reports as required.

4.3 Company/Station Captains:

4.3.1 Ensure that the members of their platoons meet the minimum training requirements of this article within the established Department Standards and time frames.

4.3.2 Ensure that Officers on the other platoons of the same company meet the training requirements for all their members.

4.3.3 Maintain and review all records and forward reports as required.

Rev. Date: February, 1997

4.4 Battalion Chiefs:

4.4.1 Battalion Chiefs are responsible for insuring members not passing the first test study and drill on the appropriate chapters of the Emergency Care and Transportation of the Sick and Injured.

4.5 Training Division:

4.5.1 The Chief of Training shall be responsible for administration of the training and shall develop, maintain and increase the EMS skills of all Department personnel.

4.6 Emergency Medical Services Division (Battalion 3):

4.6.1 Monitor, evaluate and make recommendations regarding emergency medical techniques.

4.6.2 Maintain liaison between the Training Division and the medical community.

5.0 **PROCEDURES**:

5.1 Washington State EMT Certification:

5.1.1 Upon successful completion of an Emergency Medical Technician Course, the Department of Social and Health Services shall certify those eligible graduates who have been recommended by the physician coordinator.

5.1.2 The period of certification shall be valid for three years and shall terminate on the last day of the month on the third anniversary of the completion of the course.

5.1.3 Personnel failing to successfully pass the Washington State EMT Certification Test after the initial one (1) re-test opportunity will not be eligible for State certification and shall enter into the Seattle Fire Department Emergency Medical Certification Program at the second step (see Appendix 6.2).

5.1.4 Washington State EMT re-certification of currently-certified EMT's shall be accomplished by 5.1.4.1, 5.1.4.2, or 5.1.4.3 below:

5.1.4.1 By completion of 30 hours of continuing education and successfully passing a written and practical exam.

5.1.4.2 By completing a formal course of instruction prior to the expiration date of a current certificate and successfully passing a written and practical exam.

5.1.4.3 By a combination of continuing education and formal instruction and successfully passing a written and practical examination.

5.1.5 Personnel failing to successfully pass the Washington State EMT Re-certification Test after the initial test and one (1) re-test opportunity will not be eligible for State re-certification and will enter into the Seattle Fire Department Ongoing Emergency Medical Training Program (see Appendix 6.2).

Rev. Date: February, 1997

5.1.6 Paramedics shall be automatically re-certified as Washington State Emergency Medical Technicians upon successful re-certification in their program.

5.2 Seattle Fire Department Ongoing Emergency Medical Certification:

5.2.1 Upon successful completion of the initial eighty (80) hour training program:

5.2.2 Certification shall be valid until an individual fails to pass three (3) consecutive bimonthly tests and the subsequent re-test for each bimonthly test.

5.3 Paramedic Certification:

5.3.1 Upon successful completion of the Seattle Fire Department/University of Washington Paramedic course, those recommended by the Physician Coordinator will be certified by the University of Washington as a Mobile Intensive Coronary Care Unit Paramedic (per WAC 248-15-080).

5.3.2 The period of Certification shall be valid for two years.

6.0 **APPENDIX**:

6.1 Paramedic Class Selection Standards

6.2 Washington State EMT Certification Program

Rev. Date: February, 1997

8) **Recruits may use the PAY PHONE in the north stairwell for personal calls, but only during scheduled Break periods or Lunch.**

The Department phone is for Department business and only emergency messages will be accepted on 392-1599.

9) **Fatigue clothing will be worn for all practical training periods.**

Fatigues or any part of your uniform are not to be worn when travelling to and from work. Recruits will have at least one complete change of clothes in their lockers at all times.

10) **Personnel shall practice a high standard of personal hygiene.**

All personnel shall report for Roll Call clean-shaven. Personal hygiene items are the responsibility of the individual concerned. For safety reasons, hair shall not interfere with safety protective devices, i.e.: fire helmet or breathing apparatus facepiece seal.

11) **Locker rooms, common areas, and washrooms shall be kept clean and tidy.**

All common areas within the locker/change rooms shall be kept clear of loose personal items. Floors around the showers shall be mopped after use to reduce the risks of falls.

12) **Each Recruit shall use only the locker as assigned.**

Each Recruit shall provide their own lock to secure their assigned locker. The Department does not accept any responsibility for personal items under any circumstances. All lockers will remain undamaged and shall be completely cleared out and the name tag removed from the door at the end of the course.

13) **In keeping with City of Toronto By-Laws, smoking is not permitted anywhere within the Fire Academy building and on/in department vehicles.**

Smoking is permitted on the Training Grounds in the area of the outside Draughting Pit during breaks or lunch periods. Smoking area is to remain tidy.

14) **In accordance with TORONTO FIRE SERVICES RULES & REGULATIONS and SPECIAL ORDERS BOOK, full personal protective clothing, as issued by the Fire Services, shall be worn during training and practical exercises.**

15) **Recruits shall park their personal vehicles only in those parking areas designated as RECRUIT PARKING.**

Bicycles are not to be chained or locked to any part of the Fire Academy structure (e.g. the Training Tower Fire Escape). Fire Services assumes no responsibility for privately owned vehicles parked on City property.

16) **Recruits shall enter the Fire Academy through the north-west Tower door.**

Entrance to the Academy through the front doors, Station #30, or Mechanical Maintenance Division is prohibited.

17) **Recruits will be issued a 5 m length of rope the first week.**

This rope is to be used for practising Fire Services approved knots and hitches. It will be returned at course end, cleaned, dried, and coiled.

18) **All Recruit helmets shall be marked with the reflective letter "R" indicating RECRUIT FIRE FIGHTER.**

These letters shall not be removed until notified by Fire Services Headquarters, through the Fire Academy, that the RECRUIT PROBATIONARY period has ended.

19) **All Recruits shall follow the chain of command with requests or inquiries of a personal nature.**

The Chain of Command shall be through any Training Officer or Instructor to the Course Co-ordinator, through the Co-ordinator to the Chief Training Officer, and finally, through the Chief Training Officer to the Division Chief of Professional Development and Training. The Division Chief shall be the deciding authority for any Recruit's request to address Fire Services Headquarters.

20) **Recruits shall continually be aware of any announcements or changes to timetables, routines, directions, etc.**

From time-to-time, notices, announcements, and/or timetable changes will be posted on the notice board in the lunch-room. Recruits shall routinely check the notice boards throughout the day.

21) **Recruits are responsible for the security and serviceability of all TFS items and equipment issued to them.**

Should a Recruit discover that a personal item of TFS clothing or equipment is missing or damaged in such a way that it might affect its function, it shall be immediately reported to any Training Officer.

PART 4 - CANTEEN

Every Station on each Platoon, every Division and Section has a canteen which is used to pay for the 'incidentals' of Fire Services life. This would include purchases ranging from cable TV, the private telephone, and newspaper subscriptions, to coffee, tea, milk, sugar, and so on.

The Canteen is run by the Canteen Manager who is appointed by the members of the station / division / section. The Recruit Class Canteen is non-profit and all monies received are used to the benefit of the Recruit Class.

By order of the Chief Training Officer, the Recruit Class Canteen Manager shall be the Fire Academy Canteen Manager.

OPERATING RULES OF THE CANTEEN

1. All members of the Recruit Class shall be expected to be paid-up members of the Class Canteen.

2. The Canteen dues shall be collected by the Fire Academy Canteen Manager.

3. Canteen funds shall be used to supply items such as sugar, milk, and other consumables as used by the class.

4. Canteen fees or dues shall be determined by the Recruit Class Canteen Manager in a fair and reasonable manner and shall be approved by the Chief Training Officer.

Courtesy of Toronto Fire Services.

Class Registration / Invoice

For Office Use Only

Invoice #__________

Invoice Date: __________

Last Name: ____________________ First Name: ____________________

Social Security #: ____________________

Organizational Information

Mark One of The Following:

Fire Fighter ☐ Career ☐ Volunteer

☐ Civil Emergency Mgmt.

☐ EMS

☐ Law Enforcement

☐ Other Public Employee

☐ Industry

☐ Other: ____________________

Organization: ____________________

Organization's Address: ____________________

Organization's City: ____________________

Organization's State: ____________________

Organization's Zip: ____________________

Organization's Phone#: ____________________

Course Location: ____________________

Course Title: ____________________

Course Date: ____________________

Course ID: ____________________

Course Fee: ____________________

Other Materials:

____________________ $__________

____________________ $__________

____________________ $__________

Billing Information

Bill to address if different than above:

Bill To:

Billing Address:

City: State: Zip:

☐ Please bill me.

Purchase Order # ____________________

☐ Please charge my: ☐ Master Card ☐ Discover ☐ Am. Express ☐ Visa

Name (As it Appears on the Card):

Card No.: Exp. Date:

Signature:

Release Agreement

I hereby certify that I am an active member of the above named department or organization, and that either I am personally responsible or my department / organization is responsible, for payment of any and all medical expenses or charges (including first aid treatment) incurred while attending this course. Oklahoma State University - Fire Service Training will provide instruction in the course and assumes no responsibility other than the opportunity to learn under supervision. Acceptance in this course constitutes an agreement to the conditions stated and the staff of Fire Service Training - Oklahoma State University and the state of Oklahoma are hereby relieved of liability.

Student's Signature: ____________________ Date: __________

Signature of Witness: ____________________ Date: __________

Appendix G

Structural Live Fire Training Forms

Fire departments use NFPA 1403, *Standard on Live Fire Training Evolutions,* as a guideline in adopting policies that require they carefully follow certain procedures when planning a structural burn training evolution. Included in this appendix are examples from the following departments:

- Gainesville (FL) Fire Department
- Seattle (WA) Fire Department
- Fishers (IN) Fire Department
- Maryland Fire and Rescue Institute

These departments have compiled a number of forms that they have found useful in structural fire training operations. These sample forms might be adapted for local or can serve also as a guideline to help develop new forms.

Gainesville Fire Rescue Department

AGREEMENT TO DESTROY STRUCTURE AND RELEASE

THIS AGREEMENT entered into this _____ day of _______, 19__, by and between the City of Gainesville, hereinafter called the "City," and ____________________________ called "Owners."

W I T N E S S E T H:

WHEREAS, Owners are the owners of certain property located within the City of Gainesville, which property contains a structure the Owners wish to have destroyed, and

WHEREAS, the City of Gainesville Fire Rescue Department is willing to destroy the structure as part of a training exercise for its employees.

NOW, THEREFORE, in consideration of the mutual premises and agreements herein contained and other good and valuable consideration, the parties do hereby mutually agree as follows:

1. The City of Gainesville Fire Rescue Department agrees to destroy the structure located on the property owned by the Owners, the property and structure being more particularly described as follows:

2. The Owners agree to allow the structure to be destroyed in the course of a training exercise conducted by the City of Gainesville Fire Rescue Department.

3. The Owners hereby warranty that they are in fact the sole owners of the structure and/or property described herein and, by affixation of the signatures hereunder, attest to the fact. The Owners further certify that there are no outstanding or unsatisfied mortgages, liens, claims or any other type of encumbrances on or against the above described property or structure.

4. The Owners further certify that no claim for loss under any insurance policy will be made because of damage, or because of destruction of the structure, as a result of the Fire Rescue Department's training activities at said structure. The Owners further certify that as of __________________, 19__, there is no effective insurance policy covering the structure described in paragraph one (1) under which the activities described herein by the City would constitute a claim.

Page 1 of 3

PO Box 490 • Gainesville, Florida 32601-0490 • 352.334.5078 • Fax: 352.334.2529

5. Upon completion of said destruction, the Owners agree to remove any and all debris remaining on the property. The Owners hereby acknowledge that the City's activities in razing, demolishing, and destroying the structure described in paragraph one (1) will, except as described in paragraph six (6), result in the structure being wrecked and reduced to ruin. What remains of the structure will be worthless rubble and debris, which could constitute a danger to persons entering the premises and should be expeditiously removed by the Owners.

6. If for any reason whatsoever the City of Gainesville Fire Rescue Department finds or determines that it cannot begin or complete the destruction of said structure, the Owners, upon notification of the Fire Rescue Department's inability or unwillingness to destroy or complete destruction of said structure, agree to expeditiously complete the destruction and debris removal of said structure or otherwise restore the property to such condition as to meet the minimum building and/or housing codes in effect within the government having jurisdiction.

7. The Owners understand and acknowledge their continuing control over, and liability for, damages occurring on or about the property described in Paragraph one (1) above; except during the actual training exercise authorized herein. It is specifically understood that the City's liability during the training exercise does not include the settling of ash on surrounding property. The training exercise shall be conducted between ____________________________ and _________________________. When the training is concluded, control over, and sole liability for damages occurring on or about the premises shall automatically revert to the Owners, unless notification to the contrary is made by the City.

8. The Owners agree to indemnify and save harmless the City of Gainesville, the City of Gainesville Fire Rescue Department, its officers, agents and employees from and against any and all claims, suits, actions, damages or causes of action arising out of the destruction of the structure described herein or the fact that any representation made herein was false when made, for any personal injury, loss of life, or damage to property sustained in or about the owned property by reason of the destruction of the structure located thereon except those injuries, losses or damages solely attributable to the gross negligence of the City, its agents and employees, and for and against any orders, judgments or decrees which may be entered thereon and from and against all costs, attorney's fees, expenses and liabilities incurred in or about the defense of such claims and investigation thereof.

PO Box 490 • Gainesville, Florida 32601-0490 • 352.334.5078 • Fax: 352.334.2529

9. Nothing contained in this contract shall be interpreted as a waiver of the City's sovereign immunity granted under Section 768.28, Florida Statutes.

IN WITNESS WHEREOF, we the undersigned have set our hands and seals first written above.

CITY OF GAINESVILLE, FLORIDA

By______________________________
City Manager

Approved as to form and legality:

By______________________________
City Attorney

WITNESSES: OWNERS:

______________________________ ______________________________

______________________________ ______________________________

SWORN TO AND SUBSCRIBED before me this ________________day of __________, 19__.

____
Notary Public

My Commission Expires:

____

Updated 08/24/98

PO Box 490 • Gainesville, Florida 32601-0490 • 352.334.5078 • Fax: 352.334.2529

Gainesville Fire Rescue Department

IMPORTANT NOTICE

The City of Gainesville Fire Rescue Department has a training program designed to help our firefighters maintain their skills as well as train on new equipment and fire fighting techniques. In order to do this we routinely burn vacant houses to simulate true working conditions. Since this is a common practice around the country the National Fire Protection Association (NFPA) has developed a standard for Fire Department Training Burns, NFPA 1403. This standard is designed to maintain a safe training environment and is strictly adhered to by the City of Gainesville Fire Rescue Department.

You may see a number of fire department apparatus in your neighborhood. There is nothing to be concerned about as we are running training evolutions.

Due to the smoke generated by a burning house, we request that you keep your windows and doors closed on the following date(s):

We anticipate completion by the end of the day.

We also request that you avoid parking on the street so that fire apparatus can have easy access to the building.

Sincerely,

Gainesville Fire Rescue Training Bureau

Updated 11/25/98

PO Box 490 • Gainesville, Florida 32601-0490 • 352.334.5078 • Fax: 352.334.2529

Gainesville Fire Rescue Department

Live Burn Check-Off List

		YES/NO/DATE
A.	**Develop the Plan**	
1.	Purpose____	___/___/________
2.	Objective______________________________________	___/___/________
B.	**Preburn inspection/check for the following Hazards (NFPA 1403)**	
	Utilities/Gas disconnected	___/___/________
	Flammables removed	___/___/________
	Toxic materials removed	___/___/________
	Windows and doors secured or covered	___/___/________
	Vegetation removed or trimmed	___/___/________
	Check for asbestos (by certified inspector)	___/___/________
	Powerlines in close proximity	___/___/________
	Check for holes in floors or missing steps	___/___/________
	Identify hydrants or water supply	___/___/________
	Identify traffic control needs	___/___/________

1

PO Box 490 • Gainesville, Florida 32601-0490 • 352.334.5078 • Fax: 352.334.2529

YES/NO/DATE

Check for exposures ___/___/________

C. Day before burn

Have crews in first run territory deliver notices to neighbors that could be affected by smoke ___/___/________

Arrange for traffic control support ___/___/________

Do a walk through of building to insure no hazardous conditions have occurred since initial inspection ___/___/________

Obtain necessary fire starting equipment

Fuel ___/___/________

Palates ___/___/________

Matches/Lighter ___/___/________

Notify Media if appropriate ___/___/________

D. Burn day

Determine if weather is suitable for burn ___/___/________

Discuss plan with crews ___/___/________

Appoint Safety Officer ___/___/________

Establish parking or staging for crews, including EMS ___/___/________

Do walk through with all crews ___/___/________

Ensure back-up line from second water source ___/___/________

Appoint attack and back-up crews ___/___/________

PO Box 490 • Gainesville, Florida 32601-0490 • 352.334.5078 • Fax: 352.334.2529

YES/NO/DATE

Notify dispatch of burn ___/___/________

Ensure that all safety equipment is in use ___/___/________

E. After the burn

Fire must be completely out/cold before crews leave ___/___/________

Scene shall be checked for safety hazards ___/___/________

Evaluate the activity

Were objectives met? ___/___/________

What was learned?

____

____

____

What could have been done better?

____

____

____

Document all attendants and record training hours ___/___/________

Send report to the Training Chief ___/___/________

Revised 11/25/98

PO Box 490 • Gainesville, Florida 32601-0490 • 352.334.5078 • Fax: 352.334.2529

Paul Schell, Mayor

Seattle Fire Department
James E. Sewell, Chief

April 15, 1998

John Doe
5678 Skyline Avenue S.W.
Seattle, WA 98116

Dear Mr. Doe,

This letter is a follow-up to your conversation with Captain Paul Fletcher regarding the possibility of allowing the Seattle Fire Department to use the building at 4727 - 47th Avenue S.W., Seattle, for training purposes during the period from June 15, 1998, through June 24, 1998.

With your approval, we would like to use this building for live fire training, search and rescue practice, building construction training, and ventilation. We will take responsibility for any injuries to our personnel during this training process.

The law requires that we have the following items before we use the building for training fires:

- Written permission from the owner(s) or legally delegated representative granting permission to use the building for the requested training.
- A statement in the letter that there are no outstanding loans, or any other person with a legal interest in this property. If there are others with a legal interest they must also sign the permission letter.
- A statement in the permission letter that there will be no fire insurance (covering damage to the structure) in effect on the property as of June 1, 1998.
- A document verifying ownership.
- A copy of the demolition permit.
- A copy of an asbestos survey with proof of asbestos removal (if any).
- An ***approved*** copy of the "Notice of Intent" form from the Puget Sound Air Pollution Control Agency.

This type of training is done with permission from the Puget Sound Air Pollution Control Agency, and is allowed only after the removal of all asbestos and any other heavy smoke producing materials. You are responsible for the removal of these materials with the exception that our personnel will remove roofing materials from the building if required. Also, all utilities must be disconnected back to the street by the appropriate utility provider prior to the training date.

Attached is an example of the permission letter that we must receive from you prior to our training.

If you need any additional information, or have any questions, please contact Captain Fletcher at 386-1771.

Very truly yours,

Ronald T. Hiraki
Deputy Chief of Training

RTH: pmf

301 Second Avenue South, Seattle, WA 98104-2680
Tel: (206) 386 1400, Fax: (206) 386-1412

April 24, 1998

James Sewell, Chief
Seattle Fire Department
301 Second Avenue South
Seattle, WA 98104

ATTN: Deputy Chief Ron Hiraki

Dear Chief Sewell:

The Seattle Fire Department as requested in its letter of April 15, 1998, is hereby given permission to use the building located at 4727 - 47th Avenue S.W. for training purposes.

The following is a review of our agreement regarding this building:

- You will be responsible for any injuries to your personnel.
- Your training will consist of live fires, with training in extinguishment techniques, building construction, search and rescue, and ventilation training which will include opening holes in the roof.
- You may cease training and vacate the property if it becomes necessary to demolish the building due to structural instability or for any other reason.
- As a final part of this training it is expected that a training fire will be set in the building and it will be burned to the ground.
- It is my responsibility to acquire all of the necessary permits and approvals, and to comply with them fully prior to any training fire in this building (Before June 1, 1998).
- It is my responsibility to make sure all of the utilities are properly disconnected back to the street prior to the scheduled training dates.
- I (we) certify that this property is owned free and clear (no outstanding loans) by the undersigned, that there are no underlying liens on this property, nor is there any other person with a legal interest in this property.
- I understand that this training can be discontinued at any time, and for any reason, by order of the Chief of Training. The Seattle Fire Department does not have any obligation to demolish this structure.
- I am responsible for the condition of this site after the Fire Department vacates, including all debris and any portion of the building that is left after the training is concluded, which will include the foundation, and may include other portions of the building that have not been completely consumed.
- I certify that there will be no fire insurance coverage on the property as of June 1, 1998.

Sincerely,

John Doe

City of Seattle

Paul Schell, Mayor

Seattle Fire Department
James E. Sewell, Chief

NOTICE

YOUR SEATTLE FIRE DEPARTMENT WILL BE CONDUCTING CONTROLLED TRAINING FIRES IN YOUR NEIGHBORHOOD FROM JUNE 15, 1998, THROUGH JUNE 24, 1998. THE FIRES WILL BE CONFINED TO THE BUILDING LOCATED AT 4727 - 47th AVENUE S.W.

THIS TRAINING WILL BE DONE WITH THE APPROVAL OF THE DEPARTMENT OF CONSTRUCTION AND LAND USE BY WAY OF A VALID DEMOLITION PERMIT AND WITH APPROVAL FROM THE PUGET SOUND AIR POLLUTION CONTROL AGENCY. ALL ASBESTOS AND POTENTIAL POLLUTANTS WILL HAVE BEEN REMOVED FROM THE STRUCTURE.

THE PURPOSE OF THIS TRAINING IS TO GIVE OUR NEW FIREFIGHTERS EXPERIENCE UNDER ACTUAL FIRE CONDITIONS. THE FIRES WILL INCREASE IN INTENSITY AS THE TRAINING PROGRESSES. THE BUILDING WILL BE BURNED TO THE GROUND DURING OUR LAST TRAINING EXERCISE. THIS TRAINING FOR OUR NEW FIREFIGHTERS IS NECESSARY SO THAT THEY WILL BE ABLE TO PROPERLY HANDLE STRUCTURE FIRES THAT OCCUR IN OUR CITY.

AS WE CONDUCT THESE FIRES, YOUR COOPERATION WILL BE GREATLY APPRECIATED. THE NEW FIREFIGHTERS WILL BE UNDER THE DIRECT SUPERVISION OF INSTRUCTORS FROM THE TRAINING DIVISION. ADDITIONALLY, A SAFETY OFFICER AND EXPERIENCED FIREFIGHTERS WILL BE AT THE LOCATION TO SUPPORT THE TRAINING. IF YOU DESIRE ADDITIONAL INFORMATION, PLEASE CONTACT THE SEATTLE FIRE DEPARTMENT TRAINING DIVISION AT 386-1771.

301 Second Avenue South, Seattle, WA 98104-2680
Tel: (206) 386 1400, Fax: (206) 386-1412

LESSON PLAN FOR RECRUIT TRAINING FIRES
Recruit Class #77 At 4727 - 47th Avenue S.W.

Goals	• Conduct training fires for Recruits beginning with small bed fires, and proceeding to larger well involved room fires. • Conduct a controlled burn down of the structure and protect exposures.

Objectives	Recruits under the supervision of the instructors will demonstrate: • Laying initial and back-up attack lines, then locate, confine, and extinguish fire. • Ability to perform PPV. • Ability to check for extension and perform overhaul. • Proper primary and secondary search and rescue techniques. • Ability to obtain proper suction lines. • Teamwork and coordination of tasks with other companies. • Safety and accountability in all operations.

BED FIRES	
First Engine Company	• Lay preconnect; locate, confine, and extinguish fire. Search fire room, create ventilation opening, and begin to overhaul bedding. Check wall and ceiling for possible extension. • Obtain suction from second engine company's manifold.
Second Engine Company	Simulate 1-3/4" - Manifold. Lay the handline as a back-up for first engine company.
Third Engine Company	Simulate lay of second 1-3/4" - Manifold to an alternate hydrant. Lay handline to floor above fire. Check for vertical extension of fire.
First Ladder Company	Set up positive pressure fan and provide search and rescue in the fire building.
Second Ladder Company	Bed fire operation in fire room. Overhaul and carry bedding out of building in bed fire tarp.
Team One Regular Firefighters	Interior, 1-3/4" safety line.
Team Two Regular Firefighters	Exterior, Rapid Intervention Team

Page 1

Courtesy of Seattle (WA) Fire Department

ROOM FIRE - FLOOR ONE		
First Engine Company		Lay preconnect; locate, confine and extinguish fire. Search fire room, create ventilation opening. Check wall and ceiling for possible extension. Suction from second engine company's manifold.
Second Engine Company		Simulate 1-3/4" - Manifold. Lay the handline as a back-up for first engine company.
Third Engine Company		Simulate lay of second 1-3/4" - Manifold to an alternate hydrant. Lay handline to floor above fire. Check for vertical extension of fire.
First Ladder Company		Set up positive pressure fan and provide search and rescue in the fire building.
Second Ladder Company		Provide overhaul equipment to the fire room, overhaul as directed.
Team One	**Regular Firefighters**	Interior, 1-3/4" safety line.
Team Two	**Regular Firefighters**	Exterior, Rapid Intervention Team

ROOM FIRE - FLOOR TWO		
First Engine Company		Lay preconnect; find, confine and extinguish fire. Search fire room, create ventilation opening. Check wall and ceiling for possible extension. Suction from second engine company's manifold.
Second Engine Company		Simulate 1-3/4" - Manifold. Lay the handline as a back-up for first engine company.
Third Engine Company		Simulate lay of second manifold to an alternate hydrant. Provide overhaul equipment to fire room. Overhaul where instructed.
First Ladder Company		Set up positive pressure fan and provide search and rescue in the fire building.
Second Ladder Company		Ladder the building on two sides the roof, and into one window of the second floor for secondary means of egress.
Team One	**Regular Firefighters**	Interior, 1-3/4" safety line.
Team Two	**Regular Firefighters**	Exterior, Rapid Intervention Team

Page 2

Controlled Burn of Structure		
All Recruits And Instructors		• Four 1-3/4" handlines will be placed at the corners of the building for use in controlling the fire, stopping embers, and protecting the exposures. • Two 2-1/2" handlines will be placed on each of the lee corners for additional protection and ember control if needed. • An instructor will be placed on each line to assure proper actions and to educate the recruits about aspects of exposure control, fire behavior, heat movement, and fire spread.
One	**Regular Firefighter**	Pump Operator, Engine 320
One	**Regular Firefighter**	Pump Operator, Engine 330
Team Two	**Regular Firefighters**	Staff Aid Unit

Page 3

Live Fire Training

INSTRUCTOR-IN-CHARGE

I, the undersigned, have been informed and do recognize my responsibilities as an instructor-in-charge in this live fire training exercise. These responsibilities include the following:

1. Plan and coordinate all training activites.

2. Monitor activities to ensure safe practices.

3. Inspect building integrity prior to each evolution.

4. Assign instructors:
- **Attack hoselines**
- **Backup hoselines**
- **Functional assignments**
- **Teaching assignments**

5. Brief instructors on responibilities.

6. Assign çoordinating personnel as needed:
- **Emergency medical services**
- **Communications**
- **Water supply**
- **Apparatus staging**
- **Equipment staging**
- **Breathing apparatus**
- **Personnel welfare**
- **Public relations**

7. Ensure adherance to this standard by all persons within training area.

Signature____

Date ________________________

Statement of Release:

I,__________________________________, do herein state that I am a legal representative of ___________________________________Who has legal ownership of the property at the following described location:

____

____

____

I hereby affirm that the property is free of any other ownership rights or liens.

It is herein agreed that the structures located on said property may be utilized for live fire training exercises, including demolition of the structures by fire, by the _______________________________ _______________________________and other fire service representatives as may be invited by them to participate in the training program. It is further agreed that there shall be no insurance claim filed in connection with the loss by fire of said structures involved in the training exercises.

The fire department agrees that the property owner will be held free of any liability in connection with any injury or death of any training participant or any damage to any departmental equipment involved in the use of the structure by the department. The department further agrees that any training conducted at the site will comply with all applicable federal, state, or local laws or ordinances.

The property owner agrees to hold harmless the fire department, its officers, members and representative, and the municipalities and elected officials which it serves as well as any other fire service personnel involved in the training for any damages incurred as a result of the training exercises.

______________________________________ ____________________

Signature of Fire Chief Date
or authorized representative

______________________________________ ____________________

Signature of Property Owner Date
or authorized representative

Subscribed and sworn to by me, a Notary Public in ____________________ County this __________ day of ________________, 19____ .

______________________________________ My commission expires ______________________

Signature of Notary Public Date

SAFETY OFFICER

Live Fire Training Objectives

1. Demonstrate a knowledge of department safety regulations.
2. Demonstrate the ability to maintain and complete accurate records.
3. Demonstrate the ability to identify common causes of firefighter injury.
4. Demonstrate an understanding of building construction.
5. Demonstrate an understanding of the warning signs of structural failure.
6. Demonstrate a working knowledge of temperature ranges for fire rooms.
7. Demonstrate an understanding of the warning signs of flashover.
8. Demonstrate the ability to communicate with IC and other officers.
9. Demonstrate the thorough knowledge and application of NFPA 1500
10. Demonstrate a thorough knowledge of protective clothing.

Live Fire Training

STUDENT

I, the undersigned, have been informed and do recognize my responsibilities as a student in this live fire training exercise. These responsibilities include the following:

1. Acquire prerequisite training. Provide documentation, if required.

2. Familiarize yourself with the building layout.

3. Wear approved full turnout clothing.

4. Wear approved self-contained breathing apparatus.

5. Obey all instructions and safety rules.

6. Remain with assigned group at all times.

Signature____

Date ____________________

SIGN- IN FORM

LIVE FIRE TRAINING EXERCISE AT ______________________________
(Address)

ON ______________________________
(Date)

BY SIGNING BELOW, I HEREBY STATE THAT I HAVE SUFFICIENT TRAINING FROM MY RESPECTIVE DEPARTMENT TO AT LEAST MEET THE MINIMUM REQUIREMENTS FOR THE BASIC FIREFIGHTER, TWENTY FOUR (24) HOUR COURSE OF THE STATE OF INDIANA.

NAME	DEPARTMENT	NAME	DEPARTMENT

Live Fire Training

INSTRUCTOR

I, the undersigned, have been informed and do recognize my responsibilities as an instructor in this live fire training exercise. These responsibilities include the following:

1. Monitor and supervise assigned students (no more than five)

2. Inspect students protective turn-out gear and equipment.

3. Account for all assigned students, both before and after evolutions.

4. Strive to accomplish instructional objectives.

Signature____

Date ____________________

FIRST-IN FIREFIGHTER

Live Fire Training Objectives

1. Demonstrate an understanding of personal safety.

2. Demonstrate the ability to perform basic fire attack.

3. Demonstrate the ability to perform appropriate ventilation.

4. Demonstrate the ability to perform search & rescue.

5. Demonstrate the ability to perform salvage & overhaul.

6. Demonstrate the ability to work as a team.

7. Demonstrate the ability to perfrom according to NFPA 1001, Chapter 3.

EMS / REHAB

Live Fire Training Objectives

1. Demonstrate an understanding of typical fireground injuries.

2. Demonstrate the ability to quickly triage injured personnel.

3. Demonstrate appropriate treatment of burns, cuts, smoke inhalation, etc.

4. Demonstrate knowledge of appropriate department injury forms.

5. Demonstrate the ability to communicate with IC and other officers.

6. Demonstrate an understanding of physiological effects of firefighting.

7. Demonstrate knowledge of appropriate foods and fluids for rehab.

8. Demonstrate medical screening of resting firefighters and associated documentation.

9. Demonstrate the ability to recognize signs and signals of heat exhaustion, stress, and over-exertion.

APPARATUS OPERATOR

Live Fire Training Objectives

1. Demonstrate the ability to perform pumping operations.
2. Demonstrate the ability to calculate water supply needs.
3. Demonstrate safety while performing pumping apparatus operations.
4. Demonstrate the ability to perform according to NFPA 1002, Chapter 2.
5. Demonstrate a knowledge of apparatus tools and equipment.
6. Demonstrate the ability to communicate with IC and Divisions/ Groups
7. Demonstrate the ability to calculate and supply proper discharge pressures

INCIDENT COMMANDER

Live Fire Training Objectives

1. Demonstrate the ability to perform size-up and assess incident priorities.

2. Demonstrate the ability to formulate strategic goals for an incident.

3. Demonstrate the ability to determine tactical objectives toward the accomplishment of the overall strategy.

4. Demonstrate the ability to consolidate the stategy and tactical objectives into a concise incident action plan.

5. Demonstrate the ability to form an adequate command structure for the incident.

6. Demonstrate the ability to effectively assign subordinate officers into the command structure.

7. Demonstrate the ability to communicate tactical objectives to groups, divisions, and sections of the command structure.

8. Demonstrate the ability to effectively receive feedback from subordinates.

9. Demonstrate the ability to refine strategy and tactical objectives based on information received or observed during the incident.

10. Demonstrate the ability to coordinate overall emergency activities.

11. Demonstrate the ability to coordinate activities of outside agencies.

12. Demonstrate a priority concern of personnel safety.

13. Demonstrate the ability to manage incident resources.

GROUP OFFICERS

Live Fire Training Objectives

1. Demonstrate the understanding of proper leadership.

2. Demonstrate the ability to properly assign tasks to subordinates.

3. Demonstrate the ability to accomplish tactical objectives as assigned by IC.

4. Demonstrate the ability to communicate with the incident commander and other groups and divisions.

5. Demonstrate the ability to effectively provide feed-back to the IC.

6. Demonstrate a priority concern for personnel safety.

7. Demonstrate the ability to perform according to NFPA 1021, Chapter 2.

Live Fire Training

FIRE IGNITOR

I, the undersigned, have been informed and do recognize my responsibilities as a fire ignitor in this live fire training exercise. These responsibilities include the following:

1. Coordinate all activities with safety officer and instructor-in-charge.

2. Assure that all fuels used are of a Class A nature in an amount necessary to create the desired size fire.

3. Use only combustible items with known burning characteristics.

4. Use only small amounts of uncontaminated deisel fuel or kerosene for ignition only.

5. Use of flammable liquids (NFPA 30) is prohibited.

6. Inspect building integrity prior to each evolution.

Signature____

Date ____________________

MFRI

STRUCTURAL FIREFIGHTING

BUILDING AND FIELD SITE INSPECTION FORM

1. Date of Inspection: ______________________

2. Organization requesting training: ______________________

( ) REGIONAL

( ) FIELD CLASS

( ) Co. DRILL OR OTHER

Street: ______________________

Town: ______________ County: ______________

3. Name of person making inspection:

a. M.F.R.I. Faculty: ______________________

b. Field or Training Academy Instructor/s: ______________________

c. Organization Training Rep: ______________________

4. Site location of building: ______________________

____

5. Owner of building: Name: ______________________

Street: ______________ Town: ______________

County: ______________ State: ______________ Zip: ________

Home Phone: ______________ Work Phone: ______________

6. Exposures: (Enter distances in feet in spaces below if applicable)

	Side 1	Side 2	Side 3	Side 4
a. Private Residence...........	Front ______	Left ______	Rear ______	RT ______
b. Commercial/Industrial........	Front ______	Left ______	Rear ______	RT ______
c. Institutional................	Front ______	Left ______	Rear ______	RT ______
d. Petroleum Storage...........	Front ______	Left ______	Rear ______	RT ______
e. Electrical Installations.....	Front ______	Left ______	Rear ______	RT ______
f. Electric/Telephone Wires.....	Front ______	Left ______	Rear ______	RT ______
g. Woodlands/Grasslands.........	Front ______	Left ______	Rear ______	RT ______
h. Railroad.....................	Front ______	Left ______	Rear ______	RT ______
i. Highway......................	Front ______	Left ______	Rear ______	RT ______
j. Pipeline.....................	Front ______	Left ______	Rear ______	RT ______

k. Other: ______________________

Comments/Notes: ______________________

____

7. Water Supply: (Enter location/s and distances)

a. Draft (Pond, Stream):______________________________

b. Hydrant/s: ______________________________

c. Tanker Operations: ( ) Primary Supply ( ) Secondary Supply

d. Other:______________________________

e. Road conditions for water supply:______________________________

f. Accessibility affected by weather: ( ) Yes ( ) No

g. Comments/Notes:______________________________

____

8. Vehicle Parking:

	Satisfactory	Unsatisfactory
a. Tactical Attack Vehicles................	________	________
b. M.F.R.I. Vans..........................	________	________
c. Ambulance/First Aid....................	________	________
d. Trainee Parking	________	________
e. Other: ______________________	________	________

f. Comments: ______________________________

____

____

____

9. Prevailing Wind Direction: ______________________________

10. Rough Sketch of Site (to show the following with symbols indicated)

a. Streets, Alleys, Roads, etc: ______________________________

____

b. Exposures:

PR	Private Residence	PS	Petroleum Storage
C	Commercial	H	Hydrants
I	Institutional Industrial	N	North Direction

-E- Electrical Wires

SITE DIAGRAM

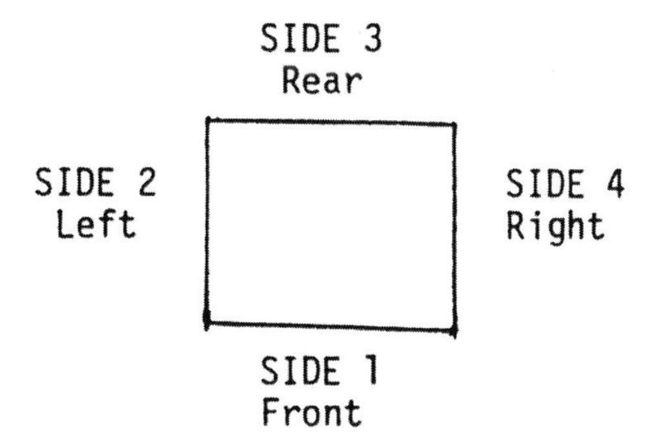

11. Rough Sketch of Building Floor Plan (Indicate "Front")

FIRST FLOOR

SECOND FLOOR

12. Building Inspection:

a. Construction: Wood Frame ___________ Exterior Cover _______________

Brick Veneer _________ Balloon Construction___________

Masonry ____________________

Other ____

Roof Type/Covering ________________________________

b. Windows...............No. 1st Fl. _____ No. 2nd Fl. _____ In Place? ______

c. Exterior Doors........No. 1st Fl. _____ No. 2nd Fl. _____ In Place? ___ __

d. Interior Doors........No. 1st Fl. _____ No. 2nd Fl. _____ In Place? ______

e. Condition of Floors...1st Fl. ______________ 2nd Fl. ____________________

f. Attic.................Type of Entrance ______ Floored ______ Walkways ______

g. Interior Stairways....Number ________________ Condition __________________

h. Cellar................Full _________ Partial ________ Crawl Space __________

i. Chimneys..............Number _____ Type of Construction __________________

i. Chimneys..............Condition _________________ Corbelled ( ) Yes ( ) No

j. Fireplaces............Number _____ Type of Construction __________________

k. Interior Walls & Ceilings:

Plaster & Wood Lath ________ Condition _________________________

Wallboard......... ________ Condition _________________________

Plaster & Metal Lath _______ Condition _________________________

Other............. ________ Condition _________________________

Patching/Sealing Required ____________________________________

Wall Paper on Walls ________ Ceilings ___________________________

l. Debris on Floor? __________________________

m. Linoleum on Floors? ________________________

n. Supporting Structures Adequate? Porches ______________________

Floors ________________________

Main Roof _____________________

Porch Roof/s __________________

Interior/Exterior Stairways __________________

13. Civilian Traffic Problem: ( ) YES ( ) NO

a. If YES, Explain:____

____

____

b. Stability of access roads and parking areas__________________________

c. Can access roads be blocked during training operations?______________

Anticipated spectator traffic ________________

d. If roads are to be blocked, by whose authority? State - County - City?

e. Who will guard the blockades?____

____

______________________________ Instructor Making Inspection

______________________________ Faculty Member Making Inspection

______________________________ Date

______________________________ Date

WRITTEN PERMITS AND DOCUMENTS, AND OTHER NOTIFICATIONS NEEDED:

	Not Required	Written Notification	Permit Received	Verbally Notified
1. Maryland Environmental Health.......	________	____________	________	________
2. Maryland Forestry..................	________	____________	________	________
3. Razing Permit......................	________		________	
4. From Building Owner:				
Permission to Burn.................	________	____________		
Proof of Cancelled Insurance........	________	____________		
OTHER: ______________________________				
______________________________	________	____________	________	________
5. From Organization Requestion Training:				
Proof of Liability Insurance obtained or in force................		____________		
6. Central Fire Dispatch...............	________	____________	________	________
7. Maryland State Police...............	________	____________	________	________
8. County Sheriff/Police...............	________	____________	________	________
9. Other ______________________________	________	____________	________	________
10. Other______________________________	________	____________	________	________
11. Other ______________________________	________	____________	________	________

NOTE: All permits and documents must be obtained 30 days prior to training date.

______________________________________ ______________________________________

MFRI

MARYLAND FIRE & RESCUE INSTITUTE
UNIVERSITY OF MARYLAND
COLLEGE PARK, MARYLAND 20742
(301) 454-2416

OWNER'S RELEASE FOR STRUCTURAL BURNING FOR TRAINING PURPOSES

I (We), the undersigned, being the owner(s) of the real property located at ______________________________

____,

in the County of ______________________________

do hereby give consent to the Maryland Fire and Rescue Institute, University of Maryland, College Park, Maryland to use the above said property for the purpose of live structural firefighting training which may include the total destruction of said building or rendering the building unuseable. I(We) certify that all insurance that I(We) have on this property has been cancelled, that no liens or encumberances apply to such property, and no claims for insurance or otherwise will be made by anyone on the above said property. Description of building to be used for training:

____

____

____

______________________________ ______________________
(PRINT) Property Owner's Name Signature

______________________________ ______________________
(PRINT) Property Owner's Name Signature

STATE OF MARYLAND
COUNTY OF:______________________

Sworn and subscribed to me before me this ____________

day of ____________________, 19__________.

____
NOTARY PUBLIC

My commission expires:

____
Date

MFRI T/A-12

BUILDING PREPARATION CHECKLIST

1. Floors made safe.............................. ________________
2. Window openings closed......................... ____ ____________
3. Loose wallpaper removed........................ ________________
4. Linoleum removed............................... ________________
5. Stairways made safe/railings in place.......... ________________
6. All chimneys undercut.......................... ________________
7. All holes in walls and ceilings patched and sealed............................. ________________
8. Openings made in end gables for observation and fire extinguishment in attic and roof areas........................ ________________
9. At least one (1) 4' x 4' hole made in each separate roof area for ventilation .... ________________
10. All inside debris cleaned-up and rooms swept................................. ________________
11. All outside debris cleared away from entrances and areas of egress.................. ________________
12. Outside grounds preparation for vehicles.. ________________
13. Cistern, cesspool, well, or other ground openings protected or filled in .... ________________
14. Check fuel tanks............................... ________________
15. Utilities disconnected:
    - Telephone.............. ________________
    - Electric............... ________________
    - Gas.................... ________________
16. Can building be secured overnight.............. ________________
17. Insect hives and toxic weeds removed........... ________________

NOTE: All repairs and preparations to building and grounds should be completed prior to start of training session to permit maximum time for learning.

____
INSTRUCTOR MAKING INSPECTION

____
DATE

____
FACULTY MEMBER MAKING INSPECTION

____
DATE

1. BRANCH	2. DIVISION/GROUP ESCAPE GROUP	ICS 204(1-82) **DIVISION/GROUP ASSIGNMENT LIST**
3. INCIDENT NAME WILDFIRE 98		4. OPERATIONAL PERIOD DATE 6/27 OR 6/28/98 TIME 0600 TO 1800

5. OPERATIONS PERSONNEL

OPERATIONS CHIEF: R. HIRAKI	**DIVISION/GROUP SUPERVISOR:** C. WILSON
BRANCH DIRECTOR: N/A	**AIR ATTACK SUPERVISOR:**

6. RESOURCES ASSIGNED THIS PERIOD

STRIKE TEAM/TASK FORCE RESOURCE DESIGNATOR	LEADER	NUMBER PERSONS	TRANS NEEDED	DROP OFF PT/TIME	PICK UP PT/TIME
PATROL 1		2		DP-4 0900	
PATROL 2		2		DP-4 0900	
PATROL 3		2		DP-2 0900	
PATROL 4		2		DP-12 0900	
ENGINE 1		3		DP-4 0900	
W T 1		3		WP-2 0900	
W T 7		3		DP-2 0900	
PATROL 5		2		TBD	
DOZER		1		TBD	
* T=TYPE					

7. CONTROL OPERATIONS

Patrols 1, 2, and 40 will be stationed along pre-burn perimeters of Branch I to control any escape fires. Div/Grp supervisor to coordinate patrol perimeters. WT-1 to manage water supply at Water Point(WP)- 2 with folding tank and portable refilling pumps. Patrols 3 and 4 to control fires in Arson area and Fire Shelter areas (Branch II). WT-7 to support Patrols 3 and 4. Division 'A' will be released and reassigned to the group upon completion of Div. 'A''s burning. Rotation of Patrols may occur throughout the drill by Escape Group Leader with approval from Operations. Intervention of suppression may be conducted to maintain safety of fire personnel, limit properyt damage, and to minimize the spread of uncontrolled fire activity. Intervention shall be intended to limit the potential of an escape fire. The Escape Group shall coordinate the continuance of firing operations and drill activities with Firing Supervisor and Operations. The Escape Group shall be the RIT.

8. SPECIAL INSTRUCTIONS

Escape fire situation: Any unit from the Escape RIT Group shall notify the Div/Grp Supervisor of an escape fire. The Div/Grp Supervisor shall notify Operations of any escape fire. The Operations Chief shall direct all firing operations to cease on the entire site. Strike Team Engines may be mobilized to assist with escape fires. The Escape RIT Group units shall support each other, as outlined in pre-incident training. The Div/Grp Supervisor shall provide a report on conditions and additional resources requested via the Operations Chief. Secondary containment lines shall be supported , as well as, escape demobilization routes. See Site Map page 3 for escape demobilization routes. Water dropping helicopters(if needed) shall obtain water from Anderson Reservoir, not the nearby water ponds! Secondary suppression support shall be under the direction of the Operations Section Chief.

9. DIVISION/GROUP COMMUNICATIONS SUMMARY

FUNCTION		FREQUENCY	SYSTEM	CHN.	FUNCTION		FREQUENCY	SYSTEM	CHN.
COMMAND	LOCAL	**153.XXX**	**YELLOW**		SUPPORT	LOCAL	**168.XXX**	**USFS TAC**	
DIV A	LOCAL	**153.XXX**	**YELLOW**		MEDICAL	LOCAL	**168.XXX**	**USFS TAC**	
ESCAPE GROUP TACTICAL		**151.XXX**	**CDF TAC 10**		GROUND TO AIR		**156.XXX**	**CAL CORD**	

PREPARED BY (RESOURCE UNIT LEADER)	APPROVED BY PLANNING SECCHIEF	DATE	TIME
PETER SELLS	BARBARA ADAMS	6/6/98	1615hrs

Incident Radio Communications Plan		1. Incident Name Wildfire 98	2. Date/Time Prepared 4/22/98 1030	3. Operational Period 0600-1800 6/27/98 or 6/28/98	
4. Basic Radio Channel Utilization					
System /Cache	**Channel**	**Function**	**Frequency**	**Assignment**	**Remarks**
OES WHITE FIRE MARS		Travel Net	154.XXX	(White 1) Travel/Coordination	Contact "Wildfire Communications"
S CLARA CO FIREMARS		Command	153.XXX	(Couty Yellow) Command Net Safety	Contact "Wildfire Communications"
OES WHITE FIREMARS		Tactical	154.XXX	(White 2) Branch I	
CDF FIREMARS		Tactical	151.XXX	(CDF Tac 6) Branch II	
OES WHITE		Tactical	154.XXX	(White 3) Branch III/Div A	
CDF FIREMARS		Tactical	151.XXX	(CDF Tac 10) Escape Group	
USFS/NIFC FIREMARS		Tactical/support	168.XXX	**Support** Logistics/Plans Mechanics	
USFS/NIFC FIREMARS		Tactical	168.XXX	Medical	
CALCORD		Tactical	156.XXX	Escape Contingency Air/Ground Medical Evacuation	Use only when directed by Command
ICS 205 4/96	5. Prepared by T. SOLBERG, Communications Unit Leader				

Medical Plan	1. Incident Name **WildFire 98**	2. Date Prepared 5/22/98	3. Time Prepared 1700	4. Operational Period 0800 to 1800 6/27/98 or 6/28/98

5. INCIDENT MEDICAL AID STATIONS

Medical Aid Station	Base	PARAMEDIC YES	PARAMEDIC NO
		XX	

6. TRANSPORTATION

A. AMBULANCE SERVICES

Name	Address	phone	PARAMEDIC YES	PARAMEDIC NO
Air Ambulance	Through Communications (911)		XX	
MFRI Life Flight	"		XX	
Life Flight II	"		XX	

B. INCIDENT AMBULANCES

Name	Address	phone	PARAMEDIC YES	PARAMEDIC NO
AMR	Through IC (request from Communications)		XX	

7. HOSPITALS

Name	Address	TRAVEL TIME AIR	TRAVEL TIME GRND	PHONE	HELIPAD YES	HELIPAD NO	BURN CENTER YES	BURN CENTER NO
St. Strickland		10	20	555-5555	XX			X
Adams Hospital		15	30	555-5555	XX		XX	
Bizjak Medical Center		15	40	555-5555	XX			X
Heyns Medical Center		15	20	555-5555	XX		XX	

8. MEDICAL EMERGENCY PROCEDURES

1. Activate Medical Plan on Command Net, through your supervisor.
2. Medical Unit Leader(MUL) will check all burn injuries for the need to be treated.
3. Transportation will be provided to move minor injuries to the aid station for treatment. AMR will provide transport to off site facilities.
4. MUL will maintain a Unit Log of all medical activity and keep the Safety Officer advised of all injuries.
5. MUL will staff and supervise the rehab unit.

San Jose Fire Department
WILDFIRE "98"
Incident Safety Message

- Follow the 10 STANDARD FIRE ORDERS
- Follow the 18 WATCH OUT SITUATIONS (checklist)
- Remember to pre-hydrate prior to activity
- Review the escape routes
- Drivers are to follow one-way signs and drive slow
- Watch each other for signs of:
  HEAT EXHAUSTION & STROKE
- Watch for snakes and ticks
- Watch for gopher holes
- Keep communications with supervisors

Look Up	Look Down	Look Around

Courtesy of San Diego (CA) Fire Department.

San Jose Fire Department
Firing Plan
for Wildfire 98

Progressive Hose Lay - Branch I:

1. One firing team will be assigned to each group in Branch I (total of 2 teams).

2. Firing teams will be led by a CDF Captain assisted by a local government member and a CDF JAC employee.

3. The Team will light fire that will create enough open flank with a slow enough rate of spread to allow Engine Companies a meaningful drill.

Teams will adhere to the following guidelines:

- Do not light head fires.
- Light fires to burn generally downhill against the wind.
- Use anchor points at all times.
- Light fire in a manner to avoid creating two large flanks.
- Attempt to minimize scorch to trees.

Division "A" Arson:

1. One CDF Engine will be assigned to Division Supervisor BC Barrett to light an extinguish his spots.

2. Several spots will be ignited and extinguished by the Team.

General Prescription Elements:

Temperature:	70-90
Relative Humidity:	20-50%
Wind Speed:	2-10 mph
Wind Direction:	Any
10-hr Fuel Moisture:	6-15%

WARNINGS: Temperature above 90, wind above 12 mph, RH below 18%.

Per Firing Supervisor BC Dave Wachtel, CDF

San Jose Fire Department
Wildfire 98 - Fire Shelter Policy

Re: Policy and procedure for conducting fire shelter deployment in conjunction with live fire exercises.

Policy

The use of fire shelter deployment in conjunction with live fire exercises may only occur after specific safety considerations have been addressed There are two methods for shelter deployment (static and dynamic). The static method will be utilized for this drill. We recognize the value provided by this type of training, but emphasize that the fire shelter is a last resort and should be utilized only after all other attempts for personnel protection have been exhausted.

Procedure

The following must be met prior to the ignition of any ground fire in conjunction with fire shelter deployment training.

Prescribed Conditions

- Deployment shall only be made in areas with light fuel loading, less than 5 ton per acre.
- Fuel shall be no more than 30 inches in height.
- Slope in deployment area shall be 10% or less.
- Two engines will be positioned in close proximity to the evolution, each having one charged hose extended and capable of applying an effective stream throughout the entire exercise area. Nozzles used shall be capable of delivering a minimum of 60 GPM.
- A safety officer or OIC other than the instructor shall be present and monitor the entire evolution. He shall have the authority to order the exercise stopped at any time.
- An experienced Torch (burn ignition person) shall be utilized at each site.

Evolution

- A detailed "tailgate" safety meeting will be held prior to each evolution. Each trainee will assure that they understand that their participation is strictly voluntary. Instructor will ensure that each trainee is in complete safety gear with shrouds down and goggles on. Trainee will be instructed to listen for the voice of their instructor who will be giving them guidance. Each trainee will be given a number that the instructor will use to direct comments to.

- Trainees will be advised of the importance of staying in the shelters until ordered out.

- Trainees will be directed to the exercise area and told to select a site to clear for deployment. Trainees will be ordered to clear an area 4 ft x 8 ft. Instructor will verify that all sites are adequately prepared. Trainees will be directed to "deploy your shelters and assume their positions."

- Trainees will be timed and advised when they have exceeded the standard of 30 seconds for deployment.

- Instructor will verify that each trainee has assumed the proper position (feet towards the fire) and that they are properly centered in the cleared area. Each trainee shall be properly secured within the shelter. Instructors will pull up on the ridge of the shelter to test trainees ability to hold the shelter in place.

- Instructor will ensure that hose lines are charged and positioned for adequate coverage of the exercise site.

- Instructor will advise trainees when ignition is imminent.

- Instructor will verify the firing Officer is notified prior to any ignition and that all safety personnel are ready. The Instructor will then order the torch to ignite the burn no more than 50 ft from the deployment site taking fire behavior into account.

- Instructor will monitor the movement of the fire and keep trainees advised of the fire status.

- After the fire has burned through the deployment area and all fire has been extinguished, instructor will advise each trainee, by number, to exit their shelter. Instructor will advise the Firing Officer that the burn has been completed.

- A post-evolution discussion will be held and each trainee will describe their experience and provide any recommendations to improving the drill.

Note to Branch Director: With two instructors per site, one instructor can have a group of trainees preparing their sites with hand tools while the other group is firing. This action is important in order to reduce reflex time and get everyone **safely** through the evolutions within the allotted time.

Appendix I

Evaluation Forms

Course Evaluation/Reaction

Please respond briefly to the following questions. Your reactions assist in making adjustments to future programs.

1. Was this course what you expected it to be? If no, why not?

2. What areas of instruction do you think could be added to the course to improve it?

3. What areas do you think could be shortened? Entirely eliminated?

4. Were the exercises and activities appropriate for the course/relevant to your job? If no, why not?

5. What did you particularly *dislike* about the course?

6. What did you particularly *like* about the course?

Course Evaluation Questionnaire
(Reaction Form)

Course Title: ______________________________ Instructor: ______________________________

Location: ______________________________ Attendance Dates: ______________________________

Directions: Mark your reaction by circling one of the following choices:

- *SA* - If you *strongly agree* with the item
- *A* - If you *agree moderately* with the item
- *D* - If you *disagree moderately* with the item
- *SD* - If you *strongly disagree* with the item
- *N* - If you have *no opinion* or the item is *not applicable*

Please read each item carefully.

1. *SA A D SD* I would take another course taught this way.
2. *SA A D SD* I did not learn or gain much by taking this course.
3. *SA A D SD* The instructor encouraged development of new viewpoints.
4. *SA A D SD* I thought that the course material seemed worthwhile.
5. *SA A D SD* The instructor(s) demonstrate knowledge of the subject.
6. *SA A D SD* The course material was too difficult.
7. *SA A D SD* I do not feel I performed as well in this course as in similar courses.
8. *SA A D SD* The course content was excellent.
9. *SA A D SD* The course format and information was good.
10. *SA A D SD* I would recommend this course to other department members.

Is this your first fire and emergency services training course? Yes___ No___

Are you (check one): Paid/career ___ Paid/on-call ___ Volunteer ___

How long have you been in the fire and emergency services? _______ years

Please state your rank or title. ______________________________

Participant Evaluation of Instruction

Instructor Name: ______________________________ Date: ______________________

Course Title: ______________________ Course Location: ______________________

Directions: Read the entire sheet BEFORE marking responses. Select the three most significant items and number them *1, 2,* and *3.* Place an *X* in the space that most accurately represents your opinion. It is not necessary to sign your name; your opinion is important so the course may be improved.

Subject Knowledge:

( )	( )	( )	( )
Very well informed	Well informed	Limited background	Poorly informed

Presentation:

( )	( )	( )	( )
Stimulating	Adequate	Routine	Dull

Attitude Toward Learners:

( )	( )	( )	( )
Very considerate	Considerate	Somewhat intolerant	Inconsiderate and rude

Explanations:

( )	( )	( )	( )
Very clear	Clear	Confused	Faulty

Composure or Manner:

( )	( )	( )	( )
Always composed	Usually composed	Easily upset	Highly insecure

Course Organization:

( )	( )	( )	( )
Well organized	Usually well-planned	Somewhat unplanned	Often disorganized

Assignments:

( )	( )	( )	( )
Very clear	Usually clear	Somewhat vague	Always vague

Examination Questions:

( )	( )	( )	( )
Clear, relevant	Adequate	Often confusing	Irrelevant and unclear

Grading Methods:

( )	( )	( )	( )
Always fair	Usually fair	Inconsistent	Biased

Time Spent Outside Class Time:

( )	( )	( )	( )
More than others	Equal to others	Less than others	Little, if any

Textbook Value:

( )	( )	( )	( )
Excellent resource	Adequate resource	Limited resource	Little, if any resource value

Attitude Towards Course:

( )	( )	( )	( )
Very favorable	Somewhat favorable	Indifferent	Negative

Instructor Rating Sheet

Instructions: Write the name of the instructor to be rated in the blank preceding the rating scale Make a check in the appropriate boxes for each subject matter and teaching area. When checking subject matter, consider, *"will the material be of value to me and my department now or in the future?"* Your check concerning teaching indicates how well you felt the instructor presented the material. Remember, material may be well presented but useless or very useful and badly presented. Use the following scale for your evaluation/reaction:

Subject Matter:
1 = of no use
2 = of limited use
3 = of some use
4 = of considerable use
5 = very useful

Teaching:
Very poor
Inadequate
Adequate
Good
Excellent

		1	2	3	4	5
______________________ (Instructor)	Subject Matter	❑	❑	❑	❑	❑
	Teaching	❑	❑	❑	❑	❑

Comments: ____

____

*****

		1	2	3	4	5
______________________ (Instructor)	Subject Matter	❑	❑	❑	❑	❑
	Teaching	❑	❑	❑	❑	❑

Comments: ____

____

*****

		1	2	3	4	5
______________________ (Instructor)	Subject Matter	❑	❑	❑	❑	❑
	Teaching	❑	❑	❑	❑	❑

Comments: ____

____

*****

		1	2	3	4	5
______________________ (Instructor)	Subject Matter	❑	❑	❑	❑	❑
	Teaching	❑	❑	❑	❑	❑

Comments: ____

____

*****

		1	2	3	4	5
______________________ (Instructor)	Subject Matter	❑	❑	❑	❑	❑
	Teaching	❑	❑	❑	❑	❑

Comments: ____

____

*****

Olympia Fire Department | Annual

PERFORMANCE EVALUATION ACTIVITY BRIEF

Member's Name	Member's Job Title	Assignment	Applicable Perf. Stds.
Type Of Activity (Incident No.)	Location Of Activity (Incident Address)	Date Of Activity	Time Of Activity

Situation* What activity did the member perform or was the member assigned to perform?

Objective* Observations What activity was observed? What evidence was there? What testimony was received?

❒ See Attachments

Assessment* What impact did the member's performance have on the operation of the Fire Department?

Plan* What actions will be taken to ensure that future activities will be positive?

❒ Commend ❒ Coach ❒ Counsel ❒ Other

Signature Of Member Completing Activity Brief	Job Title	Assignment	Date

Response Following the review of this activity, what comments does the reviewing member have?

Signature Of Member Reviewing Activity Brief	Job Title	Assignment	Date

October 1996 | Form 2.4.7.1

White: To Member's File Yellow: To Member

Courtesy of Olympia (WA) Fire Department. S.O.A.P. © Used With Permission Human Resources Systems.

COURSE ASSESSMENT FORM

SOUTH DAKOTA STATE FIRE MARSHAL'S OFFICE - FIRE TRAINING

118 WEST CAPITOL AVE. PIERRE, SD 57501 (605)773-3876

YOUR HONEST AND SINCERE EVALUATION OF THIS COURSE HELPS INSURE THAT OUR PROGRAMS ARE OF THE HIGHEST CALIBER AND THAT THEY MEET OR EXCEED YOUR TRAINING NEEDS. THANKS!

COURSE:____________________________________**DATE(S):**__________

INSTRUCTOR(S):____

RATING SCALE:
5 - OUTSTANDING
4 - MORE THAN SATISFACTORY
3 - SATISFACTORY
2 - LESS THAN SATISFACTORY
1 - POOR

MATERIALS	1. PRINTED MATERIALS WERE WELL ORGANIZED.	5 / 4 / 3 / 2 / 1
	2. PRINTED MATERIALS WERE COMPLETE.	5 / 4 / 3 / 2 / 1
	3. WERE READABLE (PRINTED WELL).	5 / 4 / 3 / 2 / 1
	4. VISUAL MATERIALS WERE RELATED TO COURSE.	5 / 4 / 3 / 2 / 1
	5. VISUAL MATERIALS WERE IN APPROPRIATE NUMBERS.	5 / 4 / 3 / 2 / 1
	6. VISUAL MATERIALS WERE OF GOOD QUALITY.	5 / 4 / 3 / 2 / 1
COURSE	7. COVERED SUBJECTS THAT YOU THOUGHT IT WOULD.	5 / 4 / 3 / 2 / 1
	8. WAS A REASONABLE LENGTH.	5 / 4 / 3 / 2 / 1
	9. CONTRIBUTED TO YOUR KNOWLEDGE AND SKILLS.	5 / 4 / 3 / 2 / 1
	10. RELATED TO YOUR NEEDS.	5 / 4 / 3 / 2 / 1
	11. WAS WORTH RECOMMENDING TO OTHERS.	5 / 4 / 3 / 2 / 1
INSTRUCTOR(S)	12. RELATED COURSE MATERIALS TO CLASS NEEDS.	5 / 4 / 3 / 2 / 1
	13. KNEW SUBJECT THOROUGHLY.	5 / 4 / 3 / 2 / 1
	14. ENCOURAGED CLASS PARTICIPATION.	5 / 4 / 3 / 2 / 1
	15. MADE COURSE REQUIRMENTS AND OBJECTIVES CLEAR.	5 / 4 / 3 / 2 / 1
	16. STAYED ON SUBJECT.	5 / 4 / 3 / 2 / 1
	17. ANSWERED QUESTIONS COMPLETELY.	5 / 4 / 3 / 2 / 1
	18. TOLERATED DIFFERENCES OF OPINION.	5 / 4 / 3 / 2 / 1
CLASSROOM	19. CONTAINED A MINIMUM NUMBER OF DISTRACTIONS.	5 / 4 / 3 / 2 / 1
	20. OVERALL, THE FACILITY WAS ACCEPTABLE.	5 / 4 / 3 / 2 / 1
SUGGESTIONS	21. HOW COULD THE INSTRUCTOR(S) IMPROVE CLASS DELIVERY?	
	22. HOW COULD THE COURSE CONTENT OR STRUCTURE BE IMPROVED?	
	23. IF YOU COULD MAKE ONE CHANGE TO THIS COURSE (OR SCHOOL), WHAT WOULD YOU ADD, SUBTRACT, OR DELETE?	

Courtesy of South Dakota State Fire Marshal's Office, Pierre, SD.

San Jose Fire Department, Bureau of Education & Training

INSTRUCTOR SURVEY

INSTRUCTIONS:

Recruits are among those who are best qualified to judge an instructor's teaching effectiveness and to offer suggestions that will improve his/her performance and promote good teaching standards. This information will not identify any student individually. Please fill out this survey, BOTH front and back, and return to the Recruit Liaison or Coordinator.

Instructor's Name:	Topic:	Date:

1. Organization & Preparation of Class Material

[A] Exceptionally well organized and prepared
[B] Consistently well organized and prepared
[C] Reasonably organized and prepared
[D] Sometimes lacks organization
[E] Disorganized or unprepared

2. Explanation of Concepts and Principles

[A] Exceptionally clear and enlightening
[B] Very clear in explanations
[C] Usually good in explanations
[D] Seldom adds to student's understanding
[E] Often confuses student's understanding

3. Ability to Create Interest in Class Material

[A] Stimulates interest to high degree
[B] Usually stimulates interest
[C] Occasionally stimulates interest
[D] Neither stimulates nor reduces interest
[E] Reduces interest

4. Apparent Knowledge of Material

[A] Exceptional
[B] Through
[C] Adequate
[D] Somewhat lacking
[E] Poor

5. Enthusiasm in Teaching Students

[A] Highly enthusiastic
[B] Generally enthusiastic
[C] Occasionally enthusiastic
[D] Shows little enthusiasm
[E] Seems to have no enthusiasm

6. Use of Examples and Illustrations

[A] Very effectively used to support materials
[B] Usually used well
[C] Used adequately
[D] Seldom used effectively
[E] Never used effectively

7. Responsiveness to Class Difficulty

[A] Extremely sensitive and responsive
[B] Usually aware and responsive
[C] Sometimes aware and responsive
[D] Responsive when asked
[E] Insensitive or unresponsive

8. Motivation of Recruits

[A] Inspires extremely strong effort
[B] Inspires strong effort
[C] Inspires adequate effort
[D] Inspires minimal effort
[E] Eliminates motivation

9. Pace of the Class

[A] Much too fast
[B] A little fast
[C] About right
[D] A little too slow
[E] Much too slow

10. Overall Rating of Instructor's Performance

[A] Excellent
[B] Very Good
[C] Good
[D] Fair
[E] Poor

Courtesy of San Jose (CA) Fire Department.

11. Comments Directed TO THE INSTRUCTOR

Please comment on the Instructor's strong and weak points of teaching.

12. Comments REGARDING CLASS

Please comment on the class content, presentation, strong and weak points, and suggested improvements.

Course Evaluation Questionnaire

OKLAHOMA STATE UNIVERSITY-FIRE SERVICE TRAINING
1723 W. TYLER STILLWATER, OK 74078-8041
PHONE: 1-405-744-5727 OR 1-800-304-5727 FAX: 1-405-744-7377

Name of Course ____________________

Location ____________________ Date ____________

Course Instructor(s) ____________________

Circle the response which best reflects your opinions regarding the following statements using the criteria:

SA - If you strongly agree with the statement.
A - If you agree with the statement.
D - If you disagree with the statement.
SD - If you strongly disagree with the statement.

SA	A	D	SD	The methods used in teaching this class were appropriate.
SA	A	D	SD	It was easy to remain attentive.
SA	A	D	SD	Much was gained by taking this class.
SA	A	D	SD	The instructor encouraged the development of new viewpoints or practical techniques.
SA	A	D	SD	The course material seemed worthwhile.
SA	A	D	SD	The instructor(s) demonstrated thorough knowledge of the subject.
SA	A	D	SD	The material was carefully explained and additional assistance was provided by the instructor(s).
SA	A	D	SD	The class was interesting.
SA	A	D	SD	The class was taught well.
SA	A	D	SD	I would like other appropriate members of my department to attend this class.

What type of agency are you affiliated with:

Fire EMS Police Industry Other ____________

If you selected Fire or EMS (above), are you: Career Volunteer Other ____________

Course Content

Please give your comments regarding the course content, subject matter, and any particular use this course has for your department or agency.

Which topic was the most beneficial? Why?

Which topic was the least beneficial? Why?

Instructor(s)

What are your comments regarding the instructor(s) for this course?

General Comments

What was your overall impression of this course?

What improvements would you suggest for this course?

Glossary

A

Accident — Sequence of unplanned or uncontrolled events that produces unintended injury, death, or property damage; the result of unsafe acts by persons who are unaware or uninformed of potential hazards and ignorant of safety policies or who fail to follow safety procedures.

Accident Investigation — Fact-finding rather than fault-finding procedures that look for causes of accidents, which leads to analyzing causes in order to prevent similar accidents.

Acquired Building (Structure) — Structure acquired by the authority having jurisdiction from a property owner for the purpose of conducting live fire training evolutions.

Administrative Law — Body of law created by an administrative agency in the form of rules, regulations, orders, and decisions to carry out regulatory powers and duties of the agency. *Also see* Law.

Affective Domain — Learning domain that involves emotions, feelings, and attitudes. *Also see* Learning Domain.

Affirmative Action Programs — Employment programs designed to remedy discriminatory practices in hiring minority group members.

AHJ — Abbreviation for Authority Having Jurisdiction.

Andragogy — Study of adult education and its methods of teaching and learning.

Application — Lesson plan component at which point the instructor provides opportunities for participants to practice or apply cognitive information to skills learned in a lesson. *Also see* Lesson Plan.

Approved — Acceptable to the authority having jurisdiction.

Assessment — Process used to find out the knowledge, skills, and abilities that a learner has; can be done by such means as observation and special assessment activities including quizzes, examinations, oral tests, and similar testing devices.

Assignment — Work that must be performed by learners outside class in order to reach a skill level, meet an objective, and/or prepare for the next lesson.

Authority Having Jurisdiction (AHJ) — Organization, office, or individual responsible for approving equipment, an installation, or a procedure.

B

Backfire — Fire set along the inner edge of a control line to consume the fuel in the path of a wildland fire.

Battery — Unlawful touching or application of force to a person.

Behavior — In psychology, any response or reaction to a stimulus, such as instruction, made by an organism such as the learner.

Behavioral Objective — Measurable and precise statement of intent that specifically describes behavior that the learner is expected to exhibit as a result of instruction. Behavioral objectives also indicate the conditions under which the behavior is to be performed (given certain equipment, a specific time period for completion, etc.) and the required standard of performance — a percentage (75 percent), a number (9 out of 10), a time constraint (within 1 minute), or an NFPA (or OSHA, SOP, etc.) requirement. The term is often interchangeably used with the terms educational objective, outcome objective, and enabling objective.

Also see Enabling Objective, Objective (1), and Performance Objective.

Below-Grade Operation — Rescue activity that occurs at the bottom or slope of an open pit or trench, for example, freeing an equipment operator from an overturned backhoe that lies on the slope of an excavation; also, per OSHA 1926.650, any open operation such as that in an open trench for footings and foundations.

Belowground Operation — Activity that occurs below the surface of the earth, for example, rescue and search operations in a mine or mine shaft; also called underground operation; per OSHA 1926.800, operation with earth cover such as mining and tunneling.

Block — Division of an occupation consisting of a group of related tasks that have some factor in common. *Also see* Unit.

Breach of Duty — Any violation or omission of a legal or moral duty; neglect or failure to fulfill in a just and proper manner the duties of an office or employment.

C

Case Study — Discussion in which a group reviews real or hypothetical events. *Also see* Discussion.

Coaching — Process in which instructors direct the skills performance of individuals by observing, evaluating, and making suggestions for improvement.

Cognitive Domain — Learning domain that emphasizes thought rather than feeling or movement and involves the learning of concepts and principles. *Also see* Learning Domain.

Cognitive Evaluation — Assessment of knowledge that shows cognitive or knowing level by requiring that learners respond appropriately to questions on various types of tests.

Communication — Two-way process of transmitting and receiving some type of message.

Comprehensive Test — Type of test typically given in the middle (midterm) or at the end (final) of instruction that measures terminal performance of program participants and whether they have achieved program objectives. *Also see* Test.

Computer-Assisted Instruction — Self-paced method of instruction in which individuals use a computer to respond to information and prompts presented on a screen.

Condition — Part of an educational objective that describes under what provisions and with what resources the learner should be able to act or fulfill the objective.

Copyright Statute — Law designed to protect the competitive advantage developed by individuals or organizations as a result of their creativity. *Also see* Law and Statute or Statutory Law.

Counseling — Advising learners or program participants on their educational progress, career opportunities, personal anxieties, or sudden crises in their lives.

Criterion (s), Criteria (pl) — Standard(s) on which a decision or judgment is based.

Criterion-Referenced Testing — Measurement of individual performance against a set standard or criterion, not against other individuals; mastery learning is the key element of criterion-referenced testing. *Also see* Mastery and Test.

Curriculum — Broad term that refers to the sequence of presentation, the content of what is taught, and the structure of ideas and activities developed to meet the learning needs of learners and achieve desired educational objectives; also the teaching and learning methods involved, how learner attainment of objectives is assessed, and the underlying theory and philosophy of education.

Curriculum Development — Using analysis, design, and evaluation to create a series of presentations that adhere to the four teaching steps (preparation, presentation, application, and evaluation) and address the learning needs of a particular audience or program. Also called Instructional Design.

Customer Service — Quality of an organization's relationship with individuals who have contact with the organization. There are internal customers such as the various levels of personnel and trainees and external customers such as other organizations and the public. Customer service is the way these individuals, personnel, and organizations are treated and their levels of satisfaction.

D

Damage — Loss, injury, or deterioration caused by the negligence, design, or accident of one person to another in respect to another person's property.

Damages — Compensation to a person for any loss, detriment, or injury whether to his or her person, property, or rights through the unlawful act, omission, or negligence of another.

Data — Facts, numbers, and information used as a basis for reasoning, discussion, or calculation.

Database — Computer software program that serves as an electronic filing cabinet and used to create forms and record and sort information. Databases can be used to develop mailing lists, organize libraries, customize telephone and fax lists, and track presentation and program outcomes.

Defamation — Publication of anything that injures the good name or reputation of a person or brings disrepute to a person. *Also see* Libel and Slander.

Demonstration — Instructional method in which the instructor shows how to perform a skill, usually in a step-by-step process.

Direct Question — Type of question or questioning method in which the instructor directs a question to a specific-named individual who is expected to respond.

Disability — According to Americans with Disabilities Act (ADA), a person has a disability if he or she has a physical or mental impairment that substantially limits a major life activity.

Discrimination (in Testing or Discrimination Index) — Measure of the extent to which any item in a test is answered more or less successfully by the testing learners who do well or poorly on a test as a whole and therefore discriminates between them. Ideally, items that do not discriminate would be omitted from revised versions of the test.

Discussion — Instructional method in which an instructor generates interaction with and among a group. There are several formats of discussion: guided, conference, case study, role-play, and brainstorming. In each type, it remains the responsibility of the instructor to steer the group discussion or activity to meet lesson objectives. *Also see* Case Study and Role-Playing.

Distance Learning — Method of instruction through which the instructors and participants rarely meet face-to-face, but communicate by correspondence, electronic mail (e-mail), radio, and television. *Also see* Open Learning.

Duty — Obligation that one has by law or contract.

E

Enabling Objective — Specific objective that learners must attain as a component of completing a behavioral objective or in order to complete certain subsequent objectives. For example, a behavioral objective requires the learner to perform cardiopulmonary resuscitation (CPR); an enabling objective toward meeting this end is to open the airway. Opening the airway is also required before moving on to other steps (checking for breathing, providing breaths, etc.). Often the standard of performance for an enabling objective is specified with the test item or instrument that evaluates achievement of the objective. The term is often interchangeably used with the terms educational objective, outcome objective and behavioral objective. *Also see* Behavioral Objective, Objective (1), and Performance Objective.

Equal Employment Opportunity Law — Law that applies to protected groups of individuals who have experienced past workplace discrimination.

Evaluation — (1) Last of the four teaching steps in which the instructor finds out whether the performance objectives have been met; (2) a process that examines the results of a presentation or program to determine whether the participants have learned the information or behaviors taught; consists of criteria, evidence, and judgment. *Also see* Formative or Process Evaluation, Lesson Plan, and Summative Evaluation.

Evolution — Set or prescribed action that results in an effective fireground activity. Also called Practical Training Evolution.

F

Facilitator — Instructor role in which he or she encourages productive interaction of group members in a manner that does not display personal expertise but channels and enhances the expertise of others.

Family Education Rights and Privacy Act — Legislation that stipulates that an individual's records are confidential and that information contained in these records may not be released without the individual's prior written consent.

Feedback — Responses that clarify and ensure that the message was received and understood.

Fire Shelter — Aluminized tent carried by firefighters offering personal protection by means of reflecting radiant heat and providing a volume of breathable air in a fire-entrapment situation.

Formative or Process Evaluation — Evaluation of a new or revised program in order to form opinions about its effects and effectiveness as it is in the process of being developed and tested (piloted). Its purpose is to gather information to help improve the program while in progress. *Also see* Evaluation (2).

G

Goal — Broad statement of educational intention that usually expresses what a training organization or instructor intends to do for the learner; contrast with the term *objective,* which states specifically what the learner will do.

Gross Negligence — Willful and wanton disregard. *Also see* Negligence.

H

High-Hazard Training — Training that involves activities that include certain known risks and potential unknown risks. Examples include evolutions or exercises in live fire suppression, hazardous materials mitigation, above- and below-grade rescue, and the use of power tools.

I

Illustration — Instructional method in which an instructor provides information coupled with visuals such as drawings, pictures, slides, transparencies, film, and models to illustrate a lecture and help clarify details or processes. Instructional outlines and key points may also be visually displayed to "illustrate" the lecture.

Immunity — Freedom or exemption from penalty, burden, or duty; special privilege.

Independent Learning — *See* Self-Directed or Independent Learning.

Individualized Instruction — Process of matching instructional methods with learning objectives and individual learning styles that enable a learner to achieve lesson objectives.

Information Sheet — Fact sheet or type of handout that provides additional background information on a topic, which supplements what is provided in the text or other course resources.

Instructional Design — *See* Curriculum Development.

Instructional Medium (s), Media (pl) — Method(s) for transmitting, sharing, or sending information or ideas; channel(s) or format(s) through which messages are transmitted between the sender and receiver.

Instructional Model — Series of steps that guide development of a program of instruction. It includes the steps of performing a needs analysis, planning the program, developing objectives, completing a task analysis, designing a lesson plan, and creating evaluation instruments.

Instructor — Individual deemed qualified by the authority having jurisdiction to deliver instruction and training in fire and emergency services.

Instructor Information — Component of the lesson plan that lists lesson resources such as personnel, texts, references, sources, instructional methods, learning activities, training locations, etc. *Also see* Lesson Plan.

J

Job — Lesson plan component that gives a short descriptive title of the information to be covered. Also called Topic. *Also see* Lesson Plan.

Job Breakdown Sheet — Form that lists a job and breaks it down into its component parts by listing the operations (doing units) and key points (knowing units). *Also see* Key Points.

Job Performance Requirement (JPR) — Statement that describes the performance required for a specific job.

JPR — Abbreviation for Job Performance Requirement.

Judge-Made Law — *See* Judiciary Law.

Judiciary Law — Law established by judicial precedent and decisions. Also called Judge-Made Law or Unwritten Law because it indicates that a decision was made by a judge. *Also see* Law.

K

Key Points — Information that must be known to perform correctly the steps in a procedure. *Also see* Job Breakdown Sheet.

Known-to-Unknown — Method of sequencing instruction so that information begins with the familiar or known and progresses to the unfamiliar or unknown while making relationships that enable learners to become familiar with the unknown.

L

Law — Rule that guides society's actions; three types: legislative, administrative, judiciary. *Also see* Administrative Law, Copyright Statute, Judiciary Law, Legislative Law, Ordinance, and Statute or Statutory Law.

Learning — Relatively permanent change in behavior that results from acquiring new information, practicing skills, or developing attitudes following some form of instruction.

Learning Contract — Formal agreement between learner and instructor that establishes an amount of work that must be finished in order to successfully complete a course.

Learning Domain — Distinct sphere or area of knowledge such as cognitive, psychomotor, or affective. *Also see* Affective Domain, Cognitive Domain, and Psychomotor Domain.

Learning Style — Learner's habitual manner of problem-solving, thinking, or learning, though the learner may not be conscious of his or her style and may adopt different styles for different learning tasks or circumstances.

Lecture — Instructional method in which an instructor provides material by telling, talking, and explaining.

Legislative Law — Law made by federal, state/province, county/parish, and city legislative bodies who have powers to make statutory laws. *Also see* Law and Statute or Statutory Law.

Lesson — *See* Presentation.

Lesson Plan — Teaching outline containing information and instructions on what will be accomplished and how during a lesson; an instructional unit that maps out a plan to cover information and skills and makes effective use of time, space, and personnel. It covers lessons that may vary in length from a few minutes to several hours. *Also see* Application, Evaluation (1), Instructor Information, Job, Level of Instruction, Preparation, Presentation, Summary, and Time Frame.

Level of Instruction — Lesson plan component that states the learning level that participants will reach by the end of the lesson and may be based on the taxonomy of learning domains or on performance of job requirements. *Also see* Lesson Plan.

Liability — All types of debts and obligations one is bound in justice to perform; a condition of being responsible for a possible or actual loss, penalty, evil, expense, or burden; a condition that creates a duty to perform an act immediately or in the future. *Also see* Vicarious Liability.

Libel — Tort; false and malicious words in print that defame a living person; written, published defamation (may vary among states/provinces; may require that other factors be proved or are present). *Also see* Defamation, Slander, and Tort.

Listening — Process of receiving, attending to, and assigning meaning to auditory stimuli; a process of steps that gives information that listeners try to understand.

Logistics — Rational calculation and reasoning used to manage the scheduling of limited materials and equipment to meet the multiple demands of training programs and instructors.

M

Manipulative Training — Lesson or exercise in a training program in which participants handle or learn to handle equipment or materials in a coordinated or skillful manner.

Maslow's Hierarchy of Needs — Theory put forth by psychologist Abraham Maslow that states that all human behavior is motivated by a drive to attain specific human needs in a progressive manner. The hierarchy begins with basic physiological needs and progresses through security, social, self-esteem, and self-actualization needs.

Mastery — High-level or nearly complete degree of proficiency in the performance of a skill based on criteria stated in objectives; in training, the ability to perform at a designated skill level, which enables the learner to progress to the next designated skill level. A mastery test checks that learners have achieved mastery (the appropriate skill level). *Also see* Criterion-Referenced Testing.

Materials — Equipment and other instructional support tools including personnel and papers that aid in teaching a lesson or course.

Mean — Term that refers to the "average" of a set of scores and is found by adding the set of scores (values) and dividing by the total number of scores. For example, if a set of scores is 98, 98, 95, 92, 92, 92, 89, 88, 87, 85, 85, 79, 75, 74, and 74, the mean or average score is 91.8.

Median — Middle score in a set of scores (values) that are arranged or ranked in size (order) from high to low. For example, if a set of scores is 98, 98, 95, 92, 92, 92, 89, 88, 87, 85, 85, 79, 75, 74, and 74, the median or middle score is 88.

Memory — Individual record or past mental and sensory experience. Some aspect of memory can be measured by the individual's ability to recall, recognize, and relearn.

Mentor — Trusted and friendly adviser or guide for someone who is new to a particular role.

Mentoring — Instructional method in which an individual, as trusted and friendly advisor or guide, sets tasks, coaches activities, and supervises progress of individuals in new learning experiences or job positions.

Message — Information or ideas transmitted and received.

Mode — Most frequent score (value) in a set of scores. For example, if a set of scores is 98, 98, 95, 92, 92, 92, 89, 88, 87, 85, 85, 79, 75, 74, and 74, the mode or most frequent score is 92.

Motivation — Arousal and maintenance of behavior directed towards a goal. Motivation usually occurs in someone who is interested in achieving some goal.

N

National Fire Protection Association (NFPA) — Nonprofit educational and technical association devoted to protecting life and property from fire by developing fire protection standards and educating the public.

Needs Analysis — Assessment of training needs that identifies the gap between what exists and what should exist; study of a selected group's needs for the purpose of providing appropriate training or equipment to satisfy those needs.

Negligence — Breach of duty where there is a responsibility to perform. *Also see* Gross Negligence, Proximate Cause, Standard of Care, and Tort.

NFPA — Abbreviation for National Fire Protection Association.

Nonverbal Cues — Messages without words often transmitted in gestures, posture or body language, eye contact, facial expression, tone of voice, or appearance.

Norm-Referenced Test — Assessment in which a learner's test performance is compared with the performance of other learners, and his or her grade is dependent on the average performance and variability of performance among those other learners. *Also see* Test.

O

Objective — (1) Purpose of a presentation or program; a step necessary to achieve a stated goal; (2) unbiased; dealing with facts of interpreting results without the distortion of personal feelings, prejudices, or interpretations. *Also see* Behavioral Objective, Enabling Objective, and Performance Objective.

Objective Test — Test or test items designed so that all qualified test developers agree on the correct answer. Test items and their answers are based on course objectives that were developed from some selected criterion or standard. Types of objective tests include multiple-choice, matching, true-false, and short answer/completion. *Also see* Test.

Observation — Actually seeing or watching a person's behavior in a natural setting. As a means of evaluation, direct observation provides very reliable information on the effects of programs.

Occupational Safety and Health Administration (OSHA) — U.S. federal agency that develops and enforces standards and regulations for occupational safety in the workplace.

Open-Ended Question — Question that requires more than a yes or no answer.

Open Learning — Form of distance learning designed so that participants attend a minimum number of classes and complete reading and writing assignments that are turned in or mailed to the instructor. *Also see* Distance Learning.

Oral Test — Type of test or form of assessment in which candidates are tested, usually individually, in face-to-face discussion with an examiner or group of examiners, often in conjunction with written and/or skill testing. *Also see* Test.

Ordinance — Local law that applies to persons and things of the local jurisdiction; a local agency act that has the force of a statute; different from law that is enacted by federal or state/provincial legislatures. *Also see* Law.

OSHA — Acronym for Occupational Safety and Health Administration.

Overhead Question — Type of question or questioning method in which an instructor asks a question of the entire class rather than just one person and to which anyone is free to respond.

P, Q

Peer Assistance — Learners who assist others (peers) in the learning process.

Percentage Score — Way of interpreting evaluation results by expressing a part of a whole in hundredths.

Performance Evaluation — Assessment of skills that shows level of learner performance or ability to perform skill steps for a particular task or job.

Performance Objective — Explicitly worded statement that specifies learners' behaviors (actions or performances), the conditions by which they will perform (what is given to them to perform), and the criteria they will meet (standards or similar measurable requirements). The objective states that the activity will be performed in some observable and measurable form. *Also see* Behavioral Objective, Enabling Objective, and Objective (1).

Performance Test — Type of test that measures a learner's ability and proficiency in performing a job or evolution that requires him or her to handle equipment or materials in a coordinated, step-by-step process. This type of test measures learner or employee achievement of a psychomotor objective and holds the test-taker to either a speed standard (timed performance) or a quality standard (minimum acceptable product or process standard) or both. *Also see* Test.

Planning Model — Organized procedure that includes the steps of analyzing, designing, developing, implementing, and evaluating instruction; a systematic approach to the design, production, evaluation, and use of a system of instruction.

Practical Training Evolution — *See* Evolution.

Preparation — Lesson plan component at which point an instructor provides preliminary information that will motivate and prepare the participants to learn. *Also see* Lesson Plan.

Prerequisite — Entry-level knowledge or abilities that a learner must have prior to starting a particular course or qualifying for a certain job or promotion. *Also see* Requisite.

Presentation — Lesson plan component at which point an instructor provides to, shares with, demonstrates to, and involves the participants in the lesson information. Also called Lesson. *Also see* Lesson Plan.

Pretest/Posttest — Test given to learners to compare knowledge or skills either before they start a training program (pretest) or after a presentation or program (posttest). Pretests check entry-level knowledge or abilities. Pretest scores are compared with posttest scores to determine progress or how much was learned in a training program. *Also see* Test.

Privacy Law — Federal and state/provincial statue that prohibits an invasion of a person's right to be left alone and also restricts access to personal information.

Private Law — Portion of the law that defines, regulates, enforces, and administers relationships among individuals, associations, and corporations. *Also see* Public Law.

Progress Test — Type of test that is used as a diagnostic device to measure the progress or improvement of learners throughout a course and helps guide instructors and participants in deciding what areas need more emphasis or learning time; typically called a quiz and may be written or oral. *Also see* Test.

Proximate Cause — That which (an act), in a natural and continuous sequence unbroken by any intervening cause, produces injury, and without which (the act) the result would not have occurred. *Also see* Negligence.

Psychomotor Domain — Learning domain that requires coordination between mind and muscle; that is, demonstrating the cognitive knowledge of riding a bike, tying a shoe, climbing a ladder, applying a splint, etc. *Also see* Learning Domain.

Public Law — Classification of law consisting of constitutional, administrative, criminal, and international law. *Also see* Private Law.

R

Ranking of Scores — Process of arranging a number of scores (values) in order from high to low; a process that makes it easy to determine median and mode.

Raw Score — Score on a test that has not yet been statistically processed to make it comparable with other scores. For example, an individual who received 38 points on a test of 40 questions has a raw score of 38. To compare the score to 100 percent, divide the raw score (38) by the total number of possible points (40) to get the percent score of 95. The raw scores and the percent scores can be compared to other learner scores (norm-referenced), or scores can be assigned to a selected mastery level based on a standard or job performance requirement, for example, 85 percent is considered mastery (criterion-referenced).

Reasonable Accommodation — Making exiting facilities readily accessible to and usable by individuals with disabilities without imposing undue hardship (significant difficulty or expense) on the operation of an organization.

References — Citations, bibliographies, and resources used in developing, planning, and researching course or lesson information.

Regulations — Rules or directives of administrative agencies that have authorization to issue them.

Relay Question — Type of question or questioning method in which an instructor returns or redirects the question back to the individual who asked the question or to the group to answer.

Reliability — Extent to which a test or test item consistently shows the same results or scores given to a set of learners on different occasions or marked by different assessors or by the same assessors on different occasions.

Requisite — Fundamental knowledge or essential skill one must have in order to perform a specific task. *Also see* Prerequisite.

Resource — Location that provides instructional training materials and support.

Rhetorical Question — Type of question or questioning method designed to stimulate thinking or to motivate participants rather than to seek a correct answer; does not necessarily require an answer.

Risk — Likelihood of suffering harm from a hazard; exposure to a hazard; the potential for failure or loss.

Role Model — Individual who others look to as an example while learning or adopting a new role or job; the part an instructor plays, the image an instructor portrays, and the actions an instructor demonstrates to learners or program participants who look to their instructor as an example.

Role-Playing — Discussion in which a group acts out various scenarios. *Also see* Discussion.

S

Safety Guidelines/Safety Plan — Rules, regulations, or policies created and/or adopted by an organization that list steps or procedures to follow that aid in reducing if not eliminating accident or injury.

Schema (s), Schemata (pl) — Refers to conceptual or knowledge structure (a mental map) in our memory system that we use to interpret or think out information that is presented to our senses by the external environment. When our senses are presented with a new object or situation, we match it against our existing knowledge or schema and act upon it based on our experience. If we don't have the appropriate knowledge or experience, we change the schema or develop additional ones. We develop more sophisticated and differentiated schemata as we gain new knowledge and encounter more experiences.

Self-Directed or Independent Learning — Method of instruction in which an individual is either given or selects a set of objectives to complete through his or her own method of learning at his or her own pace; based on an instructor's determination of the content and sequence of learning. Learners are responsible for how much they learn and how they learn it, but this type of instruction does not relieve the instructor of the responsibility of ensuring that the learner accomplishes the intended tasks and skills in appropriate formats.

Semantics — Study of meaning in words and symbols; refers to language, word meanings, and meaning changes due to context, all of which may be affected by individual background, knowledge, and experience.

Simple-to-Complex — Method of sequencing instruction so that information begins with the basic or simple information or beginning steps and progresses to the more difficult or complex information and processes.

Slander — Spoken words that damage reputation; an oral defamation; may vary among states/provinces; may require proof or presence of other factors. *Also see* Defamation, Libel, and Tort.

SOP — Abbreviation for Standard Operating Procedure.

Standard — That part of an educational objective that describes the minimum level of performance that a learner must meet in accomplishing the performance/ behavior.

Standard of Care — In law of negligence, the degree of care that a reasonably prudent person should exercise in same or similar circumstances. *Also see* Negligence.

Standard Operating Procedure (SOP) — Standard method in which an organization or department carries out a routine function; usually these procedures are written in a policies and procedures handbook.

Statute or Statutory Law — Federal or state/provincial legislative act that becomes law; prescribes conduct, defines crimes, and promotes public good and welfare. *Also see* Copyright Statute, Law, and Legislative Law.

Study Sheet — Instructional paper designed to arouse learner interest in a topic and explain specific or unique areas to study.

Subjective Test — Type of test in which a learner is free to organize, analyze, revise, redesign, or evaluate a problem based on his or her perceptions and understanding; measures higher cognitive levels than other types. Common types of subjective tests are essay and term paper. Different but equally qualified assessors will respond differently in judging the quality and characteristics of a learner's work (subjectivity) and therefore award different scores. *Also see* Test.

Summary — Lesson plan component at which point an instructor restates or reemphasizes key points with the learners by asking questions and guiding review by having them recall relationships, make comparisons, draw conclusions, etc. *Also see* Lesson Plan.

Summative Evaluation — Evaluation that sums up the effects and effectiveness of a program or teaching activity at the conclusion of a training course or session. *Also see* Evaluation (2).

T

Task — Combination of duties and jobs in an occupation that is performed regularly and requires psychomotor skills and technical information to meet occupational requirements.

Task Analysis — Systematic analysis of duties of a specific job or jobs, which identifies and describes all component tasks of that job. A task analysis enables program developers to design appropriate training for personnel and trainees who must learn certain tasks to perform a job.

Taxonomy — Classification system in which each separate class of items is given a name and contains items that are more like one another than like items in other classes. Examples are the Dewey Decimal System and Bloom's Taxonomy of Objectives for the Cognitive Domain.

Teaching — Method of giving instruction through various forms of communicating knowledge and demonstrating skill. Successful teaching causes an observable change in behavior through activities that provide opportunities for learners to demonstrate knowledge and skill and to receive feedback on progress toward expected behavior change.

Team Teaching — Instructional method in which a group of instructors cooperate in presenting information, demonstrating skills, and supervising practice of a class or several classes. A lead instructor organizes and coordinates the activities of all instructors. The instructor with the expertise in a particular topic teaches that particular topic; remaining instructors share the responsibilities of assisting with instructional details and of supervising practice of skills.

Technical Rescue — Application of special knowledge, skills and equipment to safely resolve unique and/or complex rescue situations.

Test — Any means by which the absence, presence, amount, or nature of some learner quality or ability is observed or inferred and appraised or measured. *Also see* Comprehensive Test, Criterion-Referenced Testing, Norm-Referenced Test, Objective Test, Oral Test, Performance Test, Pretest/Posttest, Progress Test, Subjective Test, and Written Test.

Test Planning Sheet — Planning form that lists and specifies the number of test items to be written and at what levels of learning in each content area. This planning sheet aids the test developer in ensuring that necessary numbers of questions are included for each objective or learning level in order to appropriately measure learning.

Time Frame — Lesson plan component that lists the estimated time it will take to teach a lesson. *Also see* Lesson Plan.

Topic — *See* Job.

Tort — Private or civil wrong or injury, including action for bad faith breach of contract, resulting from breach of duty that is based on society's expectations regarding interpersonal conduct; a violation of a duty imposed by general law upon all persons in a relationship that involves a given transaction. *Also see* Libel, Negligence, and Slander.

Training — Process of achieving proficiency through instruction and hands-on practice in the operation of equipment and systems that are expected to be used in the performance of assigned duties.

U

Unit — Division of a block within an occupation consisting of an organized grouping of tasks within that block. *Also see* Block.

Unwritten Law — *See* Judiciary Law.

V

Validity — Extent to which a test or any assessment technique measures the learner qualities (knowledge and skills) that it is meant to measure; a test that tests what was taught.

Vicarious Liability — Liability imposed on one person for the conduct of another based solely on the relationship between the two persons; indirect legal responsibility for acts of another (liability of an employer for acts of an employee). *Also see* Liability.

W

Whole-Part-Whole — Method of sequencing instruction so that information begins with the complete picture or demonstration of a skill or with an overview of all information, progresses to provide a breakdown of all steps or details on segments of information, and returns to a complete demonstration or overview in a review or summary.

Worksheet — Activity sheet that lists tasks to accomplish, guides activities, and enables learners to apply cognitive information in order to practice and develop skills.

Written Test — Test that typically assesses or measures cognitive knowledge, commonly by pencil and paper. Written tests may also be given by computer where the testing learner responds by keyboard or mouse where no pencil is used or paper is generated. *Also see* Test.

Index

O